THE SCIENCE OF PLEASURE

THE SCIENCE OF PLEASURE

Cosmos and Psyche in the Bourgeois World View

HARVIE FERGUSON

ROUTLEDGE
London and New York

First published 1990
by Routledge
11 New Fetter Lane, London EC4P 4EE
Simultaneously published in the USA and Canada
by Routledge
a division of Routledge, Chapman and Hall, Inc.
29 West 35th Street, New York, NY 10001

Typeset by LaserScript Limited, Mitcham, Surrey
Printed and bound in Great Britain by
Mackays of Chatham PLC, Chatham, Kent

British Library Cataloguing in Publication Data

Ferguson, Harvie
The science of pleasure : cosmos and psyche in the bourgeois world.
1. Man. Social behaviour. Influence of man's perception of society
I. Title
302.1'2

Library of Congress Cataloging in Publication Data

Ferguson, Harvie.
The science of pleasure : cosmos and psyche in the bourgeois world/
Harvie Ferguson.
p. cm.
Bibliography : p.
Includes index.
1. Reality–History. 2. Science–Philosophy–Social aspects–History.
3. Hedonism–History. 4. Social values–History. 5. Social
psychology–History. I. Title.
BD331.F428 1989
89-32956
CIP

ISBN 0-415-02893-0
ISBN 0-415-04458-8 (pbk)

FOR ZOË

CONTENTS

CONTENTS

ACKNOWLEDGEMENTS

My greatest intellectual debts are recorded in the Bibliography. In transcribing it from a card index I became increasingly conscious of my gratitude to the staff of Glasgow University Library, without whose unfailingly efficient and friendly service my task would have been impossible.

Many years ago Bob Young, overestimating my courage as well as my competence, encouraged me to write a thesis on the Scientific Revolution. Had I done so it would no doubt have been a very different work from that which, after many false starts and by following a rather devious path, I have now written. Yet, even more certainly, the present work would never have been written at all without the aid of his initial impetus. And if it lacks the scholarship he might feel appropriate to such a long gestation, I hope that its boldness, which owes most to his example, lives up to his initial expectations.

A number of colleagues have been kind enough to read and comment upon the manuscript. Eric Dunning and Nicholas Abercrombie made a number of helpful suggestions, as did Gianfranco Poggi who, through his teaching and writing, bears an additional and more diffuse responsibility for the final product. I would like especially to thank David Frisby for the generous gift of his time, his erudition, and his friendship. At least some of his ideas are, I trust, reflected in the pages that follow.

I would also like to thank those students who, occasionally by their persistent questioning, but more frequently by their expressions of bafflement, have forced me to express my ideas more clearly. I cannot, however, blame them or anyone else for either the content or the form of presentation of this book, which remains my sole responsibility.

Every author has impossible expectations of his or her publisher. I have indeed been fortunate. Chris Rojek of Routledge has been closely associated with the project during every stage in the preparation of the manuscript. Knowledgeable, efficient, reassuring, and interested, he is the ideal editor.

Far more than intellectual effort goes into the writing of a book. It seems appropriate therefore, though wholly inadequate, to put its last words ('Acknowledgements' being the final rather than the initial act of composition) to the unequal task of expressing my thanks to those who have also had to witness its protracted birth trauma. My wife and children, for reasons I cannot fathom, tolerated my obsession, made my writing personally worthwhile, and filled my life liberated from the word processor with genuine pleasure.

INTRODUCTION

We are all cosmologists. Few of us, of course, are scientists or can claim anything beyond the flimsiest grasp of modern physical theories of the origin and structure of the universe. We can, none the less, by virtue of our normal social activities rather than in the possession of systematic and specialized knowledge, claim to be practising cosmologists.

Such a claim, unexceptional when applied to an 'exotic' primitive society,[1] is difficult to take seriously in relation to ourselves. In our society 'everyday' social activities seem to have nothing at all to do with the abstract, technically difficult, and imaginatively esoteric speculations of the professional cosmologist. The scientist's world is highly specialized. It seems both more profound and less 'real' than the world in which we all, including the scientist, have to live.[2] If we contribute at all to the imposing edifice of modern cosmology it is surely only in the negative and uninteresting sense of sustaining in our mundane activities the social world from which it departs. We provide, in other words, the resources of time and opportunity essential to the professionalization of scientific activity. We give the cosmologist the freedom to abandon the constraints of the 'ordinary' world. Historians of science, therefore, have felt justified on 'common-sense' grounds in concentrating their efforts upon elucidating the development of modern cosmology as an 'internal' monologue. It is held, thus, that the present state of scientific theory depends entirely upon its immediately preceding state.

Appearances are deceptive. The cosmologist cannot escape, in our society any more than in any other, the imperatives of a particular way of life. Sociologists have, for the most part, accepted

1

this proposition as an article of faith. 'Society', the paramount reality, must be rediscoverable in *any* aspect of human activity. *Its* constraints and conventions must, as it were, reappear in the most recondite of theoretical endeavours. Such contentions have generally been couched within the framework of inclusive 'theories' of society and depend upon very general arguments. But where sociologists have taken a more particular interest in the development of science it has frequently been as part of a defensive gesture bestowed upon hostile critics. As a synthetic discipline, that is to say, sociologists ought to have something to say about science, but in defending a 'sociological' approach they have usually been satisfied with rather vague blandishments in favour of the social 'context' of scientific ideas. They have said little about the meaning of scientific ideas themselves.[3]

If, however, the contention that we are all cosmologists is taken seriously, then neither an 'internal' history nor a 'contextual' sociology will reveal the full significance of either scientific or non-scientific forms of thought in our society. The justification of any particular sociological approach in this matter is the insight which it allows into specific problems. Thus, rather than engage in another 'methodological' or 'theoretical' dispute, the argument will be developed substantively and by example.

The question is specific but daunting. It involves nothing less than a sociological analysis of the bourgeois world view. Clearly, within the limitations of a single volume justice can hardly be done to any aspect of such a question. Yet it is a question that hardly allows of a more modest approach. Some initial distinctions, if they do not make the task any the more manageable, might serve at least as guides to the argument that follows.

The bourgeois world view, approached in terms of its content rather than its form, is first of all a qualitative division between 'cosmos', as the order of the material world, and 'psyche', as the structure of experience. Neither can be fully understood when considered in isolation. And the relation between them can only be grasped as a specific social relation. Although distantly related to Greek philosophical, Christian eschatological, and Gnostic spiritual terms, the distinction between 'cosmos' and 'psyche' will refer here to the relatively recent and peculiarly western division between, on the one hand, an external and 'objective' order of the material universe (however interpreted), and, on the other, an internal and 'subjective' world of personal experience. The synthetic and necessarily vague term, 'bourgeois world view', will be used in a general and non-technical sense to describe the cultural traditions and social relations for which this distinction was fundamental.

Although the bourgeois world view prides itself on its scientific description of the cosmos, when taken in relation to the psyche the persistence within it of other, non-rational elements become evident. These elements on closer examination suggest the coexistence, within the bourgeois world view, of four separate but related visions of reality. These distinctions can be indicated in a number of ways, but to draw particular attention to their 'deviation' from the orthodoxy of rational science 'subjective' terms have seemed the most appropriate.

Fun, the most radically 'acosmic' of these hetorodox realities thus opposes, by virtue of its absolute inner freedom, all fixed and ordered relations. It challenges the very possibility that reality might be formed into a 'cosmos'. The difficult and changing relationship between bourgeois 'rationality' and the un-compromising 'irrationality' of fun is, therefore, the subject of Part One. Attempts to neutralize, by describing, this radically antinomian spirit are analysed in terms of the successive forms (childhood, lunacy, and the savage), which, as prototypes of chaos, serve to express and contain its subversive genius.

In addition to repressing the dangerous libertinism of fun, the bourgeois world view has to overcome a longing for *Happiness*. It attempts to do this by ascribing to the society it has overcome (feudalism) an anachronistic fascination with transcendence. Feudalism it thus regards as a symbolic reality, and this particular reconstruction of feudalism forms the subject matter of Part Two.

Scientific rationality faces challenges as it were from the future as well as the past. Modern science and modern psychology are filled with an assortment of non-classical aberrations. This new challenge to orthodoxy is all the harder to deal with in being episodic and unpredictable. Part Four, *Excitement*, attempts to pin down some of these elusive objections.

Partly for aesthetic reasons, therefore, but also to highlight a widely misconstrued relation, Part Three, which deals with the central ideas of the scientific revolution, is titled *Pleasure*.

These terms are chosen as suggestive not only of different

'subjective' dispositions but as indicative of separate formal and existential 'worlds' expressed in, and expressive of, their own social relation. They contain and reproduce a series of other distinctions. Each must be approached on its own terms. Each part therefore tries to enter as sympathetically and completely as possible into *its* particular world. An order of *signs* is replaced by a hierarchy of *symbols*, before being reduced to a system of *causes*, which is hardly formed before dissolving into a network of *images*.

The bourgeois world view is only the 'messy' interaction of these incommensurable parts. This has the unfortunate consequence of making any attempt at comprehensive description appear 'unsystematic' and arbitrary: so much is unavoidable. To discourage the rationalizing enthusiasm of the reader the distinctive character of each irreducible element is emphasized in favour of any attempt to submerge such differences within a probably fictitious 'totality'. It may nevertheless appear as if Parts Two, Three, and Four, in forming a chronological sequence, should be read as a 'history' of the bourgeois world view, to which Part One is appended as a philosophical prelude. This is, emphatically, not the intention. For reasons that emerge in the course of the argument this is an 'interpretive' rather than an 'explanatory' work; and to the extent to which it succeeds in proposing plausible interpretations it makes explanations (historical or otherwise) less appealing if not actually superfluous.

The bourgeois world view, in other words, cannot be identified with either 'science' or 'reason' however they might be defined. Nor is it founded upon a single type of social relation. It is not possible, thus, to express it in a wholly systematic and non-contradictory manner. This might as well be admitted at the outset, and has the advantage of making the author's preference for *Excitement* (intellectual and otherwise), more obvious than it ought.

Part One

FUN

The tragedy of our age is reason.

Søren Kierkegaard

In every man there is a world, a universe.

Giordano Bruno

Chapter One

THE DREAD OF CHAOS

Life in bourgeois society is contained within a series of inescapable contradictions.

No description of the human reality of capitalism can resist, nor any history of its culture avoid, the brutalizing exclusiveness engendered by almost 400 years of inconclusive struggle. Its loving polarities, subject/object, mind/matter, theory/fact, form/content, being/nothingness, exchange/use, are so many ways of rendering experience coherent by dividing it against itself. In *its* world *all* possible phenomena are categorized through the successive invocation of a universal *Either/Or*. In this, of course, bourgeois society is not unique. 'Dual organization' is the central organizing principle of many cultures.[1] And the bourgeois world cannot be defined, therefore, solely in terms of such formalisms; it must be described by reference to the entire range of social meanings embedded in the antagonistic differences specific to its ideal order.

A direct analysis of these meanings, however, tends to reproduce rather than to interpret the very categories we wish to investigate. A cautious, elliptical approach has to be adopted. Before looking more closely into the favourite antitheses of bourgeois culture, the larger implicit distinction between the possibility of *any* such ideally ordered existence (reason) and its apparently inconceivable negation (unreason), must be examined.

Bourgeois cosmology includes, that is to say, not only the familiar, rational ordering of nature and human experience, but also and simultaneously a kind of negative image of itself. It is thus something more than that self-conscious 'world view' that generations of 'theorists' have tried to make safe for reason by assiduously excluding from it everything chaotic or frivolous. It was a process of purification that never quite succeeded. *Fun* could not be completely excluded from the rational cosmos. It persisted in the diminishing but never-eliminated residue of 'unexplained' phenomena. Its awkward presence was felt more generally, within the literary tradition at least, as an uncomfortable intuition of cosmic disorder.[2] And it challenged, with growing confidence, a century of metaphysical radicalism with a number of rather obvious 'facts' of experience.

If bourgeois society was the realization of universal order, then human nature must be, in itself, the embodiment of reason. The evidence of history told against such a view, but for the *philosophe*, secure in the new cosmopolitan world of commerce and letters, history could be viewed (with some difficulty, it is true) as a perverse story of ignorance and error. Those obstacles removed, reason, nature, and society could be linked together in mutually reinforcing enlightenment.[3]

More recalcitrant than the broad sweep of narrative history, the spectacle of children, savages, and lunatics posed special difficulties for any comfortable theory of human self-improvement. Human, yet by no means reasonable, each existed alarmingly within self-enclosed worlds of their own. *They* did not share, it seemed, in the universe illuminated by Newton and Locke. A brief sketch of the *philosophe*'s response to defiant irrationality reveals the extent to which they failed to exclude from the realm of enlightenment all that was dangerously incoherent.

Simplifying a good deal, we can say that throughout the eighteenth century, especially in Scotland and France, men of letters gave themselves to the task of self-examination. As rational individuals they could, in following an introspective method, recover a universal human nature. This ambition is as evident in the magisterial coolness of Hume's *Treatise on Human Nature* as it is in the flamboyant emotionalism of Rousseau's *Confessions*. Both in fact sought to expose, systematically and unreservedly (one might almost say carelessly) the elements of a shared humanity. In this project reason came to play a dual role. It served to designate both the common criteria of the human as a species, and the method by which such an enquiry should be conducted.[4] Without reason we could not be human, and in the absence of reason we could not begin to describe the special character of humanity bestowed upon us by its possession.

In other words, during the period of confident capitalist expansion, reason was not simply a logical or intellectual faculty. It was that certainly, but it was also more than that. Reason was the synthetic unity of human nature itself, and not just one of its 'powers'. It must therefore be the common property of mankind, and could not, by being made dependent upon a technical function, become a monopoly enjoyed exclusively by the educated. Reason propelling before it a spirit of democracy must constitute the very 'frame of man'.

The divisions proper to reason (the components of human nature) underwent continuous modification throughout the eighteenth century. The simplicity of those abstract dualisms which had their origin in Descartes' project of pure thought, gave way before the claims of moral and aesthetic sensibilities.[5] Terminology varied, but some such set of distinctions, comparable to that of *intellect, emotion,* and *will,* was general. Hume's *Treatise,* for example, was divided into three books, 'Of the Under-standing', 'Of the Passions', and 'Of Morals', corresponding in an approximate fashion to such a three-fold classification. The point of such schemes was to define the absolutely irreducible and 'simple' processes common to human nature; those internal 'mechanisms' which transformed the contents of a sensory manifold into perceptions of the world.

Human nature could be defined, therefore, as an internally ordered system of relations among intellect, emotion, and will. The Enlightenment might then be characterized as the choice of intellect, as the *medium* of humanity's synthetic unity. Reason progressively came to stand both for the irresistible expansion of knowledge and understanding, and for the mechanism of coherence unique to human nature.[6] It ought perhaps to be noted that there was no necessity in such a view. In principle, either emotion or will (or some more remote 'transcendental' faculty) might have furnished such a mechanism, as indeed they were subsequently held to do. Bourgeois culture, however, even at a later date when in full possession of Schopenhauer's and Freud's alternative reconstructions of the human subject (the first from 'will', the second from 'emotion'), remained remarkably faithful to its first choice; to Man as Reason, and to Reason as Thought.[7]

Socially, these internal divisions can be related to the major spheres of collective life; to circulation, to production, and to consumption. Such a connection can, for the moment, only be suggested intuitively.[8] Production and the will 'belong together' as the energy of all humanly creative processes. Consumption and emotion, dissolving into an access of pleasure or grief, are linked as the termini of processes of exchange. And circulation, especially during the eighteenth century, evidently shares with the intellect a realm of ideal freedom. Neither in 'thought' nor in the circulation of commodities can anything be created or destroyed. Both are perfectly conserved worlds filled with objects of terrifying abstraction.[9] The epoch of merchant capital is not accidentally also the era of enlightenment.[10] Deserting systematic metaphysics, however, the *philosophe* cultivated a form of literature (varied, self-possessed, humane) that exemplified the virtues of synthetic reason rather than those of 'cold logic'.[11]

These parallel divisions allow us to grasp more securely the significance of children, lunatics, and savages as prototypically disordered lives. There was more than academic issues at stake here. It was not just a game about the limits of conceivability. It was a matter of *conduct*. Reason, it seemed, could be negated and the human world inverted. Children, lunatics, and savages were human and yet they were not rational. They were living paradoxes whose existence undermined every certainty. Each, in addition, was non-rational in a particular way, so that together they formed a strangely logical sequence; an organized multiplicity in the forms of chaos; an underground system to mock the careful elaborations of the orthodox cosmologist.[12]

Unreasonableness might be approached, then, in two rather different ways: as a 'normal' human synthesis of one or more 'inadequate' faculties, or as an inadequate synthesis of normally functioning faculties. In practice, of course, both interpretive techniques were employed in the effort to shed light on these difficult subjects.

CAPRICE

'We know nothing of childhood,' Rousseau flatly declares, setting out to exploit his own ignorance.[13] Childhood for him, as for his contemporaries, had become a cultural enigma, and children held

10

for them all the fascination of an alien species. However exaggerated or oversimplified the claims of a generation of sociologists directly linking the 'emergence' of the modern nuclear family with the rise of the bourgeoisie may be, there seems little doubt that the image of childhood has undergone significant changes during the development of capitalism.

It was specifically in bourgeois society that an association between age and dependence was established. Aristocratic youngsters of the sixteenth and seventeenth centuries could lead lives as liberated and independent as those of their parents, while during the same period economically unfree peasants, unable to marry or to establish households of their own, were held to exhibit well into their thirties those intellectual and moral characteristics that the bourgeois age came to regard as 'childish'.[14] Liberated from the necessity of labour yet excluded from the adult social world, childhood became an increasingly puzzling phenomenon. Its sequestration was justified on the grounds of children's 'immaturity' and 'helplessness', on their evident need to be 'looked after'. Yet, as a general type, childhood could not be understood solely in terms of a 'developmental' process. The non-rationality of the child was not simply the absence of 'adult' qualities which would, in due course, make their appearance. Were they, for example, completely bereft of intellect? If so, then from what source did logical faculties subsequently spring? And what of the will? In some ways children seemed possessed of a will more highly 'developed' than that of the typical adult. Clearly there was not just a change in scale here but an internal re-ordering of the basic elements of human nature.

The key to the different 'structure' of childhood subjectivity was found in the special character of the childish 'will'. This follows directly from the fact of their exclusion from the process of production, which is the social form best adapted to the expression of the will. The child's will, in being wholly 'liberated' from the necessity of labour, appears perverse, violent, unpredictable, transparent, and insincere. As a critic of bourgeois domestic indulgence as well as aristocratic indifference, therefore, Rousseau complains that 'If we did not spoil our children's wills by our blunders, their desires would be free from *caprices*'.[15] And he notes elsewhere that, more properly speaking, children cannot be said to have desires at all. They want nothing in particular, and in expressing their wishes

11

they are subject to nothing more substantial than a whim. Children altogether lack that settled self-identity which is the precondition and consequence of the will. They can *only* wish; a free and lively mobility of feeling that is directed towards nothing but the growth of their own faculties (if they are left alone), or the borrowed vanity and presumption of their elders (if they are not).

Children's actions are remote from their 'real' needs which depend for their satisfaction upon the indulgence of adults. Wishing therefore, which is the subjective corollary to a dependent relationship, is held to be typical of their 'inner' life.[16] The 'natural' activity of children thus came to be defined as play. Locke's letters to Edward Clarke are an early recognition of this fundamental proposition. Children 'must play and have play-things', he insists – a view echoed from a different perspective by Rousseau and every child psychologist since.[17]

Play supposes a world of utopian abundance. As they are ignorant of the practical necessity of labour, there is no material reality to resist children's tyrannical caprice. In play all things become possible – or rather, nothing has yet become impossible. The 'object world', variously differentiated as the toys effortlessly conjured into being by the momentary exigencies of a game, is dissolved and re-formed without limit. Play treats the 'objective' characteristics of the world as the paraphernalia of fun. And, since its endless metamorphoses are purely subjective, the 'laws of nature' can be ignored or contradicted. Who has not at some time swooped effortlessly over distant countryside, enraptured by the liberty of unaided flight?

The play world completely absorbs the child. Within it nothing is difficult. Rousseau noticed that, while playing, children frequently 'endure without complaint, hardships they would not submit to otherwise without floods of tears'.[18] Fatigue is alien to its spirit of continuous originality. Even a simple repetitive game is never tedious; the hundredth bounce and catch of a ball is as fresh and lively as the first. The child thus engaged is unaware of our world surrounding and threatening his own.

This subjective freedom is neither private nor egoistic. It cannot be 'planned' or deliberately invoked. It has merely to be granted its own possibility to 'happen'. And if it does not 'come off', children complain of 'boredom' and ask in bewilderment, 'What shall we play?'[19]

12

Such a world is sensed rather than conceptualized. 'The child's first mental experiences are purely affective, she or he is only aware of pleasure and pain', claims Rousseau, anticipating Freud, and goes on to suggest that the child takes, 'a long time to acquire the definite sensations which show him things outside himself'.[20] The senses, indeed, are originally undifferentiated, 'in the imperfect state of his sense organs he does not distinguish their several impressions; all ills produce one feeling of sorrow'.[21]

The child's wishfulness is part of a general physical, moral, and emotional mobility. Children are never still, their lives oscillate wildly between extremes that we find difficult to comprehend. Seemingly incapable of the stoicism espoused by Hume or Ferguson, 'A child has only two distinct feelings, he laughs or he cries. The child's unreasonableness is dangerously radical, his "ideas", if he has any at all, are without order or connection. It is a veritable "sleep of reason", that function of the mind which "compounded" of the rest of our faculties "is the last and choicest growth".'[22] We cannot therefore infer from speech or action anything about the 'state of mind' of the child, 'the child's sayings do not mean to him what they mean to us, the ideas he attaches to them are different'.[23]

'Who of us is philosopher enough to be able to put himself in the child's place?' asks our celebrated author, discreetly proposing himself for the honour. But if Rousseau could celebrate the absolute inner freedom contained within the bourgeois image of childhood, others (almost all others), less secure in their possession of adult reason, were more ambivalent. Viewing *caprice* (mistakenly) as recalcitrant will, they set about breaking its resistance. Ideologically opposed to personal submissiveness, the bourgeois parent none the less recognized the necessity of imposing authority upon the child. It was no longer a matter of extracting tokens of obedience or affection, but of disciplining the anarchic playfulness to which children seemed naturally disposed.[24] If children were not to be forced to work, neither could they be permitted an unlimited licence to play. Schooling, therefore, was as much a means of 'rationalizing' play as it was of 'disciplining' work.[25]

The unreasonableness of children, that is to say, consisted in a perversity of the will, so their 'education' had to take the form of 'training'. No abstract pedagogic procedure could impart to the unreasoning child the correctness of conventional behaviour, nor any system of rules be made to appear to him as a logical necessity derived from some general principle. The child could not be 'reasoned' with. The benefits of enlightenment, which was the goal of the educational process, had to be held in trust by adults rather than explicitly invoked. It was a goal reached only by way of a long detour through mindless obedience. This was a view espoused by 'progressive' educators, who sought to subvert the natural process of play in order to harness it to rational ends, as well as by the traditionally brutish.[26]

Reason was introduced to the child, in fact, not through personal example or intellectual precept, but in the organization of the classroom. The classroom operated as a small, enclosed market. Each of its members, initially regarded as equal, was differentiated through a process of exchange (marks for exercises) which established a 'rational' order among them. Differences were instituted and reinforced through continuous measurement of 'abilities', in seating arrangements, in the allocation of tasks, in the ritual of punishment.[27] A 'universal' code of rules gave rise to the distinctions of individuality, and to inequalities justifiable by an appeal to a general principle of justice (reward for effort).

The practical schooling of children was inaugurated well before the practice of developmental psychology. The separate 'reality' of childhood was given recognition directly in response to the moral and political problems that it posed. It was only much later that such a reality became the object of 'scientific' interest and the literary imagination. Bourgeois society's 'serious' interest in childhood, its fascination with the dangerous romance of play, only developed with the adoption by adult society of 'civilized' manners.[28] The peculiar status of the child became marked by the domestic boundaries of separate mealtimes and special foods;[29] a particular geography and architecture (nursery and school); and by the emergence of 'experts' (governesses, teachers, doctors) who patrolled the ambiguous boundaries of the rational world.[30] Children were dressed with sumptuary precision in clothes suddenly appropriate to their years, and encouraged to enjoy the 'freedom' of approved games. Outside the classroom, the 'official' life of the child became in practice a heavily censored version of Europe's pre-industrial popular culture. The carnival, ruthlessly suppressed by the demands of the new rationalism, thus survived in a degraded form as the variety of 'childish amusements'.[31]

The emergence of the bourgeois family, with its internal non-market relationships of affection (however idealized a picture that might be), could appear at times to threaten the coherence of the society instituted on its behalf. Might not such a family become excessively introverted and, coming to dote too much upon its offspring, forget the larger world? The bourgeois individual's double role as man and citizen created a painful conflict. As a private individual he was indulgent and loving towards his children, but as a citizen he realized the necessity of education outside the protection of the family. This conflict, a version of the more pervasive but less well-defined contest between the freedom of play and the necessity of work, was resolved in favour of the latter. Childhood became a phase. A short, disorderly period safely contained within the categorical divisions of the domestic life-cycle.

Childhood, one prototype of the world of fun, could not however be completely annihilated. For a somewhat later period (though Rousseau once again anticipates the general tendency), childhood and its play world came to hold a somewhat different significance. Everyone after all has been a child, and though fortunately we cannot accurately or completely recall the experience, its presence as unreliable recollections preserves within us an unacknowledged residue of disorder. The more the bourgeois rationalist heaped upon childhood the degraded forms of subjectivity for which he no longer had any legitimate use, the more certainly he gave way to its caprice and the more intimate he became with its image of careless freedom. Hence the fascinating ambivalence of childhood as a suppressed but unconquered 'cosmology', as a way of seeing the world so radically opposed to reason that we cannot any longer remember what it is like.

Hence, too, the emergence of the autobiographical and confessional genre that became so much a feature of post-Renaissance European literature.[32] Rousseau is better known still for his *Confessions* than for *Emile.* It is the childhood of the author, as the hidden source of his creative inspiration, that interests us. Yet even the *Confessions* would have lost their savour if they had served only to 'explain' the artistic accomplishments of their mature author, or conveyed in the process nothing but gossip that has long since lost its value as scandal. We need share nothing of the author's immediate social world, however, to remain enthralled. What we share with him, which is much more important, is the longing to rediscover a primordial experience of ourselves, and to sense the richness of the world, sensuous, unbounded, and playful, which has slipped from us. The charm, similarly, of *Swann's Way* is the glimpse it offers of our own, rather than its author's, childhood.

The attractive naïvety as well as the infuriating temper of children consisted primarily in the spectacle of unformed will. Subject to the uncontrollable perversity of wishes, they were absorbed in restless fluctuations of mood and behaviour. Their 'intellect' could be no more than an amorphous pre-conceptual 'mind', incapable of those acts of discrimination and synthesis in which genuine reason existed. Childhood existed as a perpetual wish, yet the nature of the world and of man was ordered by a rational will. In a deep and perplexing sense, then, the world of childhood must be illusory; it was at any rate an object lesson in the shortcomings of any simplistic empiricist psychology.[33]

Childhood with its play world, in knowing nothing of nature, was reason's most unselfconsciously hostile critic. It was not, however, its only critic.

BEWILDERMENT

Pre-capitalist Europe was conscious of its long history of contact with 'alien' peoples.[34] The philosophical problems posed by the 'savage' were contained, however, for some time at least, within a generous view of of human universality. The plenitude of creation demanded nothing less than the existence of all *possible* variants of human being. The variety of the savage, thus, constituted for the sixteenth and seventeenth centuries so many marvels of nature: exotic phenomena to be collected and displayed in the *cabinets de curiosités* encouraged by amateur enthusiasm for the inductive sciences. They belonged with the newly founded botanical collections and zoological gardens as evidence of God's providential abundance.[35]

Problems appeared when the variety of customs and diversity of manners seemed to outrun any credible definition of human nature. Could the brethren of Noah, within such a short period, come to live so differently? Was it simply the case that, forgetful of religion, some had fallen into barbarism?[36] More serious still, for the eighteenth-century humanist, the savage manifestly lacked reason. And, as reason and human nature were mutually defining, the common-sense approach was to regard the savage as a literally inhuman or non-human creature.

Even among the *philosophes* such a view became popular. Lord Monboddo became prominent in the wide-ranging, serious, and inconclusive debates over the precise nature of recently discovered *orang-utangs*.[37] The idea that they might be a specific subspecies of the human seemed plausible to those who held it to have been already adequately demonstrated that in the same area there lived tribes bereft of all the 'arts of life: lacking in speech, the social conventions of marriage, domestic settlement, the rudiments of cooking, or the most primitive forms of exchange'.[38] In such a condition, regulated only by the periodic fluctuations of natural appetites, no moral or aesthetic distinctions were possible and *a fortiori* no form of reason could be sustained. Cannibalism, for example, existed among such peoples because no action had yet been defined as disgusting.

More moderate commentators agreed that savages lived in a state of perpetual intellectual confusion. Lord Kames, for example, who enjoyed considerable authority upon such matters, believed all savages 'capable of higher powers', and detected in reports of their customs a kind of rudimentary religion, a 'sense of Deity'. Yet so impoverished were their mental faculties that, lacking a sufficiently developed language, they were incapable of giving shape, in the form of explicit 'beliefs', to their diffuse feelings. About nature they were, not surprisingly, held to be 'grossly ignorant of cause and effect'.[39]

The savage lacked the most rudimentary of *concepts*. His language being restricted to the names of concrete and particular things, he was unable to carry out the elementary operations of comparison, generalization, or analysis.[40] As a consequence, he remained innocent of himself as an individual and as a moral being. His behaviour was inconsistent, his actions unreliable. The initial friendliness, for example, with which Columbus was met by the Caribs turned quickly and inexplicably to hostility. The treachery of the savage became legendary. It was not, however, malice (a rationally comprehensible motive) but thoughtlessness that made them wholly untrustworthy.

Savages had not really succeeded in separating themselves from nature, which was the first step in bringing life under the control of intellect. Their instincts, which were in the circumstances more or less adequate to the purpose, guided all their behaviour. Indeed, they had survived only by a quirk of nature. Living in lands of natural abundance they had no need to labour to procure their daily subsistence. That 'We possess already all that is necessary for our existence' points out the Tahitian native.[41] Unmoved by necessity, they could not make the initial effort of the mind which subsequently revealed itself as the first stage in the progress of the arts and sciences; the foundation of all human self-improvement.

Unresponsive to the prompting of reason the savage could be enslaved but not employed.[42] He might even be domesticated and, with training, become a loyal servant; but he could not, given his lack of calculative ability and insecure self-identity, survive if released on to the market-place. To become the beneficiary of reason the savage must therefore, first of all, be deprived of his natural livelihood. The colonial trader and the missionary justified one another.[43] In introducing the savage to the life of reason he must first be made to feel the necessity of work and the exchange of commodities.

Throughout the eighteenth century, innumerable tours were also conducted to imaginary islands whose fabulous inhabitants satisfied an insatiable demand for the exotic and threw into yet bolder relief the technical, moral, and commercial advantages of civilized life.[44]

The simplicity and brutishness of savages prompted, in addition, a number of conjectural 'histories' of the development of the civilized state. The authors of the best-known of such accounts divided man's putative history into 'epochs', arranged chronologically and systematically (narrative and system happily coinciding) according to the relative 'progress' of reason which was the accomplishment of each.[45] And even those scorning complacent histories, such as Rousseau in his prize-winning *Discourse on the Sciences and the Arts*, or those, such as Swift and Diderot, who used their hypothetical voyages of discovery as vehicles for satirical and critical intentions, concurred in viewing the savage as a 'natural' and *therefore* thoughtless being.

18

Rousseau subsequently developed his ideas on the nature of savage man in his important essay, *A Discourse on Inequality*. This is not an attempt to write an empirical history of human society; such an undertaking would be impossible if for no other reason than the absence of the materials upon which it might be based. It is, rather, an analytic description, suggested by the spectacle of contemporary savage life, of man's 'original' condition.

His method, he feels, avoids the major difficulty of the well-known 'contract' school who, in invoking a 'state of nature', succeed only in hypostatizing the less visible but no less conventional institutions of their own society: 'All those philosophers talking ceaselessly of need, greed, oppression, desire and pride have transported into a state of nature concepts formed in society.'[46] As clearly as he had seen the difficulty in describing the real experience of childhood, Rousseau takes seriously the gulf between ourselves and a presumptive 'original' state of man. Society (Reason) cannot be deduced from itself; it must spring from something other than itself. Rousseau therefore suggests a 'mechanism' through which society might have come into existence. This hinges on two related principles whose interaction led 'blindly' to the establishment of social life. 'The first gives us an ardent interest in our own wellbeing and our own preservation, the second inspires in us a natural aversion to seeing any other sentient being perish or suffer, especially if it is of our own kind.'[47] From these principles, which Rousseau takes to be self-evident, he deduces the possibility, though not the necessity, of private property, the division of labour, the development of language, and the birth of reason. 'In this way we are not obliged', he comments, 'to make a man a philosopher before we can make him a man.'[48] Reason does not 'develop' of its own accord as if by some inner necessity, it springs to life as an adjunct of passion: 'we seek to know only because we desire to enjoy; and it is impossible to conceive a man who had neither desires nor fears giving himself the trouble of reasoning.'[49] In spite of its heretical implication (that civilized man tended to an 'unnatural' suppression of passion in favour of intellect), Rousseau's vision of original man enjoying a 'solitary and idle' existence that demanded nothing of him beyond instinctual responses, seemed far from implausible.

The contemporary savage can be understood, then, not as an 'original' man, but as a being none the less in many ways closer to

19

that primordial condition than that of civilized man. His intellect is little developed, and the urge to self-improvement has not yet taken hold of his passions. His needs are simple, his wants easily satisfied. He lives *immediately*, incapable of the act of abstraction which so complicates our relation to the world: 'his soul, which nothing disturbs, dwells only in the sensation of its present existence'.[50] He cannot envisage anything beyond the 'end of the day', which is the 'extent of the foresight of the Caribbean Indian'.[51]

Sunk within himself, as absorbed as the child at play, 'everything appears to remove the savage man both from the temptation to quit the savage condition and from the means of doing so'.[52] The original impersonal 'happiness' of man is something unrecognizable to us; it is a world anterior to the metaphysical niceties upon which our own particular existence has been stretched.

Rejecting Rousseau's method of reconstruction, Adam Ferguson, at almost the same time, described the society of 'rude nations' in rather similar terms. A sober and reserved *philosophe*, he also rejected Monboddo's popular natural history. 'Men have always appeared among animals as a distinct and superior race', he notes, 'he is, in short a man in every condition; and we can learn nothing of his nature from the analogy of other animals.'[53] The distinctions among men are exclusively social and cultural, but between man in a 'rude' condition and the more developed 'barbarian' there is what amounts to a qualitative distinction, a classificatory difference even more evident in the later emergence of 'civil society' as a uniquely sophisticated way of life.

The social relations of the savage resemble 'more the suggestion of instinct, than the invention of reason'. And while the condition of 'rudeness' is not to be taken for a 'state of nature', it recalls the 'nascent society' to which some of the most imaginative pages of A Discourse on Inequality had been devoted.

Drawing on somewhat unreliable accounts of North American and Caribbean Indians, Ferguson tried to establish the general characteristics of those societies in which there is 'little attention to property'. Subsistence is there a daily renewed process of hunting and gathering. It is the difficulty rather than the ease in securing a livelihood directly from the wild that, he argues, prevents the emergence of the institution of private property and the intellectual elaboration that goes with it. The savage acts in direct response to his appetite, without calculation or co-operation. And since property depends upon a 'method of defining possession' and a 'habit of acting with a view to distant objects', he concurs with Rousseau in describing the savage world as an immediate and overwhelming reality confined within a timeless present.[54]

Knowing nothing of the exchange of goods (or the circulation of ideas), savage relations are formed from the 'commerce of affection'. Incapable of the intellectual detachment of cynicism or hypocrisy, their sentiments are open and honest. A gift, for example, is always the expression of a pure act of kindness: 'they delight in them, but do not consider them as a matter of obligation'.[55] Ferguson's admiration of their benevolence, fortitude, and skill in warfare does not blind him, however, to the 'childish imbecility', of their 'grovelling and mean superstition'.[56] Prompted by appetite alone they 'go in pursuit of no general principles',[57] are incapable of understanding their own surroundings, and fall victim to groundless fears.

The intellectual feebleness of the savage is the most striking feature of his unreasonableness. His will, in consequence, is weak and amorphous. When not pressed by an immediate need, savages do not even exercise themselves in play. 'Their aversion to every sort of employment', remarks Ferguson, 'makes them pass a great part of their time in idleness or sleep.'[58]

Neither children nor savages, it was held, have the possibility of *mastering* themselves. They were beyond reason. The former, in being excluded from labour, acted wishfully; the latter, subsisting directly from nature without labour, existed in a state of bewilderment.

DERANGEMENT

The nearness of lunacy made the *philosophe* uneasy. The child and savage could be excused on the grounds of 'immaturity' from the responsibilities of reason. But the adult who became a lunatic did so in open defiance of his already established nature. In turning his back on reason the madman created a glaring contradiction. Nature could not act against itself.[59] No self-possessed individual could deny himself by plunging voluntarily into the abyss of chaos; yet madness could not be other than unreasonable.

This had not always been the case any more than it had always been the case that children or savages had been seen in the image of disorder. Foucault's pioneering work draws attention to the inconspicuous origins of such an image and, in charting the emergence of new visions of lunacy, reveals the history also of a specific conception of reason.[60] The antithesis of reason and madness, Foucault argues, is founded upon the imperative order of the market conceived as a set of *logical* relations. The mad 'distinguished themselves by their inability to work and to follow the rhythms of collective life'.[61] The ability to labour becomes a badge of reasonableness. Those unable or unwilling to work are absorbed into the residual category of unreason and consigned to the safety of the asylum.

The seductive rationalism of the market is not restricted, however, to the organization of work; it embraces the social process as a whole. Foucault imputes to eighteenth-century writers a view of madness (as disordered intellect), which is characteristic in fact of a somewhat later period.[62] The *philosophe* viewed madness, first and foremost, as *derangement* of the affections or passions. This was a view quite consistent with the significance they accorded the passions in the 'moral economy' of human nature. [63] There were other and better examples of intellectual confusion and perverse will. The character of madness, then, should be sought first in the sphere of consumption (the social domain of the affections), rather than in that of production or circulation.

Diderot, one of the supreme polymaths of the century, provides for us in *Rameau's Nephew* a portrait of the contemporary madman.[64] Its central character is more than eccentric but is clearly not without intellectual gifts. His conversation indeed sparkles with cleverness. What is immediately striking, however, is his physical appearance and manner: 'At times he is gaunt like somebody in the last stages of consumption . . . . A month later he is sleek and plump.'[65] This variability is not the outcome of confusion or personal neglect, for he 'thinks of nothing but himself', and like a savage 'lives for the day'.[66] His behaviour is disconnected from the rhythm of nature felt by normal people as the regular periodicity of appetites. A chaos of wants is evident in his completely unpredictable behaviour. Unkempt one day, he appears luxuriously attired the next. He hardly sleeps twice in the same place. He pays no attention to the *value* of things. And beyond himself, 'the rest of the world is not worth a pin'.[67] It is a form of irrationality which begins, in other words, with disordered consumption, in a superficial and accidental relation to the object world.

Samuel Johnson provides a comparable and equally lucid exposition in his biography of that 'strange, gifted psychopath', Richard Savage.[68] 'His mind was in an uncommon degree vigorous and active', we are told, 'his judgment was accurate, his apprehension quick.'[69] The madness that afflicted him had nothing to do with perturbed thought, his intellectual powers were acute, and 'his attention never deserted him'.[70] It was, rather, 'an irregular and dissipated manner of life', already indicative of unruly appetites, that led him to become the 'slave to every passion that happened to be excited by the presence of its object'.[71]

Professional observers adopted a similar standpoint. Tuke, for example, went so far as to 'conceive that mind is incapable of injury or destruction',[72] claiming with Pinel that 'passions are the most frequent causes of mental aberrations'.[73] Affections as the 'inwardness' of acts of consumption might logically be disordered in two rather different ways. This gives to lunacy its basic division between melancholia, a disease of under-consumption, and mania, the frenzy of excess.

The melancholic suffers from passivity, lassitude, and lack of interest in the world. Since Elizabethan times, melancholia had been the affliction of the sensitive, the cultivated, and the scholarly. As a 'character type', it already enjoyed a rich cultural heritage, which Burton was able to draw upon in fashioning his medico-philosophical *Anatomy*.[74] The old humoral psychology, however, had undergone a profound change. It was replaced first by vague 'animal spirits' flowing through fine channels in the nerves, and subsequently by a more strictly mechanical image of organic fibres vibrating under various states of 'tension'. The melancholic are, literally, too relaxed: their nervous fibres, weak and flaccid, refuse the stimulation of the senses. The outside world, in consequence, is rendered vague and insignificant.

The nerve fibres of the maniac, on the contrary, stretched to breaking point, are set violently in motion at the slightest external stimulus. The delirium typical of the maniac is a consequence of the distorting and magnifying effects of these taut, vibrating nerves.[75] The tension of the nerves regulates appetite. Nervous afflictions could be treated, therefore, by physical means. This had been the orthodox view since Thomas Willis had linked hysteria to disturbances in the brain and 'nervous stock'.[76] The melancholic requires the world to be brought close to him, he needs the 'stimulation' of physical extremes. Alternating hot and cold baths was a good starting point. He could also benefit from a rich diet, narcotic tonics, and the distraction of music and lively company.[77] The violent gyrations imparted by a variety of specially designed pieces of apparatus might also prove beneficial. The 'raving' lunatic, conversely, needs to be constrained, secluded, isolated, and calmed. The 'vice or fault of the brain' was difficult to control and much ingenuity was spent in pursuit of effective measures. In this regard 'management did much more than medicine', and a period of calm allowing the 'natural' tension of the nerve fibres to be restored: 'confinement alone is often times sufficient'.[78]

There was no question here of 'suppressing' the emotions or affections. Each 'faculty' contributed to human nature its essential element. The emotional life, therefore, required cultivation rather than simple restraint. The significance of aesthetics in the Scottish Enlightenment finds its counterpart here in the adoption of specific codes of civilized 'manners'. The mode of experiencing bodily sensations itself became an aspect of 'cultivated' feelings.[79] And it is in just those forms (ordered, calm, predictable, friendly, and in everything moderate) that we discover the civic virtues of consumer rationality.

In setting the boundaries of rational consumption, melancholia and mania were, for the most part, associated with different ranks of men.[80] Melancholia remained an affliction of the privileged. Their material wants satisfied, they withdrew from society and enjoyed the company of their own imagination. Johnson, continually struggling against the temptation to melancholia, was saved by the insecurity of having to earn a living.[81] Mania, as an uncouth display of emotionality, correspondingly found its place among the pauper lunatics who came to litter the asylums of the nineteenth century. There were of course exceptions. Waves of a more fashionable frenzy swept the bourgeoisie from time to time. Hale, writing in 1720 and commenting on the consequences of the South Sea Bubble, draws attention to the numerous respectable people 'whose heads were turned by the immense riches which fortune had suddenly thrown their way', releasing in them a 'force of insatiable avarice' that quickly succeeded in 'destroying the rational faculties'.[82] A loss of cultivation was alluded to more cryptically by Pinel, who notes that 'The storms of the revolution stirred up corresponding tempests in the passions of men.'

Madness was not a matter of quantitative extremes alone. The affections were perverse as well as inflamed or flaccid. The lunatic consumes, immoderately, things that are worthless. Johnson, fearing that 'all power of fancy over reason is a degree of insanity',[83] conceives of a melancholic as a person who consumes only himself emotionally, who lives upon 'fancy'. Then, 'fictions begin to operate as realities, false opinions fasten upon the mind, and life passes in dreams of rapture and anguish'.[84] The maniac, on the other hand, has no emotional interior, he is completely 'open' and unmannered. His extravagant passions, expressing unbounded appetite, betray a complete lack of discrimination. He tries to consume the entire world, bewildered, his rage can only alight upon a succession of trivialities.[85]

Self-control was admittedly difficult to achieve. Madness, an ever-present possibility to us all, could be tolerated only by being confined. A hundred years of profiteering demonstrated the superior efficiency of a system of 'moral management' within the asylum.[86] The introduction of 'enlightened' treatment was little more in fact than a technical improvement in the art of incarceration. 'Therapy' was slow to be introduced and was rarely practised with conviction. The minute regulation of daily life within the safety of a 'retreat' ideally provided a Sallean 'schooling' of the passions. But while the child had first to submit to the authority of a master as a precondition of instruction, the doctor ought to forge a relationship of equality with his patient. He must trust in the humanity of the madman. This was an 'extreme' but consistent view, most frequently practised by way of some 'safe' and tentative gesture.[87]

During the eighteenth century, the prototypes of unreason were organized conceptually as caprice, bewilderment, and derangement; politically as education, slavery, and confinement; commercially as the school, the plantation, and the asylum. Each excluded from the general social process those incapable of acting in conformity with reason. And each was defined by inadequacy or perversion in one of those specific faculties whose synthesis normally constituted human nature. An imperfection in one faculty thus corrupted the others. The child was, to a degree, bewildered and deranged as well as capricious; but he was so *because* of his overwhelming propensity to playfulness. The general unreasonableness of the savage and lunatic stemmed similarly from their particular and respective deficiencies.

Within their secure enclosures the unreasonable were made the subjects of small Utopian societies.[88] Sheltered from the harsher reality of capitalism, they enjoyed a special kind of 'humanitarian' protection. The slave and the madman, like the child, was sufficiently 'helpless' to require constant supervision. They needed the perpetual care of professional custodians. It was an ideological opportunity that proved irresistible.[89]

They were contained but not forgotten. The fascination with unruly passion and the exotic, disordered intellect was undiminished by the growth of 'enlightenment'. Bedlam, for over a century, was a place of popular entertainment, an attraction as popular as any literary Bougainville and as morally ambiguous as the growth of a sentimental attachment to childhood.[90]

United as it was in the acceptance of the commercial world and the image of man's rational nature it supported, the enlightened bourgeoisie could not suppress an urge to look upon the world in its primordial nakedness. There was something of a sense of loss in their relation to the varieties of unreason, a hint of regret that gradually intensified into a longing, infrequently and timidly expressed, to renounce civilization. *Fun*, as reason's antithesis, became the subject matter of the new sciences of child psychology, anthropology, and psychiatry. It was through them that reason could be cleansed of its impurities, but it was also through them that a *frisson* of contact with the cosmology of fun could still be felt.[91]

Chapter Two

THE FEARFUL COSMOGRAPHER

The close and difficult relationship between Hume and Rousseau exposes the inherent instability of reason in the Age of Enlightenment.[1] Aspiring to the broadest freedom within the republic of letters, they pushed beyond the boundaries that more cautious *philosophes* had accepted as the limits recommended by nature to human conduct. But while Hume remained secure and at ease within an everyday world that his intellect had time and again completely undermined, Rousseau, the less radical metaphysician, could not resist the temptation to experience directly, as will and passion, the world's unreasonable aspect. Rousseau's emotionalism, his 'terrifying eloquence', testify to an existence just beyond the reach of conventional reason; an existence Hume could only grasp intellectually. Each clinging to the conventions discarded by the other, their mutual admiration as writers could not withstand the shock of a personal encounter.

Rousseau and Hume, in fact, from the perspective of a later period, become twin critics of the Enlightenment.[2] If reason were a genuine synthesis of human faculties, then it must be a goal beyond the reach of practical life. Most people, however, including most *philosophes*, were less exacting and less disturbed. The harmonious interrelation of will, intellect, and emotion was largely taken for granted. The subversive genius at the heart of the Enlightenment was ignored, and their writings, during the nineteenth century, became unfashionable.

The distinction between reason and unreason became simplified in accordance with new social imperatives. Reason, at once more specific and more practical, became defined as an exclusively intellectual function. It was not, however, simply another term for the intellect. Reason was the *instrumental* use of intellect; a practical intelligence. It was only in pursuit of a practical goal that thought became reason. More specifically, it was the form given to thought in the process of our gaining mastery over the natural and social world. Reason was both the precondition and consequence of our power to subdue nature; a power represented primarily by science.[3]

In this context the shrinkage in the scope of reason is associated with a growing awareness of the underlying productive mechanism of social life.[4] Capitalism is viewed more as the *means* of producing commodities and less as the universal system of their exchange. Reason, modelled upon this process of production, becomes the most general 'means' at our disposal. Distinguishing itself from the complex of 'civic virtues', it takes on a hard, unyielding aspect. As the social logic of production, it reduces all forms of unreasonableness to equivalent instances of the irrational. Thus, while many eighteenth-century writers might for the sake of literary embellishment draw comparisons between, say, children and savages, nineteenth-century scientists saw the development of child psychology, anthropology, and psychiatry as genuinely cognate disciplines.

This was given formal and somewhat belated recognition in Ernst Haeckel's 'biogenetic law'.[5] The conceptual condensation which allowed madness, for example, to be viewed as ontogenetic and phylogenetic 'regression', however, had taken place a good deal earlier. And, more generally, the theory of 'development' of which such views were a part was enunciated well in advance of Darwin's demonstration of its specific scientific validity.

Auguste Comte's *Cours de philosophie positive* might be taken, then, as one of the earliest, and certainly the most systematically ambitious, expressions of the new point of view.[6] It remains the most comprehensive attempt to render history intelligible through the use of 'intellectual evolution as the preponderant principle'.[7] Written during the 1830s, and delivered first in the form of private lectures, Comte assembled what he took to be the materials requisite to an inductive demonstration of his celebrated 'law' of the three stages. At the outset, he makes explicit the assumption which justifies merging into a single category of the irrational all previous varieties of the unreasonable. 'The point of departure of the individual and of the race being the same', he claims, 'the phases of the mind of man correspond to the epochs of the mind of the race.'[8] Both are to be measured and judged by the standard of the 'positive sciences'.

It is a standard established by the totality of human historical development. In opposition, therefore, to his Enlightenment predecessors, Comte insists upon the inherently progressive character of religion. The 'theological stage' through which all forms of thought must pass is not viewed negatively. It is not simply an obstacle to the attainment of a rational truth but a necessary stage in the development of a science that will ultimately free itself from finalistic prejudices. There is indeed nothing in the past which is not in some sense 'progressive'. This dogmatism is a consequence of Comte's conviction that sociology, in becoming a positive science, can express the entire history of humanity in terms of invariable laws. All discernible differences in modes of thought and ways of living must be reducible to elements within an unbroken series. All events must find an appropriate place within the continuous process that finally delivers reason into the world.[9]

In resting his entire 'philosophy' upon an historical classification of the sciences, Comte forcefully expresses the intellectualist vision that was to dominate the rest of the century. He insists that, even at the earliest stages of development, man requires the world to be conceptually ordered. And, in constructing such an order, Comte assumes the primitive 'must begin by supposing himself to be the centre of all things'.[10] Bewildered by the world of appearances, archaic man can organize the world only upon the basis of his own inner experience. His knowledge is exclusively knowledge of himself. It is his internal states and feelings, particularly the fear generated by his impotence in the face of nature, that prompts his first stumbling efforts in conceptualization. The most primitive form of thought is thus a direct projection of subjectivity upon the external world. This fetishism is the initial disposition towards the world of things.[11] Behind every natural phenomenon, each event or occurrence, there is held to stand some spiritual entity as its 'cause'. Fetishism 'allowed free exercise to the tendency of our nature by which Man conceives of all external bodies as animated by a life, analogous to his own, with differences of mere intensity'.[12] Subjective 'feelings', in other words, were held to account (directly) for human experience and (projectively) for the natural world.

The fundamental difference, therefore, between the savage and ourselves is not that we possess reason where he has none, but that our reason has been 'liberated' from the feelings in which it was originally entangled.[13] Even in the earliest times, man sought the causes of things. The savage reasoned, however, on the basis of false assumptions and untested judgements of reality. Only very slowly could the rational faculty establish its supremacy. That human history is the story of this gradual ascendancy Comte has no doubt. For him, it is self-evident that the '*development* which brings after it the *improvement* in human society' is a process of subjecting 'all our passions to rules *imposed* by an ever-strengthening intelligence'.[14]

This 'development' is not an unequivocal blessing. 'Savages, like children', he notes, 'are not subject to much *ennui* while their physical activity, which alone is of any importance to them, is not interfered with.'[15] The lassitude of reason is, none the less, much to be preferred to the terrors of fetishism. Comte purges himself of all sentimental attachment to the primitive. The primordial world has no value of its own. It exists only as an initial, faltering step towards the accomplishment of the rational control over every aspect of life. The underlying interconnectedness of nature making itself felt, as it were intuitively, through man's improving adaptation to the world, gradually expresses itself as a *system* of spiritual forces. When fully developed the powerful intellectual apparatus of the theological stage mutates into a metaphysical extravagance, the preparatory interlude to the wholly disenchanted positive stage. The prehistory of positivism is in fact reduced to a series of inadequate but inescapable anticipations of the final triumph of science.

ASSOCIATIONS

Victorian social thought, in calling on the scientific evidence of evolution, gave fresh impetus to what remained a broadly Comtean version of positivism.[16] Increasing emphasis was placed, however, on the character of the savage cosmos itself, rather than the narrative of reason's ineluctable progress beyond it. Since all such descriptions were, in practice, reconstructions by western academic rationalists of the unreliable testimony of missionaries and travellers, the result is liable to tell us more about the image of the cosmos the bourgeois scholar felt compelled to renounce than it is to inform us directly of the primitive world view.

An aura of impropriety clung to even the most austere of these works. In dwelling on the irrationalities of primitive life, a way was opened to discuss everything that enlightened opinion and good manners had already settled. It was therefore the ethnographic detail and not the theoretical reflections they contained that made such books popular.[17] Their neglected analyses were, notwithstanding, of considerable significance. If reason is defined instrumentally as efficiency, and in evolutionary terms as adaptation, then the unreasonable becomes irrational and can be nothing other than a form of maladjusted thought – an ineffectual and therefore misguided science.

Sir Edward Tylor's *Primitive Culture*, published in 1871, best illustrates these tendencies. As confidently as Comte, he holds to the 'scientific naturalism' characteristic of the period.[18] The present state of the 'industrial arts' in western Europe and America could be adopted without argument as the standard against which the 'level' of development of any other society might be judged.[19]

At various points, Tylor also makes an open comparison between the 'stage of thought' typical of the savage and of the child in modern advanced society. Children's games, for example, 'keep up the record of primitive warlike arts'.[20] And in children we see the same facility in the imitative function of speech that had proved 'so important in the formation of language'.[21] There is, in addition, a similar fascination with riddles, games of chance, and, most significantly of all, in a propensity to magical thought.[22]

The child, the savage, and our archaic ancestors are alike in their thought world being *mythic*; 'in our childhood', he assures us, referring to our collective childhood in archaic society as well as our personal infancy, 'we dwelt at the very gates of the realm of myth'.[23] And myth is an outgrowth of the same basic movement of projection that Comte had defined as fetishism. 'First and foremost among the causes which transfigure into myths the facts of daily experience, is the belief in the animation of all nature, rising at its highest pitch to personification.'[24]

This is for Tylor, however, an aspect of a broad philosophy of nature rather than a 'mechanical' reflex. The animism typical of children and savages springs directly from primordial intuition of nature. Incapable of reducing the world to an ordered set of relations, they yet strive to discover within it a meaning deeper than appearances. By the operation of an unanalysable intuition, proper names are given to natural phenomena; the sun and moon, for example, become the living embodiment of sexual difference. It is a cosmology 'deeper than metaphor'.[25] Against Max Müller and the dominant philological school, Tylor argues that, far from betraying an elementary linguistic confusion, the universality of such a vision of the world (the diurnal romance of the sun and moon, the living monster of the rainbow, the sky populated with ancestor spirits condensed into points of light) springs from an original and 'deep consciousness of nature'.[26] It involves 'a direct comparison of object with object and action with action', resting 'on a basis of real and sensible analogy'.[27] Myth, indeed, is more profound than poetry, which survives as its linguistic echo. And, in passages reminiscent of Rousseau, he argues that language itself begins in myth, in the adaptation of imitative sounds to proper names projected into the cosmos as the 'grammar of nature'. Through it, the savage does not simply communicate; he draws himself into the 'reflection of a mythic world'.[28]

The romantic and visionary side of Tylor's work has been somewhat neglected, yet there is little in Lang that is not taken directly from the master.[29] *Primitive Culture* is none the less fundamental, unlike Rousseau's *Discourse.* The Victorian moralist restrains and finally suppresses the initial tendency towards romanticism. Magical thought and the myths grown from them may be formed intuitively, but from the perspective of our own controlled reflection we can detect the associative mechanism which in fact guides savage thought. The association of ideas – that is to say, a principle central to our own rationality – covertly organizes the savage cosmos. It is 'a faculty which lies at the very foundation of human reason', and, he adds soberly, 'in no small degree human unreason also'.[30] Once formed, the most superficial associations tend to persist, 'frozen' into superstitious and magical beliefs. Tylor exposes the contemporary popularity of spiritualism, for example, as just such a vestigial belief.

The savage finds connections everywhere. It is all too easy to discover 'similarities' among things. The network of symbolic relationships expands alarmingly and quite unsystematically. New links are continuously generated, drawing the savage into a richly interconnected and interactive world. It is the association between experiences in dreams, reveries, and imagination, on the one hand, and the events of the 'real' world, on the other, that prove particularly powerful and pernicious. These metaphorical relations are taken, as are all relations of similitude, to be causally effective.

Reality and its representation is confused. The process of association, the psychological mechanism which carries the potential of reason, remains at an unconscious and therefore uncorrected level. It acts in an uncontrolled fashion as an intermediary between an original intuition of nature, and the completed mythological world within which everything stands transformed. Thus, 'any idea shaped and made current by mythic fancy may at once acquire all the definiteness of a fact'.[31] The underlying process remains opaque, and causal relations, uncritically inferred from every associative bond, are understood as spiritual forces. The cosmos thus becomes a web of immaterial forces giving rise, at various unpredictable points, to the sensible transformations of the physical world. Given that it is a 'theory' resting on nothing but accidental relations, the intimate cosmology of the savage is a kind of spiritual terrorism. The savage is helpless in the face of nature. Divining relations within it, he represents these unsystematically as the controlling spiritual forces to which he also is subject. Unpredictability is hypostatized; yet his security depends upon regularity. The savage creates for himself an uncomfortable cosmos; a reality more frightening than the 'raw' nature from which it sprang.

Tylor's work became the foundation, in conception and method, for a generation and more of ethnographic compilations. His ideas were continually discussed, often distorted, and rarely surpassed.[32] Victorian intellectualism found ample confirmation within the generous proportions of *Primitive Culture* before losing itself completely in the luxuriant growth of *The Golden Bough.* Victory over the German philological school secured, Frazer could go on to develop Tylor's method on a global scale.[33] First published in 1900, *The Golden Bough* was subjected to a continuous process of revision and enlargement. Not content with reviewing ethnographic literature from all parts of the savage world, Frazer arranged his sources 'developmentally' in such a way as to lead effortlessly into the more complex mythological field of ancient

Near Eastern civilizations and their contemporary degraded remnants that clung still to European soil in the form of peasant superstitions.

Frazer asserts the unity of human thought not only in terms of a single underlying mechanism (association) common to primitive and to advanced peoples, but also historically in terms of its content. 'The melancholy cry of the Egyptian reapers', he tells us in a typical generalization, '... down to Roman times could be heard year after year sounding across the fields. Similar cries ... were also heard in the harvest fields of Western Asia .... And to this day Devonshire reapers utter cries of the same sort'[34] – cries, that is, to the corn god, who might still be fleetingly sighted making his way across the edge of half-cut fields into wooded areas. For people 'unable to discriminate clearly between words and things', the natural world could take on a living appearance; it must certainly be experienced as a profoundly different place from the world in which we lived.[35] Nature, for them, was not a system of fixed relations. It was constantly, and quite literally, changing shape. The corn-spirit, for example, might appear and reappear in the guise of any number of apparently different beings. The savage cosmos was possessed of such an extraordinary liberty of metamorphosis that 'magical change of shape seems perfectly credible'.[36]

Frazer considerably elaborates on Tylor's basic insight into the origins of magical thought. The association of ideas can follow either of two fundamental paths. Similitude, which reduces to the simple idea that 'like produces like', is the foundation of 'mimetic' magic; and continuity, which assumes that things once in contact remain 'linked' by invisible forces, flowers into chains of 'sympathetic' magic. Both principles are slavishly followed as the invariable indicators of efficient causality, binding the world into an unbroken but chaotic network of determinism. It is to the latter that Frazer traces the spirit of scientific enquiry, 'Whenever sympathetic magic occurs in its pure unadulterated form, it assumes that in nature one event follows another necessarily and invariably without the intervention of any spiritual agency. Thus its fundamental conception is *identical* with that of modern science.'[37]

It is a conviction of determinism, however, in the absence of a suitable conception of nature. It is applied universally, to every accident of immediate experience, in complete ignorance 'of the nature of the particular laws which govern its succession'.[38] In attempting to render the complexity of empirical reality directly into a science, magic, rather as Comte had complained of political economy, 'systematises anarchy'.[39]

Again echoing Comte, Frazer goes on to suppose that this purely intellectual error was itself the stimulus to progress: 'the shrewder intelligences must in time have come to perceive that magical ceremonies and incantations did not really affect the results which they were designed to produce'.[40] As a technology, magic was a failure; the mere repetition, symbolically or actually, of specific events or circumstances associated with some desirable state of affairs was held to be sufficient for the reappearance of such a state. The coincidence, of course, is not generally repeatable, and it is this failure, Frazer suggests, which prompts the development of religion.[41]

Frazer goes to some lengths to distinguish the separate intellectual mechanisms and motives at the roots of religion and magic. Religion is not just the 'application' of magical thought to specific problems of life. Magic, he had pointed out, is in fact bereft of spiritual forces so that Tylor's discovery of animism was in reality a description of the most primitive form of religion rather than the beginnings of science. 'Our primitive philosopher must have been sadly perplexed', he remarks, 'by the impotence of his magical technique.'[42] This failure must be due to the *de facto* control of the world by beings more powerful than man, 'beings, like himself, but far stronger, who, unseen themselves, directed its course and brought about all the varied series of events which he had hitherto believed to be dependent on his own magic'.[43] Magical practices therefore gave way to religious rites – to rituals, that is, designed to influence by paying tribute to those spiritual beings responsible for nature.

In subsequently concentrating upon the development of religious beliefs and practices, Frazer renders the simultaneous development of 'rationality' a somewhat mysterious process. Is the growth of reason a slow liberation at the hands of the unusually insightful, or a series of happy accidents consequent upon misguided technologies? The latter possibility is suggested in relation, for example, to the Matlock Islanders who, conscious of the practical advantages of reliable navigational aids, '*incidentally* stumbled upon the elements of another science, which is astronomy', without fully understanding its rational principles.[44]

The rationalist rejection of magic can also be seen as an intellectualization of *its* world; a means of maintaining 'at a distance' our relation with its primordial irrationality. By tracing its distance in evolutionary terms from our own scientific concepts, a chain of continuity was established, lending empirical weight to the abstract similitude said to link modern rationality with the most primitive mode of thought. It was itself, in other words, a piece of intellectual 'magic', which served to legitimate the tangible closeness of the vanished cosmos, allowing us once again to sense the 'real and substantial bond' between words and things. Among 'civilized' people, it was not only the peasants who 'in their hearts [could] never really abandon their old superstitions'.[45]

Frazer was only the most prominent proponent of a widely held viewpoint.[46] Through a *theory* of animism, the Victorian rationalist momentarily escaped into the cosmology of fun. It is a universe within which all things become contiguous and the similitudes uniting all things spring spontaneously to light. The only law of change within such a cosmos is the unrestrained freedom of metamorphosis. Wholly enmeshed in boundless associations, the 'subject' is drawn into 'nature'. All the distinctions proper to his individuated ego are obliterated. The separateness of the human body is barely recognized, the distinction between 'inside' and 'outside' barely applies. 'Subjective' dispositions cease to be the primary experiences, upon which is modelled a picture of the cosmos, and become instead the real forces of nature.

Frazer and most of his contemporaries, however, shrank from the implications of their theories. They avoided direct contact with the cosmos of fun, preferring in the mechanism of association and the theory of animism a glimpse of chaos from the safe perspective of an undisturbed ego. This half-heartedness is precisely the weakness of their 'theory'. From Comte to Frazer (and beyond), an individuated 'bourgeois' consciousness lurks just beneath the surface of ethnography. The foundation of animism, the *projection* of subjectivity on to nature, assumes an elemental separation of 'subject' from 'object' which the evolutionary perspective had itself denied to those societies in which it was made to play a leading role. The whole process of projection, that is to say, is incomprehensible unless we assume that the savage, as well as his theoretical observer, is possessed of a typically modern ego.

It is as if the social theorist, suddenly finding himself in the midst of a primitive society, assumes that everyone else there has similarly arrived, as if in a time machine, just a minute or so before him. Bereft, but not forgetful, of civilization, they assess their predicament still encumbered by many of the psychological features of 'advanced' cultures. The world therefore becomes a frighteningly dangerous place. Without the control over nature exercised by instrumental reason, ethnographer and savage live in a condition of perpetual insecurity. Nature appears ferocious and unpredictable.[47] Magic develops spontaneously in a vain attempt to avert periodic catastrophe, illness, and hardship. Having failed, religious rites are instigated to enlist the aid of those responsible for forces beyond human control.

The conception of reason as the instrument of nature's subjugation and of the irrational as intellectual confusion, brings together the previously separated prototypes of unreasonableness within a single category. Madness, which had been viewed as a disorder of the affections, thus became a disease of the mind whose 'symptoms' might have been culled from the ethnographic literature.

Baron Ernst von Feuchtersleben's *Principles of Medical Psychology*, published in Vienna in 1847 and generally regarded as the first modern psychiatric text, is not, therefore, just a 'medicalization' of previous notions. He defines 'health' as 'the perfect adaptation of the body, without injury to its integrity, to the *purposes of the mind*'.[48] As man's 'material nature is not wholly material', madness is quickly transformed into a disease of the *mind*. Bodily appetites are effective only when represented to the mind as desire. And the mind itself, to remain healthy, must establish its own internal order as self-conscious reason.[49]

Just as Tylor, therefore, criticized popular spiritualism, Feuchtersleben attacked an earlier fashionable interest in 'magnetic' states.[50] Such altered states are 'not a more exalted but a more fettered state of mind'. No longer an instrument of self-conscious reason, the mind, like the body, falls under 'self-delusions'.[51] It destroys the 'free self-regulation' which alone is genuine rationality.

The common general cause of madness is traced to 'the erroneous combination of manifold ideas often united with the patient's own inclinations without his being aware of the error, or being able to overcome it'.[52] The patient begins to 'see' causal relations where none exists, and, in consequence, his perceptual world becomes distorted. The process, if unchecked, ends in individual cases of 'animism'. Examples of such 'mental alienation' abound.[53] Thus, the madman 'is not called mad because his brain is over-excited but because he judges and acts absurdly'.[54] Once the process of association becomes disordered, 'fancy rules without control' and reality, as the savage myth world, becomes 'a waking dream'. Quoting Reil with approval he argues that 'Fools have no ruling idea . . . . Besides their general craziness there is a remarkable weakness of all the powers of the mind, especially of the judgement.'[55] It is just in this intellectual feebleness that Feuchtersleben recognizes the 'regression' typical of madmen, 'amusing themselves and playing pranks like children'.[56]

The madman was then, like the savage, a poor scientist. He formed hypothetical connections among things without ever subjecting them to the test of experiment. The subsequent errors were never corrected and became magnified, ultimately into a wholly delusional reality. That uncontrolled association was responsible for individual cases of lunacy became an orthodox doctrine of the new science of psychiatry.[57] It survived to receive its most elaborate exposition in Eugene Bleuler's authoritative text *Dementia Praecox*. It was in this work that Bleuler introduced the term 'schizophrenia' to describe a loose grouping of symptoms characterized by a disorganization of the associative function.[58] The train of associations may lose its continuity, or be subject to periods of relentless spontaneity, to pathological 'blocking', or the 'flight of ideas'. Thoughts are not related and organized through a hierarchy of logical concepts, but are led from one to another by way of superficial and accidental relations: 'a senseless compulsion to associate may replace thinking proper'.[59] Thought is reduced to 'an automatic compulsion, quite independent of its content'.[60]

The same irrationality (the maladaptive absence of self-conscious control over the process of association) defined the world of childhood. Inspired by the different works of Darwin and Froebel, there emerged a specific genre of works on the psychology of children.[61] First-hand observations of children, usually the author's children, were adduced as evidence of the child's undeveloped powers of thought. The child was conceptualized as an undifferentiated sensorium undergoing a gradual process of organization into specific functions. The entire process was viewed from the perspective of the maturing intellect; it was its gradual discovery of logic which most clearly expressed the underlying process of development observable in all aspects of child life. A leading spokesman outlines the general approach: 'The infant is contemplated in the process of gaining command over himself. His sense organs gradually become available for perception; his muscles become controllable by his will. Each new acquisition becomes in turn an instrument of further progress.'[62]

The fact that children 'naturally' grew up gave to these works a somewhat different tone from ethnographic or psychiatric texts. The optimism of associationist psychology remained unclouded. The many arbitrary and 'mistaken' connections formed in the mind of children, their characteristically 'mythological' fantasies, their confusion over the limits of reality, were reported as the charming details enlivening otherwise sober works. Intellectual development, the central process of maturation, was guaranteed as the natural unfolding of an immanent power of thought.

The uncompromising sensuousness of the play world, celebrated by Rousseau and Schiller, was reduced, therefore, to a sensori-motor 'practice' for adult life.[63] The new pedagogy similarly subverted the play world by harnessing it to the goal of self-development.[64] Play, from being beyond reason, became the most powerful of educative technologies. This 'developmental' approach distorts the work of the most perceptive of pre-Freudian child psychologists. James Sully, for example, begins his *Studies of Childhood* with a perceptive analysis of play as the 'imaginative transformation of objects . . . the sheer liveliness of wanton activity'. But he was unable to free himself from intellectualist prejudices. The child, he insists, 'when at play is possessed of an *idea*'.[65] Play is the ill-formed medium more or less adequate for the expression of immature ideas. It is reduced to being an initial 'stage' of development whose significance rests solely upon its relation to later, more adequate, conceptions. Sully's original deeply felt chapter on play, like Tylor's introduction to the mythic world, is rapidly forgotten in the overwhelming onrush of 'rational' analysis: 'We are learning at last that the inventive fantasy of a child, prodigal as it is of delightful illusions, is also a valuable contribution to the sober work of thought.'[66] Play, in becoming self-conscious as thought, becomes 'like that of primitive races . . . a crude attempt at a connected system'.[67] At first 'a picturesque fancy, and a crude attempt at explanation', it rapidly gravitates towards 'sober reality'. [68]

Child, savage, and madman belong momentarily together as instances of the irrational. They are beings who have failed, or have not yet learned, to subjugate nature to human intentions, intentions to intellect, and intellect to science.

IDENTITIES

The mystery of growing up remained and could not, by an appeal to the biogenetic law, be ignored or avoided.[69] What governed the succession of stages in the ontogenetic development of thought? Did not the conception of 'stages' lead ultimately, in a formal sense at least, to the rejection of any notion of *continuous* maturation? Such difficulties (together with circumstances to be examined later) combined to promote a *Gestalt* approach to child psychology. In its nascent form, this is already evident in Comparyé's pioneering book, and it constitutes the real advance of Stern's work, both of which were written well before Bühler's theoretically more sophisticated examination of the issues involved.[70]

The attempt to grasp childhood subjectivity as a whole, and to understand its principles of organization from within, was not accompanied by any fundamental revision of the broadly functionalist and intellectualist standpoint from which the analysis was conducted. 'Development' thus becomes a series of discontinuous jumps and rearrangements; a progressive shifting of interrelated elements through which the familiar adult 'faculties' become crystallized. Such schemata were ordered, however, according to a certain vision of intellectual 'progress': an order established by an unswerving commitment to the norm of adult rationality as self-possessed instrumentality. What might easily be regarded as a fundamental change in the history of psychology is a much less decisive break with the past than it at first appears. Reason remains the *telos* of development, and is still conceived as the *instrument* of adaptation to the world.

The change in method and approach, however, is real enough and by no means restricted to child psychology. Piaget's description of 'syncretic' thinking in children, for example, echoes

40

Bleuler's depiction of the 'autistic' intellect in schizophrenia and Lévy-Bruhl's evocation of the 'pre-logical' mentality of the primitive. The irrational here is not simply the mistaken result of reasoning logically upon faulty assumptions; it is a 'mode' of thought which is in itself incoherent by our standards.

In Piaget's early work, as in Sully's, the child's world is both a play world and a world of thought. It is not, however, composed only of intellectual fallacies. We cannot understand the child's thought if we assume that it is nothing but a feeble version of our own mode of reasoning. The child's utterances, thus, should not be viewed as efforts to articulate feelings or intentions, but should be seen rather as part of the material resources of play.[71] Children of four or five talk together but do not really converse. Their speech is not 'adapted to reality', and 'creates for itself a dream world of imagination'.[72] Childhood thought, which is hardly thought at all, so different is it from our own, is fundamentally 'egocentric'. Adult intelligence and egocentric thinking, 'represent two different forms of reasoning, and we even say without paradox, two different forms of logic'.[73] It is a thought world arising spontaneously on the basis of primitive distinctions rather than being constructed on the foundation of conceptual classification. Piaget argues that children do not articulate their egocentric world to any great extent because they do not believe it is necessary to do so. For them, thought is as 'real' as anything else. And, not being distinguished from physical reality, it is therefore 'obvious' to all. Every child thus 'has his own world of hypotheses and solutions which he has never communicated to anyone.'[74] Young children live in a strangely lucid world. Like schizophrenics they believe themselves to be 'transparent' to others.

Logically relations remain virtually unknown. Egocentric thought jumps intuitively from premise to conclusion by making use of general schemata of imagery and analogy. 'The child', Piaget points out, 'does not compare perceptions but perceives comparisons.'[75] For the purpose of forming perceptual wholes *identities* rather than associations have to be established.

In childhood, that is to say, reality is not constructed from associations. The child does not venture general concepts upon the basis of chance contiguities or similitudes. He lives, rather, as part of a complex subjective synthesis within which everything is

41

immediately interconnected. New, unfamiliar objects may not arouse curiosity, therefore, because all possible objects are defined in advance as aspects of one, single, known world.[76] This 'syncretism' is just another aspect of the 'realism' children ascribe to their subjectivity.

Piaget, in spite of all this, is not prepared to abandon his general biological functionalism. 'There is nothing unintelligent in these schema,' he claims.[77] The 'condensation' and 'displacement' of child thought may be reminiscent of dreams rather than self-conscious intellect, but it is to the latter that these intellectual perversities is somehow 'aimed'.

Psychiatric studies similarly had challenged intellectualist assumptions. Bleuler, beginning his analysis of the schizophrenic with a conventional account of his confused and distorted associations, went on to explore the characteristic symptoms as expressions of 'autistic' or 'dereistic' thinking. It is a 'caricature' of reality, a 'compulsive' form of thinking which appears to the patient as completely 'objective'. His thoughts are assimilated to the work of nature. They are no longer due simply to an 'accidental insufficiency of logic', but are the consequence of an 'inner need'.[78] Delusions, he points out, as if directly challenging Frazer, 'have their psychological analogy not in error but in belief.'[79] It is the power of their personal (mistaken) convictions that creates the world anew for them, making of 'reality' a simulacrum of their ideas.

A new ethnographic perspective completes this reformed intellectualism. Lucien Lévy-Bruhl thus makes conceptual indifference the central feature of 'primitive mentality'. The primitive, of course, distinguishes innumerable 'objects' in the everyday world: if he did not, he would not survive. His mental world, none the less, sustains a good deal of the preconceptual innocence of the child. Although his world is filled with an overwhelming abundance of concrete differences, he is able to suppress abstractly all particularity in favour of an immaterial homogeneity that binds all things together. The primitive intuits a 'mystic reality' which, including himself within it, allows him to 'participate' directly in the fullness of the cosmos. Its underlying unity is 'felt rather than represented'.[80] Every conceivable thing 'participates in the same essential nature, or in the same ensemble of qualities'.[81] The intellectual ambition of the primitive is represented for us by Spinoza rather than Aristotle. The task of division and classification, where it takes place at all, is subordinated to the task of revealing in any phenomena the degree of the 'soul' possessed by its particularity.

To grasp the thought world of irrational subjectivity we must 'imagine a being, knowing nothing of the distinction between mind and body'.[82] Within such a world, animistic 'projection' is the basic form of cognition. And as the noumenal essence of things is directly perceived as at times both threatening and benevolent, the task of knowledge is to probe nature for its moral secrets. 'Everything happens as though nature were the outcome, or rather the reflection of a mental activity whose reasons or intentions the child is always trying to find out.'[83] Each object is related, not directly to another, but to the quantity of *mana* or *imunu* it contains.[84]

The psychotic, alone in his cosmos, discovers a world he does not understand and cannot control. He adopts animistic 'theories' as a measure of self-defence. The *necessity* of primitive thought becomes clearer in this context. The psychotic is identical with a world that threatens him with indescribable torments. As the world is contained within himself he must, if he is to survive, 'project' it into some other domain.

The simultaneous subjectivity/objectivity of experience is viewed by the outside observer as a hideously mistaken 'omnipotence of thought'. The psychotic 'introjects' the cosmos and then 'projects' its fearful, threatening aspect in a single, unanalysable movement. The most remote regions of space and time become intimately personalized. Equally, portions of the body may become 'detached' and appear as lumps of inert matter. Less ominously, in young children, the discovery that parts of the body respond directly to his wishes conveys the magical charm of 'action at a distance'. In the infant's smile on watching himself manipulate his own limbs, 'one has the impression of the joy felt by a god in directing from a distance the movement of the stars'.[85]

Through the logic of identities, the 'appurtenances' of the individual (secretions, hair, nail clippings, shadow, footprints, and so on) possess 'magical' attributes. They retain, directly rather than merely by association, the undiminished essence of the person. The 'action at a distance' typical of sympathetic magic is not a (fallacious) logical manipulation so much as an attempt at direct physical coercion. The distance is illusory. To manipulate the 'part' is in fact no different from operating with the whole. The *Deschagga* mother thus hides nail clippings and hair stolen from her sleeping child so that he will not stray, during the day, from the village.[86]

It is hardly surprising, then, that the names of things are also the things themselves. The word 'sun' contains the quality of hotness. 'What are names for?' Piaget asks obscurely of one of his young philosophers: 'They are what you can see when you look at things', comes the prompt reply.[87] Some, without the benefit of Anselm's guidance, claim that God in creating things simultaneously planted their names in men's heads. In like fashion, a man and his shadow, or the psychotic and his spiritual persecutors, are not conjoined by a mechanical association of ideas, so much as separately express the underlying unity of a single substance.

Intentionality and purpose, as soon as they can be felt as subjective dispositions, thus become nature's organizing principles. The sun and the moon follow the young child about. They move, therefore they are alive. And their movements are governed, as are the child's, by purposes given in the nature of living things themselves. The sun exists 'in order' to keep us warm and give us light. Rain falls 'because' we need water to quench our thirst.[88] Nature exists by virtue of the moral finalism manifest in the fact of our own existence. Life, which is the essence of things, exists to preserve the harmonious relations among its various aspects. The *Deschagga,* for example, conduct precise rituals in relation to bees, a species abundant in the human spiritual essence, in order to gain access to the reservoir of their own inner being, and in doing so reaffirm their nature as human beings.[89]

All things, appearing ideally as momentary aspects of an undivided substance, become interchangeable. Man, plant, and animal undergo endless spontaneous transformations. Every creature is potentially deceptive. A frail body may house a dangerous animal. A man might be a metamorphosed tiger. A child, intuiting such transformations, believes himself capable of flight. The psychiatrist becomes a 'fleetingly improvised figure' concealing the divine presence intent upon the destruction of his final victim.[90] These relations, once again, are not the outcome of distant metaphorical connections; they are aspects, rather, of an immediately sensed reality. It is only the existence of such a reality, indeed, that allows analogies subsequently to be drawn.

This is an intimate cosmos. Everything is close at hand because, directly sensed, it cannot extend beyond the range of the senses themselves. The child sees the moon resting in the branches of a nearby tree. Nothing can reach higher than the top of the tallest visible building.[91] The psychotic's intimacy with the cosmos grows more exalted. Expanding himself, he literally explores once distant regions of space. He too can reach up a hand and touch the stars. Time is no more a barrier to the senses than space. The ancestors are only our contemporaries who happen to exist in a somewhat different way from the living.[92]

The primal thought world is a veritable 'orgy of identification'.[93] It is constructed, formally, by the substitution of parts for wholes, rather than through the association of pre-existing elements. Identical predicates are transformed into identical subjects. 'The Virgin Mary was a virgin, I am a virgin, therefore I am the Virgin Mary,' claims one modern schizophrenic.[94] And Bleuler tells us of a man who claimed to be Switzerland; 'Switzerland loves freedom, I love freedom, I am Switzerland.'[95]

The formal 'similitudes' linking the world of children, primitives, and psychotics as forms of primal thought are the focus of a comprehensive 'morphological' study by Werner.[96] And the stress on 'intellectual development' recedes in favour of bringing to light the subtlety, depth, and internal coherence of each evolutionary 'stage'.[97]

Common to both associationist and identificatory forms of intellectualism is the view of primal subjectivity as responsive primarily to fear of the 'external' world. Reason is the 'proper' instrument of adaptation to the world; irrational beings who do not possess the intellectual 'maturity' adequate to the tasks of life are therefore conscious of the world as a threatening and terrible place. The practical exigencies of life are 'assimilated' to their own irrational view of the cosmos. Kelsen, for example, holding genuine causal thinking to be 'utterly beyond the grasp of primitive man', argues that nature is first conceptualized finalistically as retribution.[98] Revenge, he supposes, is a simple and easily understood motive, physical hardships are therefore viewed as the punishments for moral transgression.[99]

Associationism was overthrown, in part at least, because it assumed an 'advanced' individuated ego where none could be expected to exist. Lévy-Bruhl's great merit was to show that the savage did not project his ego upon the natural world for the simple reason that he had not yet discovered his ego.[100] Yet the process of identification appears often to rest upon a similarly individualistic account of 'motives' as the content of consciousness. Why should fear and retribution be assumed to be more 'primitive' than any other feeling? The child's anxieties indeed ought not to be confused with adult fears. They do not yet know of what they should be afraid.[101] And psychotics, Bleuler remarks, will describe the most horrifying delusions in a calm and 'unnaturally' detached manner.[102] More generally, since it is the unpredictable which is held to be a particular source of terror, the primitive can only be expected to recognize such a category of events as deviations from a view of nature as a continuously functioning and integrated order – the very conception, that is, denied to him by the intellectualist tradition.

The bourgeois world, of course, as a rational order, ought already to have transcended fear. The intellect, as the instrumental mastery of nature, ought to have superseded the magical or religious techniques of coping with threats to life. Yet fear, like play and superstition, persists. The fear that is said to motivate the primal cosmos is really a bourgeois unease 'projected' upon it as an unwelcome anachronism inexplicably persisting within the midst of *its* world.

Chapter Three

THE ROMANCE OF SIGNS

Has reason abandoned the relation between means and ends? Surely not. It is, none the less, tempting to argue that increasingly throughout the twentieth century the self-conscious identification of reason with rationality, and rationality with instrumental efficiency, has been, if not broken, then certainly weakened.

Our society, of course, is just as dedicated to the production of commodities, and our lives just as constrained by this necessity, as before. Yet in establishing its openly acknowledged and unchallenged supremacy over the social world, the entire process of production has been rendered strangely invisible. In being revealed as the most fundamental of social relationships, production has been stripped of any lingering enchantment.[1] It is *only* the most fundamental of social relations; important but uninteresting.[2] Reduced to being the *ultimate* cause of life's startling diversity, production relations are constituted as a brute, insensitive force.[3]

It is the extraordinary 'objectivity' of the modern production process which makes it both more fundamental and less significant than ever before. It provides, underneath all the interesting superficialities of modern culture, its own dull account of things; an explanation, appealed to on the point of conversational exhaustion only 'in the last resort'.[4]

Production no longer depends – or appears no longer to depend – upon a voluntaristic commitment on the part of all those whose activities are in fact required for its daily reconstruction. Demanding no more than passive assent it persists rather from neglect than from sedulous activity.[5] It is as if the goal of mechanical efficiency (the reciprocal interdependence of rationality and production) had been achieved once and for all, releasing the subject into a new 'liberated' existence outside of all exterior necessity.[6]

This is just another way of expressing the critical understanding of capitalist society as so many forms of 'alienation'.[7] It is within this perspective that a specifically 'modern' notion of reason emerges. 'Modernist' revisions have not so much transcended as viewed from a different angle the classical bourgeois notion of reason. Bourgeois rationality could be seen as just another 'version' of the process of alienation.[8] Society's vital processes, however, having been wholly 'objectified' as the direct expression of *natural* laws, its members could be allowed an unlimited inner, 'subjective' freedom of thought and feeling. 'Ideology' has consequently become oddly insignificant. This is not to say that 'values' have ceased to play a meaningful part in people's lives or that modern culture, in being radically secularized, has been simultaneously emasculated.[9] All such considerations have, however, been trivialized into 'aesthetic' categories, or the spiritual turmoil of private uneasiness.[10] Subjectivity would hardly have been liberated had it been otherwise.

Reason is no doubt still organically linked to the process of production. But it cannot any longer be defined as instrumentality. The new freedom of subjectivity has led to a certain blurring of once well-defined boundaries. Once it has been 'detached' from production, subjectivity is able to discover, or rediscover, a rational image of itself in hitherto neglected or degraded forms of life. Having passed, in turn, from the sphere of circulation to the sphere of production, reason (partially) evaporates into the sphere of consumption. The distinction between reason and unreason consequently becomes more difficult to sustain. Both undergo a thorough relativization.[11]

The primordial forms of experience untouched by reason have not simply been absorbed into general philosophical discourse; they have secured certain metaphysical privileges. Children, primitives, and the insane have preoccupied modern writers to an unprecedented extent. Their fascination has intensified. From being equivalent examples to a common irrationality, they have once again drawn apart and established around themselves a dense world of unique experience. They have become the centres of cosmologies unlike but related to one another and to our own.

What had been 'merely' the irrational is revealed as the 'human'; a fullness and purity of thought and feeling that reason had led us to neglect. Here are located those 'unalienated' worlds out of which our own was born, and from which it has become severed. Their thought worlds, therefore, could not be the antithesis of reason; rather, they constituted the plenitude from which the life of reason must itself be replenished. 'Irrational' beings, thus having a value of their own, were quite suddenly discovered to have 'rights', and to be due the respect we demanded for ourselves.[12]

Now that reason, in being stamped upon the world as man's 'second nature', has achieved an enviable security, man himself has been freed of the responsibility of taking it altogether seriously. The timidity of the Victorian theorist lay in his belief that the 'real' world still depended in some way upon his own efforts; upon his ability to think and act rationally. Now we are able to relax. Liberated by impotence we can begin to dream again. But we have almost forgotten how. No matter, the primitive lives nearby on the diminishing margins of the civilized world, the insane can be rescued from their hiding places and subjected to yet another tormenting examination, or we can simply recall something of our own early childhood. These primordial worlds are inexhaustible. And closeness to them can still shock our imagination back to life.

Reason has ceased to be an instrument of human perfection, or a tool of adaptation. It seeks dominion, not over nature, but over a realm truly its own. It is master of a purely subjective domain which exists for consciousness as a variety of *signs*. Neither synthesis nor causality are of much importance here. Those are the goals of reason's previous incarnation when, still ambitious to define a reality 'other' than itself, it was constituted as a 'mechanism' linked directly to nature.[13] Reason as a system of signs, however, is composed, as it were, from its own substance. It no longer points beyond itself. It seeks only the perfection of an inner order, matching by analogy rather than reflection the crystallized structure of the modern production process. The sign, as pure relation (humanity), has no material reality of its own and remains indifferent to the medium through which it expresses itself.[14] It is an elementary particle of human subjectivity, a quantum of meaning, and as such can be defined only in relation to other such elements.[15]

The modern conception of reason, then, seeks to perfect itself in the image of man as a consumer of signs.[16] It becomes the systematization of play, and in maintaining a strictly arbitrary relationship to the object world, finally frees thought from the dead weight of matter.

The social world, that is to say, as a process of production has become 'frozen' into fixed relationships 'independent of our will'. The human can be realized, therefore, only subjectively in the absolute freedom of aesthetic categories. The division between reason and unreason gives way to the gradation from the human to the social. No longer an instrument of calculation, reason presides over a reconstructed play world of signs. Its task is to illuminate the inner meaning of these signs by revealing the systematic relations they maintain one with another. The human sciences are dissolved in this process into the specialized techniques of *interpretation*.

Reason loses its clarity. A victim of crumbling dichotomies, it recognizes itself in the midst of its opposites. The primordial worlds of childhood or insanity or the primitive are equally rational and non-rational. The irrational, strictly speaking, has almost ceased to exist. Its only meaning now can be that of a 'meaningless' act; the contradiction of an empty signifier.[17] The semiotic view of reason begins with a clear distinction between the 'meaningless' and the 'inconsequential'.[18] It is, in fact, precisely in the trivial debris of consciousness and action that the modern psychologist and anthropologist have found the special signs of the human. It is just among those cultural objects, uncluttered with older rationalist associations and freed from the prejudices of historicizing mythologies and utilitarian banalities, that the fullness of reason can rediscover itself. It is, then, outside the realm of everyday necessity that reason can observe itself as the power of signification.

There is no condescension in extending humanity to previously stigmatized cases of the irrational; we do so entirely upon our own behalf. We hope to borrow back from them the unlimited freedom of signification in which we think our lives consist.[19] We have completely inverted the judgement of the Enlightenment. The primordial world is a semiotic wonderland. Linguistics therefore becomes the first of the human sciences, the discipline to which all others must turn for their method.[20] And it is in Ferdinand de

Saussure's *Course in General Linguistics* that we find one of the first clear statements of the modern conception of reason. The long trek through historical philology comes abruptly to a halt.[21] It is a meaningless story. Language is best understood synchronously as an interconnected system of signs, each of which is constructed arbitrarily and maintained by convention. Language resists 'reduction' to explanation in terms of the materials from which it is composed. It is a reality *sui generis* and, as such, furnishes reason with its model.

SIGNATURES

Within the order of signs constitutive of bourgeois culture, those defining personal identity have become particularly problematic. Bourgeois society presupposes a set of relationships among individuals, each one of which is conceived as an integrated 'ego'. The 'ego' is the function of reason within a personal context and operates by an apparently infinitely expanding memory. Memory and personal identity are simply different aspects of the ego's function. So it was with the problems clustering around the memory that a distinctively 'modern' psychology began.

Every act of memory has become guided by the overwhelming necessity of sustaining the 'self' as an imaginable entity; an entity which is in fact nothing but the operation of repeated acts of recollection. In this way the world is reconstructed from a unique viewpoint. 'There is no perception which is not full of memories,' Bergson insists;[22] so that, inverting the direction of the rational faculty, he can claim that 'our consciousness of the present is already memory'.[23]

Classical rationalism had insisted upon a practically demonstrative reason. Modern consciousness, aware of its own impotence, seeks to test itself inwardly. Kierkegaard, before Freud or Proust, thoroughly explored the baffling psychology of 'advanced' consciousness and constructed an autobiographical work, *Either/Or*, as a testament to its paradoxes.[24] Yet, in spite of himself, his writings turned into dialectical masterpieces that led, unpredictably but inescapably, towards a revised version of traditional religious salvation.

A purely secular psychology had to make do with memory alone. But, unaided, memory could not pierce the mystery of our own origin. The foundation of our identity remained a secret to ourselves, and the frustrating amnesia obscuring our early experiences only heightened the conviction of its overwhelming but undefined significance.[25] Nor could observational studies of other children adequately substitute for this failure of recollection. Other children (or even our own, for that matter), as they are reintegrated into bourgeois society as consumers (and are dignified therefore as at least quasi-rational beings), cease to exert their charm.[26] The simple fact of childhood amnesia, once reason had settled upon memory rather than foresight as its vehicle, became tormenting. Why did our active, rational memory fail to stretch back to the very origin of ourselves? It was not simply the distance it had to traverse. It could only be because our initial experience of the world was so profoundly different from that of the mature 'ego' that our memory could not grasp its 'otherness'. Our initial experience of ourselves was forgotten as a corollary to the growth of reason within the personalized psyche.[27] The primordial world was, in other words, literally and profoundly irrational; a pure sensuousness from which nothing could 'escape' into the disembodied world of signification. Such sensuous density could not simply wither away through neglect or distance in time. Its being 'forgotten', was a clear indication of the active role of memory in creating the ego from protean subjectivity. Reason began with an act of repression.[28]

Repressed, forgotten, but not annihilated, childhood lies still within us, revealing itself from time to time in small inexplicable 'accidents', or momentary, incomprehensible pleasures. Distributed over our body, expressed in sudden clumsiness or local paralysis, in tiny rebellions against our will, in unpredictable sensations of dissolution, there is a curious, mute form of personal memory.[29] Superficially no more than bodily jokes, these parapraxes, like dreams or minor neurotic 'symptoms', seem quite meaningless. But properly approached, with a relaxed, uncalculating consciousness, such tiny irrationalities spread within us a rich network of 'associative' connections.[30] Spontaneous recollections, which Proust noticed never failed to announce themselves in an anticipatory and quite generalized thrill of pleasure, are not quite meaningless. This is memory of another sort. It does not so much lead us back intellectually (and falsely) through an unbroken sequence of events to a presumptive origin, as reveal to us, with the vividness of an original perception, fragments of our buried past.[31] Bergson, as clearly as Freud or Proust, distinguishes between these fundamentally different memory-images, between the rational ego and the primordial world of pure sensuousness. 'The first, conquered by effort, remains dependent upon our will; the second, entirely spontaneous, is as capricious in reproducing as it is faithful in preserving.'[32]

Freud is often credited with 'demonstrating' the persistence of the 'irrational' impulses in normal and pathological behaviour; in fact he achieved something much more important. In succeeding in *interpreting* dreams, symptoms, parapraxes, he shows the manner in which the residues of our spontaneous recollection are formed into systems of signs. He thus extends a rationalizing principle over the entire content of the psyche.

Freud's initial approach to the problem was somewhat 'mechanistic'. In a famous phrase from the 'Preliminary Communication' to the *Studies in Hysteria* he claims that hysterics 'suffer mainly from reminiscences'.[33] A spontaneous recollection which could not be expressed as a rational memory is 'converted' into a physical symptom. This is almost always the result of some initial 'traumatic' event, the circumstances of which prohibited the discharge of a powerful affect which had been aroused. Rage, or shame, or sexual excitement, for example, which it would have been inappropriate to express openly, was actively and consciously suppressed.[34] The event was 'forgotten', but another occasion, perhaps years later, which bore some superficial 'associative' resemblance to the suppressed original, recalled the undischarged emotion. Now, however, as circumstances are never more than remotely similar, the originally 'natural' response no longer 'makes sense'. The affect is now discharged, however, in its 'converted' form as a physical symptom. The 'choice' of symptom is always extremely cunning. It both conceals and reveals. It is a physical 'representation' of the original event, but in using an alien medium expresses itself cryptically. Symptoms, as coded messages rather than conscious memories, always require interpretation.[35] They exist somewhere between rational, abstract memory and spontaneous, sensuous recollection. It is only incidentally that an individual may 'fall ill' from them.

Freud's analytic work, and especially his self-analysis, convinced him that such symptoms were always unconscious recollections of childhood events. For a time he favoured a 'seduction' theory, which he quickly abandoned, partly on empirical grounds but more compellingly in response to a deepening insight into his own method of interpretation.[36] Hysterical symptoms were not reminiscences of actual events, but of fantasies; wishes which it was just as necessary and just as difficult to suppress.[37] Symptoms, that is, recalled suppressed wishes by exploiting the conscious memory's associative network.

In all this there was little that was specific to the neurotic. The wishes preserved in symptoms were the common inheritance of us all. Their suppression, however, was usually so effective that when Freud reminded us of them many were unwilling to accept their reality.[38] The neurotic's troublesome reminiscences were the result of inadequate forgetting, and analysis was designed as an aid to forgetting more than as a technique of recollection. By revealing the meaningful connection between the symptom and the wish it partially expressed, the symptom could be made to vanish. This, however, was a temporary relief. A purely intellectual method could not serve to annihilate the wish which remained, temporarily thwarted, to pour itself into a new deformity. Just as in a joke, analysis destroys its effectiveness but does not eradicate the idea contained within it.[39] The aim of analysis became, therefore, a comprehensive re-education of the heart, rather than a mechanical removal of specific symptoms. A 'cure' could be effected only through transference and counter-transference;[40] that is, by a 'repetition' of the process of growing up through which the patient could finally forget those wishes which had survived his first attempt to do so.

Symptoms, then, do not join two events distant in time so much as connect two separate and continuously present realms of meaning. Like dreams they point simultaneously in different directions. In referring to some contemporary situation, they take on the guise of memory and address themselves to the conscious, rational ego; but in carrying within themselves the remains of older wishes, they preserve a psychic life that knew nothing whatever of the ego. To be interpreted, a symptom must be 'read' as a sign of this 'primary process'. Freud is at pains to point out that such a primary process is not to be identified with simple organic functioning. Sensuousness cannot, any more than memory or reflection, be grasped biologically.[41] It should be understood, rather, as wishes. The wish is a kind of approximation, tolerable to the ego, of the unfathomable primordial psyche. The primary process is *known* only as the unlimited mobility of the wish. In place of the fixity of the self, preserved in deliberative acts of memory, there is the absolute inner freedom, the 'polymorph perversity', of uncorrupted libido. It is the shock of this freedom that we feel as the thrill of spontaneous recollection.[42] Less intensely, but more regularly and predictably, it is in dreams that we re-experience the world of *fun*.

Reason breaks into this freedom, selecting through a process of progressive inhibition specific aims and objects to which wishes might be attached. It is misleading to see this process as the superseding of an original 'pleasure principle' by a 'reality principle'. No such simple dualism is implied by Freud.[43] Reality (which can hardly be a *principle*) is *created* from the primary process; it is a specific 'selection' from its infinite possibilities, a part of *its* whole. Nor should the process of inhibition which defines reality be regarded as the 'repression of pleasure'. It is fun, not pleasure, which must be forgotten. Pleasure indeed is the reward we gain for undergoing the process of inhibition.[44] As a relation of the ego, pleasure must await the loss of childhood to our conscious memory. Then, when we can no longer enjoy the immediacy of the wish, we taste its socialized afterthought as pleasure. Bourgeois culture, in Freud's view, being directed towards the psychic completion of the ego, is built upon the *pursuit* of pleasure.

The renunciation of fun is the renunciation, but not destruction, of the 'primary process'. In a deeper sense, pleasure may be seen as its continuation in an 'adult' form. The inhibition and differentiation of the primary process, its continuous fragmentation through the 'developmental' stages of psychic organization, and its establishment of the opposition between the ego and its 'object world', preserves an echo of itself. Pleasure is itself 'symptomatic' of primary narcissism, and all its forms derive from a common source. Pleasure is yet another form – in fact, the most general form possible – of recollection. The neurotic symptom is a physical perversion, a recollection of a forbidden wish. But normal pleasure is qualitatively the same; no pleasure is wholly 'innocent'.[45] In satisfying legitimate desires, the ego is in fact 'borrowing' psychic satisfactions rooted in the primary

not so sure about this

process. All pleasure leads back (whatever its function in consciousness) to the primordial, undivided world. Hence, once again, the necessity for disguise, censorship, and amnesia. Legitimate pleasure must be separated, arbitrarily but categorically, from the forbidden wishes of the primary process. If it were not, we would continually fall back into infancy.

If the connection between pleasure and fun were actually severed, the real aim of conscious psychic life would be lost. To disguise itself as 'innocent' pleasure, the primary process must make use of a richer set of structural principles than those allowed by Tylor or Frazer as constitutive of 'magical' thought. Processes of 'condensation' and 'displacement' conceal the real origin of pleasure.[46] These processes are 'playful' transformations of the 'rational' sign system of narrative memory. 'Condensation' joins together elements usually separated along what Saussure terms the 'syntagmatic' or linear axis of speech. In dreams, for example, events and places distant from each other in time and space can appear contiguous. And by 'displacement' any sign can be replaced by another lying along its 'associative' axis.[47]

Dreams, jokes, symptoms, parapraxes, are the different materials from which the same system of signification is constructed. Each effects in its own way the transformation of fun into pleasure, of primary narcissism into object love. In doing so, each threatens a dangerous proximity to our own childhood and its subversive freedom from the ego.

Childhood has become a personal 'signature', the hidden origin of ourselves. Within us it still wishes, inexhaustibly, for what it can never have. Happily, the advanced economies need more than the specific desires of the pleasure-seeking ego to sustain their passion for commodities; they need the insincerity of the wish. A little more of the child, therefore, can be allowed to reappear, and in our relation to the commodity world even a certain 'playfulness' can be encouraged.[48]

GESTURES

It is only superficially that the recollection of our childhood charms us. However much we might be attracted by the *idea* of intimate self-knowledge, we take care to shield ourselves from the primary process. Direct contact with the primordial experience of ourselves is an abyss of sensuousness from which we might never return. We protect ourselves therefore by interposing between ourselves and the uncompromising radicalism of fun a complex string of personal signifiers.

These signifiers serve also to protect us from the equally uncompromising abstraction of reason. <u>Those who cannot or will not experience themselves in terms of such images we call insane.</u> Here more than anywhere the historical opposition between reason and unreason has been in practice clouded, yet nowhere else has the conceptual language of its outmoded dualities persisted with such tenacity. The *opposition* between reason and madness has none the less finally succumbed to the terrifying conventionality of the distinction between sanity and insanity.[49]

The social logic of consumption, which is the general context of modern views of reason, cannot afford to be neatly prescriptive. As use values are held to be subjective and *therefore* arbitrary, it must begin by acknowledging a realm of conventionality. A particular but constantly shifting boundary of tolerable eccentricity is defined. This is a classification which cannot yield in advance to the establishment of a fixed 'rational' norm. It has to be allowed the general freedom of subjectivity in responding to 'fashion'.[50]

However insecure the boundary, beyond it insanity luminously beckons. In Kandinsky's circles, Artaud's physiognomic essays, Nietzsche's aphoristic genius, art, literature, and philosophy all suddenly side with madness as the realm of truth.[51] They share with the insane a vision of life deprived of its personal mythologies.

The tradition of Enlightenment has been finally and completely abandoned. For the eighteenth century, to be mad was to be under the sway of illusion; now it is just in the conviction of a certain kind of conventional illusion that we can claim to be sane. We cannot defend this sanity by a sincere appeal to 'reason'; it has deserted to the other side. In constituting itself as a system of signs 'detached' from the impurities of immediate experience, reason discovers in itself an analogy to the 'primary process'. Reason, that is to say, as pure 'mediation', enjoys the unlimited freedom of transformation among arbitrary signifiers. Being wholly abstract, reason accepts no practical limit upon the range of its internal self-reference.[52] And, as pure 'relation', it avoids the compromises and contradictions of empirical reality. Sanity exists, however, in tolerating the contradictions, inconsistencies, and incompleteness which has been expunged from the life of reason. Our 'signatures' are composed from just such impoverished and imperfect materials. Insanity, therefore, is as much a temptation to reason as a resurgence of the primary process. Both tendencies are visible in the abundance of modern psychopathology.

In using their inner freedom, perversely, to refuse the consolation of 'signatures', the insane become transparent to us.[53] Defenceless before the world, they act as passive recording devices of all its most fundamental processes. There is an appalling directness in the gesture of the insane. It is the truthfulness of their symptoms which frightens us, their helplessness as signifiers.[54]

The neurotic, burdened with partially discarded wishes, is too honest to accept the cunning of conscious memory and too demanding to be satisfied with the intermittent pleasure of spontaneous recollection. The assumption of continuity, implicit in the operation of conscious memory, is too great a leap of faith for the neurotic.[55] The neurotic tries to live without the illusions of time. Experience is decomposed into discrete moments accidentally occurring as a linear series. Each moment might be the last. None carries the promise of a successor which, should it materialize, might do so in some unimaginable way. Equally, however, as their lives are only 'gestures' to the truth, the neurotic tries to coexist with a conventional world in which he cannot believe. There is no escape to the playfulness of the instant. The work of inhibition proves irreversible. Instead, therefore, of a release into the atemporal paradise of fun (the primary process), the neurotic suffers the torment of anxiety, which is simply a fear of time. Anxiety manifests itself in the 'freezing' of movement. It is an inability to act. Each moment, heavy with doubt and possibility, threatens both to appear or not to appear. The neurotic's defensive gestures, the ritualization of behaviour, symptomatic obsessive acts, endless preparations for actions which never take place, prolong the present beyond its 'normal' duration.[56] Anxiety, like play, is open before the world of infinite possibilities. But whereas in play each momentary metamorphosis is without consequence, in anxiety each instant becomes an absolutely decisive choice. Reason is helpless; only the biographical fiction of an extended 'self', projected into the future by a reflex of the will, can guide the subject through such fearful discontinuities.[57]

'Hovering above existence', the neurotic in a sense retains an ideal humanity. Refusing to become one person rather than another, he contains, crammed into the anxiety of each timeless moment, the unalienated essence of endless possible identities.[58] The neurotic, to put it crudely, cannot make up his mind; or, more accurately, tries too hard to make up his mind. In attempting to 'think ahead', the neurotic suffers from a surfeit of reason as well as an excess of sensuousness.[59] He cannot 'realize' himself in spontaneous action because his 'self' exists as a kind of volatilized essence desperately leafing through a catalogue of its own future. In seeking to be led by reason, the neurotic comes to a standstill, unable to decide upon the correct path.[60]

The neurotic cannot rid himself of childhood. Endless metamorphoses, interiorized and made anxious, circulate within him. Attempts to 'solve' the problem (avoidance rituals, obsession, hysterical symptoms), rather than crystallizing from the flux of subjectivity a fixed personal identity simply make him more 'nervous'. It is tempting to interpret these signs as an appeal to be 'looked after'. But there is no hypocrisy here. Neurotic helplessness is more a measure of seriousness than irresponsibility.[61]

The neurotic is all terrified openness, unable to pick his way through the overwhelming complexity of the object world. For the psychotic the moment of choice lies irretrievably in the past. Everything is settled and complete. He must set about conforming the object world to his decision. He has traversed the entire length of the road upon which the neurotic cannot set out. He has become the unique individual which is said to be the goal of rational self-development. He has an absolutely clear and determined identity which 'reality' must vindicate. The psychotic withdrawal from the world is a logical transformation of the neurotic's anxious sign-system.

The psychotic 'illness' is primarily a disease of space. The literature reveals a truly formidable variety of examples.[62] The object world is dissolved into a plastic medium from which can be created, effortlessly, an entire cosmos to confirm and threaten the psychotic's chosen identity.[63] Not simply unique among other unique beings, the psychotic leads a solitary existence. He is the *only* individual, the sole survivor of a cosmic catastrophe. Empirical reality is a deceptive appearance populated by the 'fleetingly improvised' creatures 'miracled' up by his enemy.[64] He is continually threatened by the world he creates, which appears to him as the macabre invention of a demiurge. Spatial relations are arbitrarily transformed. He finds himself stretched across vast reaches of space. Distant stars are felt as the pores of his own skin.[65] But he might just as easily shrink to nothing. The interior of his body becomes a laboratory of hideous experimentation. It is metamorphosed into a series of mechanized contraptions. Schreber's description of 'miracles' perpetrated on his body is the most ample of modern pornographies. All those distinctions normal to the developed ego, self/other, inside/outside, body/world, melt away.[66] If *he* is wholly 'objectified' and fixed, then all else must be 'subjective', malleable, and transitory. The world is dangerous because it is never still; each contact with it threatens the frozen personality of the psychotic. Space itself is dangerous and must be contained. Where the neurotic seeks safety in the abolition of time, the psychotic, fearful of everything other and therefore beyond himself, annihilates extension. He takes the cosmos into himself and attempts physically to master it.[67] Nothing should be 'left over', no place remain uncolonized by the psychotic's expanding soul.[68] But realizing that he cannot succeed, he fears that the cosmos will master him, that he will be 'absorbed' by it, that already every other human being has been sucked into some hideous machine of destruction.

The fear of time and the fear of space constitute the fundamental axes of psychopathology, the signs of insanity.[69] In this respect confirming the judgement of the Enlightenment, the unreasonableness of the insane is manifest in disturbed consumption. The neurotic is too anxious to consume. He cannot bear the doubt of selection. He wants everything and has nothing. The psychotic, having already swallowed the cosmos, can find nothing else to consume and becomes a voracious anorexic.

Gabel ingeniously argues that these opposing tendencies can be readily conceived as respectively an underestimation, and an overestimation of the level of alienation characteristic of 'normal' social relations.[70] We cannot tolerate the truth of capitalism. We resist, psychologically, the fact of our alienation into 'objective' relations and live instead under the illusion of personality.

The neurotic goes too far in this refusal. He insists upon the real individual humanity of everyone he sees. He cannot act in a partial or fragmented fashion. He cannot accept the facility of stereotypes. His is a disease of sensitivity. Burdened with the duty of authenticity in a wholly unalienated world, he is overwhelmed by its plenitude. The psychotic, on the contrary, does not resist enough. In accepting the present reality of alienation, he refuses to accept the comfort of an imagined past. His life is absorbed into the general process of production. As the last human survivor, he realizes his predicament when it is already too late, and shrinks from a world whose touch would transform him into a lifeless commodity. Neither can tolerate the superficial inconsistencies of sanity.

Our normal personality is 'opaque'. It reaches towards the 'primary process', retracing its own path of 'development' by an indirect route. It exists in the small delusions of a personal 'signature'. By comparison, the 'gestures' of the insane are 'transparent'. Insanity, then, like childhood, comes to enjoy a privileged status, not as some exotic deviation, but as an exemplary instance of the life of reason.

REPRESENTATION

Signatures are 'rational' illusions, gestures 'rational' disillusions. In the age of consumption, reason spills over from the totality of the productive process (instrumentality) to reappear, fragmented and transitory, in the most 'perverse' forms of subjectivity.

The radicalism once espoused by reason becomes softened by innumerable relativizing tendencies.[71] The extremism of any view of life as an integrated totality is found only in the limiting cases of 'sensuous immediacy' (childhood) or 'abstraction' (insanity). Our view of the primitive has similarly been revised: from the antithesis to the epitome of rationality. Signatures and gestures are but fragments of modern culture, isolated forms of the new subjectivity of reason. The primitive, however, as a type 'beyond' our own society, provides us with an image of reason which can be both modern and holistic. Here are recombined into actual living matter the disparate tendencies of sensuousness and abstraction, the simultaneous and opposing temptations of the cosmology of fun.

Reason consists only in the relation among signs; and signs are indifferent to their material 'carriers'. There can be no society without signs and no thought without difference. Durkheim realized this very clearly and, abandoning his initial picture of simple societies as undivided unities, inaugurated the modern view of the primitive with an arbitrary act of classification.[72] The very notion of society implies division.[73] We will never arrive at an understanding of the rational foundation of society by an appeal to some alien form of 'necessity'. Reason, as a system of signs, must be understood from within; society comprehended from its very beginning as a reality *sui generis*. In claiming therefore that 'the classification of things reproduces the classification of men', Durkheim implies a common principle of division that is necessary to both.[74] It was this common principle which provided, he believed, an explanation of the universality of primitive religion.

The Elementary Forms of the Religious Life is, if nothing else, an important statement of the modern conception of reason. The necessity which reason and society share begins in an artificial act of separation: in the differentiation of the sacred and the profane. All societies, in order to exist at all, create and express an internal order. This order has no basis other than its own conventions: rules which cannot be deduced from the laws of nature or the principle of utility.[75] Durkheim claims in *The Elementary Forms* that, whatever their content, all conventions presuppose an original distinction, arbitrary *and* universal, which is the foundation of society.[76] It is the *first* division which creates the cosmos. All objects must be either sacred or profane: 'there exists no other example of two categories of things so profoundly differentiated or so radically opposed to one another.'[77] It is a transparently subjective distinction, since 'the sacred character assumed by an object is not implied in the intrinsic properties of this latter; *it is added to them*'.[78] The sacred is defined by a specific attitude and by rituals of avoidance which sets apart a specific category of objects from everyday use.

The sacred exists as the most general of social conventions. It is a 'pure' rule. It draws an arbitrary line and insists upon its being observed. The sacred is the *possibility* of society and evades any more specific definition in terms of its purpose or consequences. It is an empty rule, a demonstration of man's power to create social life.[79] In positing society, the distinction between sacred and profane also rouses reason to life. The social order, as conventionality, is rational not by virtue of any ideological judgement but because reason and society are composed of the same system of *representations*.

The sacred/profane distinction, as the first system of classification, is the principle of representation itself. Durkheim's students developed this insight in a variety of contexts.[80] Again, it was less a case of claiming that primitive thought 'modelled' itself upon a primordial social opposition than the discovery at every turn of divisions 'participating' in the arbitrary exclusiveness of the sacred. However 'obvious' or 'natural' they might appear to us, classificatory distinctions, such as left/right, living/dead, male/female, depend upon the continuous application of a social rule. Reason, liberated from nature, must be capable of rediscovering itself in *every* social relation.

The peculiarity of primitive thought lies not in its lack of differentiation as such, but in the non-specialized nature of its signs. 'Thought' does not constitute for primitive society a 'category' separate from any other relation.[81] In like fashion, the 'gift' is not a specific economic relation, or 'sacrifice' its peculiar religious rite.[82] The totality of society is represented in each domain of its activity, all conjoined through a common 'spiritual matter'.[83]

Magic is not then a kind of dissolution of society into a mass of uncontrolled identities. In apprehending *mana* and its cognate forms, the primitive generates a complex series of relations, each propagating, as it were, the conventional order of his society. The same conventions are discovered – in each division of the natural world, in every possible arrangement of its parts. Magic is neither inadequate science nor impoverished religion, but a form of perception. Its aim is neither explanation nor understanding, but representation.[84]

Magic renders the social process into a system of signs.[85] In pursuing this approach, the tradition of Durkheimian sociology slowly undermined its own foundation. If all representations in primitive society are versions of the same thing, the most mundane of acts becomes laden with the 'sacred' signs of social totality. The 'sociological' interpretation of the distinction originally offered by Durkheim (that the sacred represents the totality of society where the mundane does not) is lost.[86] In primitive society *all* is totality, therefore everything is sacred and the radical exclusiveness upon which all distinction was based dissolves into nothing.

Lévi-Strauss, sensing this difficulty, returns beyond Durkheim to Rousseau in search of the difference immanent in social life. In society there is nothing which is not a sign. Representation therefore is not a specific 'faculty' of social life to be grasped functionally in terms of its consequential 'solidarity'.[87] Durkheim's fundamental insight is embraced more radically. All social life is sacred, and its various aspects, which can be described broadly as a series of exchanges, are ordered *grammatically* as transformations of each other. The rules, implicit and explicit – for example, in the preparation and consumption of food – are just as basic to our understanding of primitive society as are the conventions ordering kinship, or hunting, or warfare. They are, indeed, the same conventions, variously 'embedded' in practical activities. Just as Freud could reveal the variety of childhood wishes in the arbitrary diversity of dreams, jokes, and parapraxes, so Lévi-Strauss seeks to uncover the 'savage mind' from the insignificant details of primitive social organization.[88]

Primitive cosmology is therefore 'lived' as much as it is 'thought'. The demand for intellectual order and coherence, for rationality, is just as imperative as it is for the bourgeois theorist, but the signs through which it is expressed serve simultaneously as the descriptive labels of the sensory world. The differentiation of animal species might be used, thus, to represent certain differences among the groups within a society.[89] These totemic relations are 'structural' rather than 'symbolic'. There is no direct relation between a specific species and a social group. A particular animal is not 'adopted' out of admiration for some specific physical or moral attribute imputed to it. Nor is totemism simply indicative of primitive 'confusion' over the boundary between the human and the animal. Totemic species, rather, form a system through which the order of society can be expressed.

It would be misleading, however, to regard totemism as only that. It is a system of representations rather than the represen-tation of a system in some sense other than itself. Lévi-Strauss never returns to Durkheim's functionalism. Totemism 'expresses' social relations only, as it were, accidentally. It is because social relations *are themselves* systems of representations that such analogies appear compelling. It is in both being constituted as orders of signs that they can be said to represent each other. The relations among separate domains of social exchange (food, women, stories) do not therefore replicate one another metaphorically, but are linked according to rules of grammatical transformation.[90] This formal linguistic rationality of representations is not 'added' to their primary function of designation and discrimination. For Lévi-Strauss it appears indeed that grammatical correctness was the first constraint upon human action.

Lévi-Strauss none the less dissociates himself from any such formalism.[91] Representations are always representations of something, and must always be considered in relation to some specific content. Like dreams, they can only be interpreted in relation to a complex ethnographic context.[92] The 'latent content' of all sign systems, however, is the power of representation itself. All specific signs embody the general distinction between nature and culture. Nature is transformed into signifiers. As a realm of necessity, nature is 'useful' primarily in providing 'reason' with a series of empirical differences (species, gender, season, and so on) which can stand as the 'phonemic' elements of culture.[93] 'Social structure', therefore, as the grammar of social life, 'has nothing to do with empirical reality'.[94]

The structural rules implicit in representations are as well disguised as is the primary process in our 'signatures'.[95] Lévi-Strauss turns consequently, like Freud, to what appears to be the most unpromisingly 'irrational' of signs (myth and ritual) to demonstrate the generality and coherence of such rules. Divested of its spurious narrative form, primitive myth is revealed as a universal form of self-consciousness. In analysing its complex internal relations, the anthropologist, as much as the psychoanalyst, is engaged in the pursuit of self-knowledge. Myth operates according to a logic immanent in the human power of representation. The commentary upon the myth itself becomes part of the anthropologist's material.[96] His own analysis does not escape the magic circle of reflexive signifiers. Myth 'absorbs' human thought in much the same way that play absorbs human activity. From its perspective, bourgeois science becomes yet one more transformative pattern within which to inscribe its code.[97] In being liberated from material constraint, reason first takes on the characteristics of 'subjectivity' (relativism, malleability, plasticity, the marks of consumption), before reconstructing itself as the formal relations which make subjectivity possible.[98] Reason proclaims, then, as its only necessity, a principle of inner freedom.

The subject as well as the object become predicated upon

reason. The exemplary types of unreasonableness are transformed in consequence into the most free and therefore the most truthful of signifiers. As the hidden foundation of individuality, as the frank inhumanity of social relations, as the universal power to generate meaning, signatures, gestures, and representations are three orders of signs 'in but not of' bourgeois society. They lie uneasily along the margins of the productive order. We often wish they might refer exclusively to a 'mythological' world beyond our reach. Each is in its own way quite dreadful, at once attracting and repelling the ego with their purity.

THE COSMIC BODY

'Fun' is the dissolution of all concepts and categories;[1] a subversive germ we cannot live without. In its absolute freedom, no principle is acknowledged or served other than the possibility of its own unrestrained mobility. It therefore exists as the limit to the known world. Yet we feel as a palpable reality its continuing presence in the very form and structure with which we lend order to experience.[2]

The bourgeois order, like any conventional order, is built first upon the renunciation of fun: in fact upon a stern rejection of the plenitude of *its* possibilities in favour of the creation, in reality, of a single conceptual and practical world. In bourgeois society, the contrast between this reality and its dissolution as fun has, for the most part, been conceptualized as the opposition between reason and unreason. Fun, however, evading all linguistic designation, playfully insinuates itself into the life of reason itself.

In the development of rational cosmology, fun has been more than a boundary of logical speculation. It has nurtured its own vision of the world. The cosmology of fun, paradoxical as the notion might appear, has remained integral (by negation, exclusion, division), to the bourgeois world view and appears time and again as reason's shadowy companion.

As a 'mode of thought', fun refuses to leave the body. It 'begins with the mouth' and does not flatter itself by 'escaping' into some other, more ethereal medium.[3] The body, as appetite, mechanism, or memory-image is its exclusive locus. Its cosmos is directly felt. Sensuousness is not simply the empirical 'basis' upon which its concepts are erected, it is itself the cosmos.[4]

The 'thinking body' is, with the establishment of capitalism, an evident contradiction. The body is here defined first of all as the source of unruly perversities, the origin of all unreasonableness. It is in a regrettable lack of self-possession that the irrationality of fun is manifest. Then, as reason is identified with instrumentality, the cosmology of fun takes shape as a universal 'undeveloped' consciousness. But as the initial point of a linear process of change, fun can be separated from reason only by an arbitrary act of classification. A process of 'rationalization' must then be present from the outset and be immanent in the variety of sensuous metamorphoses. The 'modern' forms of subjectivity, finally, have carried fun into the very heart of reason. The frankness, concreteness, and lack of constraint typical (but not definitive) of fun suddenly appear in a new light favourably contrasted to the 'frozen' artificiality of reason founded upon the ineluctable order of production.

In terms of its 'object-relations' rather than its 'consciousness', fun is unlimited playfulness. Its unfettered movement creates from the empirical world a theatre of dreams. There is nothing 'beyond' its own transformations to resist the prodigality of its own creativity. Its world is the momentary creation of an interior whim. It is action outside of the constraint of labour, which, as the 'reasoned' use of the body to serve its own 'needs', is dedicated to a single reality.

The cosmic body enjoys unlimited freedom without 'evaporating' into abstraction. As sensuous immediacy it performs 'intellectual' operations by metamorphosis. It is a direct sensing of internal relations we can only grasp conceptually. For a long time, 'advanced' rationalist thought found the primordial world of fun somewhat distasteful. 'Primitive thought' in all its guises appeared obsessively 'stuck' in physicality; in widespread mythologies about the origin of the world as a process of birth,[5] as neurotic wishfulness, or psychotic delusion, as the primary process. But now, as reason's double, it enjoys a new respectability. The body as a concretion of signs is a thinking mechanism, the first and most fundamental site of reason's architectonics.[6] In the differentiation of bodily feelings we feel the impulse of reason. The categorical opposition with which we began – mind/body, subject/object, theory/fact – begin to disintegrate. Each term is recognizable in the other. Metamorphosis is reborn. And just for a moment we believe that the primordial world is once again within our reach.

Fun, however, cannot be reached by rational action or reflection. It cannot become the aim of social life. The moment we deliberately seek out the infinitizing movement from sensuousness to abstraction and back again it dissolves into nothingness.[7] To make the attempt would be to suppose that fun is no more than a differentiated segment of a larger world of experience. But this contradicts the only thing we might 'reasonably' know of fun: that it is infinite. Our world is already contained within *its* countless possibilities.

Responding to a fugitive presence, a new style of art, the 'little phrase' of a sonata, a face glimpsed against a backdrop of sea,[8] we pursue, in order to make it our own, the object in which it seems embodied. We are disappointed. Fun is not to be mastered by deliberate appropriation. It remains tantalizingly beyond our grasp. Fascinated, we cannot stop analysing its litter of abandoned forms; a process which, playing host to the spirit inadvisedly sought in objects themselves, occasionally springs to life.[9]

Part Two

HAPPINESS

The universe, which does not exist through itself, cannot exist from out of its own nature.

St Anselm

It is God himself who loves himself in us.

William of St Thierry

Chapter Four

HIERARCHY

The bourgeois world view is not constituted by a simple act of rejection. The cosmic body (*fun*), with its absolute freedom of metamorphoses, must certainly be renounced in favour of the order of relations we know separately as 'cosmos' and 'psyche', but this might be said (in some form) of any complex society. The distinctive character of the bourgeois order must be defined additionally in relation to its own past, in relation to the society it has 'overcome'. This society had in its turn also been built upon the renunciation of fun. Its techniques of repression, however, were quite different, and it is largely in comparison to them that the unique rationality of capitalism becomes evident.

During the greater part of the period of the development of capitalism, the 'Middle Ages' was viewed unsympathetically as a prolonged but ultimately insignificant interruption in the continuity of western history. The two great periods of 'ancient' and 'modern' culture were separated and joined by an alien intrusion belonging properly to neither. And while the 'Dark Ages' might have been reserved as a term descriptive of the 'barbarism' immediately consequent upon the collapse of the Roman Empire, European civilization at any time prior to the Renaissance was seen as hardly less steeped in ignorance and superstition.[1] There was nothing of the innocence of fun in the irrationalism of such societies. In place of a primitive but 'progressive' system of magic there was a dogmatically absurd religion forced upon an unbelieving but uncritical populace, itself caught up in the greater absurdity of superstition. The two were mutually supportive. The power of the Church was a brutal fact no one cared to justify. It did not touch the hearts or the minds of the

73

masses, who tolerated its supremacy just so long as they were left in peace to take what comfort they could from flimsy and incoherent 'popular beliefs'. So that when the Church was at last moved to 'Christianize' its nominal congregation, its only hope of success lay by way of the Inquisition.[2]

This, of course, is a caricature of eighteenth- and some nineteenth-century historical writing. It none the less remains the case that a coherent account of 'pre-capitalist' society had to wait upon the maturity of rational theories of capitalism itself. Thus, during the seventeenth century the term *féodalité* was known only to lawyers.[3] And even when it became more widely used to designate a particular form of society it carried with it the pejorative connotation of an *ancien régime* riddled with abuses.[4] We can therefore avoid the task of reviewing the changing conception of feudalism; it is a term that belongs exclusively to the period of high capitalism and its aftermath.

THE PRINCIPLE OF FEUDAL SOCIAL RELATIONS

Otto Hintze, neatly summing up the prejudices of German historical scholarship, claimed feudalism to be 'a system brought about by the lack of rational institutional arrangements'.[5] If society was not generated and maintained through an appeal to the power of reason vested in each individual, how was it ordered? The answer lay in the development of two particular types of 'non-rational' social relation: vassalage and lordship.

Vassalage was a relation of mutual service and respect contracted among individuals of similar social standing. Its origins have been traced to the practice among warrior knights within the Frankish kingdoms of uniting under a freely chosen military leader. The *Gefolgschaft* was thus a band of retainers linked through bonds of personal loyalty and subordination to a leader.[6]

In periods of endemic warfare, the mutual benefit of such relations was obvious. It was under the Carolingians, however, particularly Pepin II and Charles Martel, that such practices became organized into a system of vassalage.[7] Here personal subordination (*commendation*) was directly linked to the granting of land and goods (*benefice*) thus ensuring that obligations of military service could be discharged. With the grant of a *fief* the vassal was not only enabled to meet his obligations of service and protect his own interests more readily than before, but he also came under the direct protection of 'immunities' granted by his lord in relation to that particular benefice.[8]

Relations of vassalage were held in high esteem. In spite of involving a 'reciprocity' of unequal obligations, they were regarded as being quite free of compulsion.[9] In becoming 'the man of another man' the vassal, in somewhat diminished form, took on the personal qualities of greatness emanating from his lord. 'I will love what thou lovest: I will hate what thou hatest,' ran an Anglo-Saxon oath of commendation.[10] Indeed, the law came in time to deal with disputes between fathers and sons, as if the fathers 'were the lords and the sons their men, bound to them by the rite of homage'.[11] Similarly, angels were sometimes represented as the 'theigns' of God. And the common attitude of prayer became that of *commendatio*, hands closed tight over the chest rather than open and outstretched.[12]

The vassal was not so much the instrument of his superior's will as the expression of his person. He did not therefore require the protection of legal rights held independently from his lord. The lord held 'immediate and direct power over the vassal', constrained 'solely by the notion of what was incompatible with the dignity of a free man'. However, as the vassal's dignity was inseparable from his lord's, no further regulation was deemed necessary.[13]

By the tenth century, in many areas of western Europe, the advantages of vassalage were much sought after. In order to exploit his fief, a vassal frequently became in his turn lord to a lesser vassal. At the same time, his lord might seek the security of a more powerful magnate than himself. Great men established ties of vassalage with many dependants, and inversely, the same person could become vassal to more than one lord.[14] Overlordship and multiple vassalage greatly complicated political society so that 'there began to be built up a vast system of personal relationships whose intersecting threads ran from one level of the social structure to another'.[15] It should be remembered, however, that vassalage was an institutional arrangement 'peculiar to the upper classes who were characterised above all by the profession of arms and the exercise of command'.[16]

The ideology of vassalage was invoked also to justify a quite different type of personal dependence: that exercised by the fief-holder over those unconditionally bound to the land within his benefice. These *seigneurial* relations were non-contractual, materially exploitative, and unregulated by common adherence to a code of 'honour' or 'dignity'.

Seigneurial rights were more ancient than vassalage. Manorial lordship in some form was continuous in western Europe from the late Roman Empire. The *villa* and *latifundia* were transformed, often becoming the property of the Church before being parcelled out as benefices through which the ecclesiastical hierarchy could extend its control over the secular elite by creating its own system of vassalage.[17] Throughout these changes, however, the lower orders remained the passive objects of rule: unfree and tied to a specific community.

The local lord enclosed the community in his own system of 'justice'. There was in principle no limitation upon the use of his power, though in practice his freedom became restricted by the gradual accretion of customary practices which came in time to bear the legitimacy of 'traditional rights'.[18]

The feudal pyramid was in fact two superimposed structures. At its apex was a ruling group of politically free individuals organized into loose hierarchies through voluntary and mutual personal relations. Beneath them lay the mass of the populace, contained within communities unconditionally subordinated to the rule of a feudal superior. Bloch points out that the essential difference between bondman and free lay in the separate system of justice to which each might appeal: 'On the one hand, there were the true subjects of the king to whom was extended, at all times, the protection of the courts; on the other, there was the mass of the peasantry, largely abandoned to the jurisdiction of the lord of the manor.'[19]

Even theoretically, the feudal hierarchy was never conceived as a single, unified, and coherent order; and in practice, of course, it is often difficult to discern even the most rudimentary principles of vassalage or lordship in operation.[20] The centrifugal tendencies of such a system were counteracted, though not always successfully, by the two major centralizing institutions of feudalism: the monarchy and the Church.

Kingship during the earlier period of feudalism was simply the political expression of military superiority.[21] Success depended to a large extent on the king's ability to establish powers of lordship over demesnes greater than any of his subjects, and therefore the possibility of creating, through relations of vassalage, the most extensive network of military support.[22] The king's position of supremacy created unique problems. If every man should be 'the man of another man', whose man was the king? To what authority did the king's command appeal? The hierarchical principle could not extend upward simply to terminate in the practical control of a mundane individual. The king must therefore possess or express extraordinary characteristics and kingship must itself be mysteriously shrouded in 'routinised charisma'.[23] During the twelfth century, when feudalism was well established throughout western Europe, the king was conceived indeed as a *persona mixta.* Like the hierarchy of the Church, he was touched by the divine presence; 'Spirit "leaped" into the terrestrial king at the moment of his consecration.'[24] He became 'another man', in nature surpassing all others; a gift 'attributed to him as an effluence of his consecration and function'.[25] It was specifically the ceremony of coronation, most dramatically in the case of Charlemagne on Christmas Day, AD800, which thus elevated him.

'Liturgical kingship' in the later feudal period became somewhat overshadowed by legalism, without however obscuring the fundamental principle that 'The power of the king is the power of God. This power, namely is God's by nature, and the king's by grace.'[26] This is not to say, of course, that the king could rule only by the explicit and continuing support of the Church. Feudalism, in most cases, was not a genuinely theocratic state.[27] In being consecrated, however, the king was able to justify his position as political leader and, more than that, *claim* an inexhaustible source of authority. Both emperor and pope claimed a divine inspiration, both acted as mediators between God and man, both realized, within the limits of possibility, a perfect nature. It was indeed just because of this theoretical requirement for purity that periodic reform movements were instigated in both Church and state.[28]

The interdependence of Church and monarchy was given formal recognition by the mid-eleventh century in the functionalist theory of the 'three orders' of feudal society. 'The City of God, which is believed to be one, is divided into three. Some pray, others fight, others again work. These three co-existing orders could not suffer separation. The services rendered by one permit the world of the other two.'[29] These distinctions, *oratores,* *bellatores,* and *agricolatores,* was itself, Duby suggests, a revival and synthesis of several different classificatory divisions current in the Carolingian period.[30] As a self-conscious 'ideology' it appears to have originated among the more senior French bishops who, together with the monarch, advanced it as a means of 'Christianizing' and controlling the knightly class.

The 'horizontal' as well as the 'vertical' links within the ruling group became, in consequence, more clearly defined. It was not just at the mystical peak of the hierarchy that a symbiosis of secular and ecclesiastical authority was established. At every level, complex interrelations based upon an exchange of spiritual services (the protection of prayer, the forgiveness of sins, the promise of salvation) for material and judicial protection (immunities, gifts of land, feudal obligation), were established.[31] Few were willing to risk death without remission of sin. Yet everyone must hold themselves in a state of perpetual readiness; death was a commonplace, but none the less alarming for all that.[32] Even the knight, who had conquered his secular fear, was seized of a spiritual terror that could be calmed only by the intercession of a religious expert. A continuous process of cleansing and forgiveness justified the 'privileges' of the clergy. Prayers from the lips of the vocationally devout were more likely to be effective than the hasty devotions of those brought daily by the secular defence of Christendom to the unavoidable commission of sin.[33] Ultimate salvation was assured by the magnificence of gifts and endowments to church or monastery. To die within the holy confines of either made doubly certain. Many thus arranged, well in advance, to spend their last days as monks; and not surprisingly (eternity in bliss was being offered in exchange), 'no man could hope to take the frock on his deathbed unless he had something substantial to give for it'.[34] The lower orders, who were less exposed to the danger of major sins, in order to satisfy their superstitious craving could, for the gift of labour and produce, gaze upon holy relics, and hope.[35]

These functional relations were often the focus of fierce conflicts, more often because the protagonists believed that a hypothetical reciprocity had been evaded or disturbed than from a conviction that the ideal system was in itself unjust. Sanctity, courage, patience; social virtues were no more homogenous than any other human characteristic. Man and his attributes were distributed functionally over the entire range of social activities, and ordered hierarchically in relation to a transcendental norm.

The feudal hierarchy cannot be viewed therefore as a 'continuous variable'. Those at its apex did not simply possess more of some infinitely divisible 'good' than those at its base. It was additionally, and more significantly, an order established among separate 'essences'. The serf was a different kind of being from his lord. The universal promise of salvation did not extend to all the bond of a common *human* nature. The 'culture' of feudalism was similarly dispersed. There were no common 'national' languages uniting its scattered elements. Latin, often of a rudimentary sort, expressed the freedom and mobility of the elite, while dialect and strong regional variation reinforced the localism of the lower orders. The social world was extraordinarily varied. The market was a local affair. Money, where it was used at all, was equally limited geographically, being no more than a local means, rather than a general medium, of exchange.[36] The date, even the time of day, were matters of local usage and local dispute.[37] Sumptuary laws regulated the appearance of different groups in an outward show of feudalism's *hylomorphic* order.

THE CHAIN OF BEING

The physical world in its entirety, the visible cosmos, was also constituted as a feudal hierarchy. It was indeed a more complete, unambiguous, and perfect instance of such an order. The gradations of being into which the universe was organized were seen as the system of natural subordination which had been imperfectly copied in human society.

The cosmos, that is to say, was more perfectly and completely ordered than was the social world. It was composed of a concentric series of crystalline spheres, the fixed unmoving core of which was the earth. Each successive sphere, moving outward from the centre, was physically closer to perfection, and had impressed within it bodies graded in the 'likeness' they bore their creator. The moon thus, its surface visibly scarred and imperfect, was none the less carried effortlessly and recurrently in a perfectly circular path, participating to that extent in a degree of perfection denied the earth. The planets ranged beyond were so many steps towards the sublime. The 'space' containing perfect physical objects gave way to celestial regions 'containing' the purer forms of angels and the Trinity.[38] The entire nested structure had motion imparted to it by the 'action' of the *primum mobile*, the sphere of Divine Being which was the transcending terminus to the chain of being ranged beneath it.

The interconnected movements of the visible heavens were complex and somewhat baffling. Each sphere moved in the only conceivable manner appropriate to physically perfect forms: uniform, circular motion. The transparent substance of each sphere communicated by direct contact its motion to the sphere immediately 'beneath' it. By varying the number and the relative viscosity of the jelly-like substance of each sphere, quite sophisticated physical descriptions became possible. Authorities, all of whom accepted this general scheme, disagreed over the number and precise qualities of each sphere. At the inception of the feudal era, few ventured to revise Bede's modest estimate of seven. His scholastic successors, however, by a process of cosmic subinfeudation, increased the number of steps in the hierarchy to fifty and more.[39]

This picture of the physical universe was not original with feudalism. Similar hierarchical systems had been commonplace since classical times, and medieval writers, particularly following John Scotus Erigena's translation of the Neoplatonic writings of Pseudo-Dionysius, borrowed freely from older sources.[40] During the feudal period, however, such schemes took on fresh moral and religious meaning. As a description of the cosmos, the chain of being was not so much the representation of a single physical structure as a 'polysemic' symbol within which the material world found its proper 'place'.[41]

The earth, the physical centre and therefore most distant point from the *primum mobile*, was man's temporary abode. Expelled from Eden by the reckless exercise of his freedom, man was condemned to exist in the midst of decay and corruption. His own physical annihilation constantly before him, he inhabited a world which, physically stationary, knew only the inner movement of generation and death. The material world was itself a reminder of the transitoriness of life, each glance heavenward, equally, an irresistible intuition of eternity. This was not just a matter of 'reading into' its physical structure 'extraneous' moral significance. Sin was *physically* part of the universe. The four elements composing the sublunary world, a promiscuous flux of earth, water, air, and fire, were contrasted unfavourably with the simple, immutable essence of the superlunary spheres. Both separately

and in combination, the earthly elements were subject to degeneration, so that all material terrestrial reality participated in the
sensuousness of continuous transformation.

The cosmos was divided hierarchically, not into regions of
space, but into the localities of *place*.[42] And each place constituted
an indivisible unity of moral and physical qualities. There were no
general 'laws of nature' unifying and linking together the various
places of the cosmos; such a notion was contrary to the very idea of
place. Its coherence sprang, rather, from the single unconditional
act of benevolence from which it was created. The cosmos is the
overflowing abundance of God's love.[43] Its divisions expressed the
variety of possible degrees in which He could express Himself. All
existence therefore shared to some extent in a divine essence, but
did so unequally, falling naturally into an unbroken hierarchy,
within which each being was dependent upon that immediately
superior to it.[44]

Man was not the least significant of God's creatures. Sharing his
earthly place were species even less exalted, those lacking a
rational, or even a sensitive soul; creatures and plants corrupted by
man's sinfulness.[45] The sphere of the earth therefore marked a
particularly significant gradation in the chain of being. Within it,
everything was tainted and corrupt, all its transitory physical forms
the temporary remission of disintegrative processes. Beyond it, all
material bodies were perfectly regular spheres, smooth conden-
sations of quintessential matter, carried in stately procession upon
their regular course for ever. Here there was no change or decay,
no undisciplined metamorphosis, only the timeless recurrence of
pure movement.[46]

Space was extended but limited. It was not void. Its internal
differentiation was a precise separation of *qualities*. Space and
object were defined as a mutually shared essence. The physical and
moral characteristics of matter belonged together, therefore, as
the bundle of properties 'natural' to some particular 'place'.

Qualitative divisions of space were made at every level and on
every scale. 'Christian topographers', for example, drew maps that
proclaimed a more profound truth than might have been guessed
from the accidental distribution of land and sea. On the authority
of Ezekiel, Jerusalem was placed 'in the midst of nations', the
spiritual and therefore also the geographical centre of the world.[47]

Pilgrimages were thus spiritual odysseys as much as physical journeys; the one implied the other. Movement from one place to another was simultaneously a change in the moral 'qualities' of the traveller.[48] The religious orders prior to the fourteenth century withdrew from the world, physically, by the enclosure of a wall, if nothing else, and spiritually, in part at least to satisfy the logical requirements of the theory of place. They needed to construct a place uncontaminated by human wickedness. A place appropriate to a holy 'relic' which, suffusing its aura of holiness throughout their community, would act like a beacon to pilgrims.

From the cosmic design of space we move effortlessly to the human significance of place. At every point such transitions were effected. The medieval cosmos, to put it concisely, was a feudal hierarchy. It was the ideal prototype of its social order. The idea of dependence, of a nested structure of subordinate forms, of a theoretically continuous chain of being within which every object found its natural place, of a self-inflicted gulf between, on the one hand, celestial beings invested with the dignity and freedom of their own motion and, on the other, sublunary mortals clinging wretchedly in their bondage to the 'customs of the manor', were images shared by 'society' and 'nature'.

It was a structure reproduced in the visible form of the cathedral. Above all a holy *place*, the cathedral enclosed, as it were on a minute scale, the entire creation. 'Chartres is medieval thought in visible form, with no essential element lacking,' Mâle tells us.[49] Its alignment, the obligatory symbolism of its figures in glass and stone, the position, grouping, and symmetry of its parts, the arithmetical proportions realized in its structure combined to make of it a realistic 'model' of the universe as a whole.

At the mid-point in the chain of being, man, the cathedral builder, was himself the most subtle and accurate replica of the cosmic hierarchy. He 'discovers in himself an analogue of the universe'.[50] The macrocosm is reproduced in the body, the microcosm, which replicates its hierarchical structure. The head thus represented the sky, the stomach the seas and the feet the earth. As a sensuous being, the human body was a mixed nature composed of the same four elements as every other form of degraded corporeality. Man contained within himself, additionally, in imitation of the most fundamental of cosmic divisions, a 'fifth' essence; a rational soul which, more perfect than earthly elements, was the 'sphere' of his privileged freedom.

Man was a point within which all the forces of the cosmos were gathered. Spoiling creation he yet retained a special significance within it. He occupied a privileged position, a place, from which the entire cosmos became visible. This is not another version of the cosmology of fun. The *immediate* experience of the body is not in itself a revelation of cosmic design. The cosmos cannot be 'read off' directly from the somatic engram. Nor are macrocosm and microcosm transformations of some more abstract rational 'structure' buried separately within each. The microcosm is, rather, a structure which informs man's rational soul *analogically* of the macrocosm, which cannot be known directly. The body is a 'natural symbol' of the cosmic hierarchy which exists as its essential prototype. Hence, when Alan of Lille claims that 'All created things exist for us as a book, a picture, and a mirror', he implies that alone among created earthly beings man has the potential for knowledge, a capacity fulfilled if we learn how to look and read.[51] And if man was a mirror of nature, nature was itself, 'a mirror in which man can contemplate the image of God'.[52] The entire universe does not lie, as it does in the case of the psychotic or the infant, wholly within the subject. What does exist within the subject is a picture of its own place (the body) which offers a means of understanding the larger structure of which it is *both* a replica and a part.

Just as there is no space which is not simultaneously a place, with all the moral and physical qualities appropriate to it, so there was no time which stood apart from the drama of creation. All the divisions of time possessed real and essential qualities. Augustine had already separated – duplicating the divisions of space and society – the City of God, whose time with the movement of the heavenly bodies was eternal, from the Earthly City, with its linear, degenerative, and secular time. Yet, since the promise of salvation was held out to all, earthly time must be superimposed upon something more profound; and just as specific places might become charged with spirituality, salvation promised the possibility of sacred time.

A series of analogies linked separate systems of time. The most fundamental relation was between events in the Old and New Testaments, a correspondence which was carried forward to the present as a series of repetitions or re-enactments of an original story.[53] The divisions of time followed the invariant order of Creation, reproduced in the ages of man and the ages of the world, whose final epoch, the millennium, was close at hand.[54]

Secular time was equally a divisible substance whose qualities varied with place and social relation. The time of the questing knight was very different from that of the monk or peasant.[55] This is true of course of our own society, but within the feudal hierarchy it was not a matter of a variable 'consciousness' of time detached from its 'objective' order, so much as a different ordering of the world in accordance with time's inherently variable qualities. Time was not in the least abstract. Its specific local characteristics could not be generalized. It struck no one as odd that day and night were divided into equal numbers of hours, so that in consequence the duration of the hour varied seasonally. Or that the bells announcing the canonical hours were never synchronized, even within earshot.[56]

There were no difficulties, therefore, over the 'literal' reading of Genesis. The six days of Creation marked the divisions of God's time, not a fixed duration on some human scale. Succession and duration, as with spatial order, were always viewed *symbolically*. The intellect sought to penetrate their secrets by isolating some particular instance and combining it with others through a general process of subordination.

The hierarchical structure of nature, as of human society, is a symbolic order. It is not just that each element in the visible chain of being has a place of its own within the hierarchical scheme, but each has to be understood as the physical emblem of a more essential reality that lies concealed within it. All empirical order is a mere copy of the hidden relations of Creation. Nature is best conceived as an order of 'motivated signs'. There is nothing arbitrary in the relation between signifier and signified. It is through a shared essence that a symbol comes into being and proclaims its fixed meaning. The bestiary as much as the stained-glass window conformed to an involuntary iconography.[57]

This is a cosmology opposed to fun. Everything is motionless, fixed, and immovable. The cathedral, with its complex inner divisions, defines an exact and unchanging order.[58] There is no possibility of metamorphoses; every form is possessed of its own inner, unique value. The observer within its midst is made aware as a result not of individual objects alone but of an entire system of

84

meaning. Drawn to some particular object or event the intellect, the imagination, and the senses are immediately connected to another, and another formed into an ascending scale. In being drawn into an object, the human subject, as it were, passes straight through it and fastens on to the immaterial presence it symbolizes. The intellect, then, is continually led away from reflection upon the physical world. It soars, as do cathedral walls, above and beyond material existence in search of its authentic transcendental object.

Any discernible bundle of qualities thus contains a number of different meanings. Jerusalem, for example, is in the historical or factual sense the town in Palestine. In an allegorical sense it is the Church Militant and a potent symbol of the unification of Christendom against its external enemies. Topologically, it refers to the Christian soul, as the 'place' most completely integrated with the moral and ethical teaching of the Church. It also means the Celestial Jerusalem, the terminus of human history and every connected system of symbols. It is this final 'anagogic' form that defines, additionally, the organized structure of dependency in which the various symbolic 'levels' are held.[59]

Hierarchical relations stand behind and condition the structure of appearances. Both empirical and symbolic relations point towards a transcending reality, whose necessary being is the precondition of all existence. Where the internal relations of fun can be pictured as a labyrinth of horizontal lines belonging to the same 'plane' of experience, the symbolic realm must be conceived as a set of vertical lines that lead beyond themselves to an absolutely unconditioned reality.

THE SYMBOLIC ORDER

The fundamental reality of feudalism, then, was an invisible and immaterial reality in relation to which the empirical world stood as its symbol. 'The world is a book written by the hand of God in which every creature is a word charged with meaning', such that 'in reading nature [we] read the thoughts of God.'[60] This was so much taken for granted that, during the medieval period, there was no general term for 'symbol' or 'symbolic'. The clerical designation *symbolum* was occasionally used in the sense of an article of faith, as in *symbolum Nicea.*[61] And Augustine had defined the more specialized notions of *signum* as *figura, imago,* and *allegoria* to describe the various 'correspondences' between human reason and the inner nature of things.[62] The task of the intellect was to recognize the similitudes obtaining between the world of appearances and the 'real' world that stood within and beyond it.

The real world being inexhaustible, its physical symbols could be endlessly replicated. The apparently endless multiplication, for example, of 'holy relics' need not shock us. What mattered was that the piece of wood, or the length of twisted metal, 'contained' the essence it symbolized – not that, physically, it was cut from the actual cross or had once, in fact, pierced Christ's hand. We are dealing here with faith rather than credulity, with a particular vision of the cosmos which 'consecrated' such remains. It is understandable, similarly, that medieval lawyers and scholars would not simply copy documents but would 'mould' them to some symbolic purpose,[63] or that litigation should be contested by symbolic devices.[64]

The symbolic hierarchy of feudalism only gradually became fully Christianized. In an important study, Le Goff has shown that the central relation of feudalism, vassalage, was established through a ceremony which drew many symbolic elements from the traditional institutions of kinship found among the pre-feudal Germanic tribes.[65] The entire ritual comprised three separate, related acts; homage, faith, and investiture.

Homage was inaugurated in a formal declaration of the wish to become a vassal. The subordinate places his joined hands between the hands of the superior which close over them in a 'polysemic' gesture expressive of 'instruction, defense, judgment and protection'.[66] The homage is not, however, a unilateral act of submission. It is an act of reciprocity. In clasping the vassal's hands the lord already displays a much greater degree of shared dignity than his less 'feudalized' Spanish counterpart, who allowed his intending vassal only to kiss his left hand. The homage is completed, indeed, by an open expression of equality, the exchange of a kiss. A mutual oath is then sworn on the Bible or some saintly relic and the vassal is publicly declared the lord's 'man of mouth and hand'. Le Goff emphasizes this formula in relation to the significance of the body (microcosm) in medieval ritual. The investiture is completed by the lord conveying to his vassal an object representative of their new relationship and its mutual obligations. A large number of possible objects were available, most of them connected fairly obviously with the vassal's new privileges in relation to a fief. Du Cange lists no less than ninety-nine, which Le Goff classifies according to their domain of reference.[67] Most commonly, their mutually supportive connection to the land was invoked (a clod of earth, branch of a tree). Alternatively, the investiture might be by a bodily gesture or 'appurtenance' symbolic of their social position (by the finger, toe, hand, hair, belt, gloves, and so on). More rarely, though not uncommonly, if an ecclesiastical community was involved, some specific 'socio-professional' symbol conveyed new duties and obligations (by bells, keys, or books; or by sword or lance; or in more lowly cases by knife, pitchfork, or sharpened pike).

These complex exchanges carefully regulated the establishment of new relationships. The two participants were quite precisely defined within the symbolic system, and through its ritual inequality is both expressed and overcome. Exchange, if it is to be genuine, must be between equals. But the essence of vassalage (and of the feudal relations modelled upon it) is inequality. Hence the necessity of elaborate symbolic manipulation, including the designation of a special place (a church or great hall) in which to perform the ceremony as a public spectacle.

Such rituals attracted, over a period, a penumbra of Christian symbolism. The religious element merged with the pagan before establishing its iconographical dominance.[68] This was less a direct borrowing of ecclesiastical authority than a consequence of invoking symbolic forms (at first their own) which made transparent an ideal hierarchy of essences in relation to which the Church came to have a special claim to expertise, and whose own forms then became the natural idiom in which to express everything profound. Subtle differences, however, remained. The *osculum* of fidelity was neither the liturgical kiss nor the kiss of peace, but just as effectively crystallized the entire structure of feudalism.[69] And as certainly as the more explicitly religious ceremony of dubbing, it captured and expressed its spirit.[70]

The sustaining of reciprocal but unequal relations is the fundamental political problem of feudalism. It was dealt with in two different ways. First, as in the ceremony of vassalage, inequality is denied through the creation of a purely symbolic realm within which genuine and spontaneous reciprocity can be practised. Or second, as in the theory of the three orders, differences in power and wealth are 'converted' into qualitative and essential differences between ranks of men. Thus, just as the superlunary has dominion over the sublunary world, or man has dominion over other earthly species, so a nobility ought to exercise its dominion over a dependent peasantry. Inequality, that is to say, may be ignored in order that exchange can take place, or exchange is denied in order that inequality can be justified and sustained.

The symbolic order was not an 'ideal' we should strive to realize; it was already fully realized. The symbolic order was a weak imitation in physical form of the order of reality itself. This reality, hinted at in imperfect actuality, could be reconstructed more fully through the operation of its privileged 'mirror', the human mind.[71] The fundamental nature of feudalism is therefore exposed most fully in examples of its 'abstract' reasoning. Hence theology, even more than law, is the real 'science' of feudalism. In it we can find, free of historical complication, an analysis of its distinguishing social relations.

St Anselm's is the first and most profound theological exploration of feudalism. His first work, the *Monologion*, written in 1076, was not, he claimed, a religious book. It was originally titled *An Example of Meditating About the Rational Basis of Faith*, and this accurately describes its intent. Anselm insists that in it he is putting forward no new doctrine. It consists exclusively in rational reflection and 'nothing at all in the meditation would be argued on scriptural authority'.[72] He wanted, rather, to demonstrate that by reason alone (*sola ratione*) we would be led to the knowledge of God we already enjoyed by the privilege of revelation and authority. The nature, far less the existence, of God is not a matter of dispute. But by 'disputing with himself', His fundamental truth can be grasped in a different and equally legitimate way. Reason, of course, cannot replace faith; but it can follow in its path. It cannot pierce the mystery of God's transcending otherness, but in perfecting its own method it clarifies the image of God appropriate to the human intellect.[73]

Anselm begins with the characteristic observation that 'there is something that is the best, the greatest, the highest of all existing things'[74] – a statement which, whether taken for granted or deduced from more primitive assumptions, makes sense only in the context of a hierarchical society. Existence, and not only its

accidental qualifications, is itself ordered by successive subordination and dependence. Anselm illustrates his contention by
appealing to psychological rather than natural 'facts'. Everyone,
he points out, seeks what they believe to be good or beneficial to
themselves. In doing so they seek a multiplicity of particular
things. This heterogeneity should not obscure the fact, however,
that each is being sought in relation to its 'goodness' as the
common measure of all attractive qualities. 'Goods' are commensurate because dispersed in its varied manifestations is something
identical in relation to each. Goods can thus be ordered according
to the quantity of 'goodness' each contains. Moreover, 'that
through which all goods are good is itself a great good'.[75] The
heterogeneity of the good leads inevitably to the notion that, in
addition to specific instances of beings, 'good through something
other than what they are', there must also exist being which is
'good through itself'.[76]

It is 'greatness' which defines the specifically Christian-feudal
conception of existence. The demonstration of supreme and
necessary goodness could be found in Plato. Anselm, however,
identifies the Christian God with *being* itself, rather than with
goodness.[77] Relative greatness, therefore, is another term for the
degree of being which allows some specific thing to exist. Every
particular thing exists, like goodness, only by 'borrowing' from the
essence of a supremely existing being: 'There is a Nature which
exists through itself, which is the highest of all existing things, and
through which exists whatever is.'[78]

'Greatness' is a noun rather than an adjective. The world owes
its existence to this essence which replicates itself, with
diminishing intensity, throughout the cosmos. We cannot help
comparing the value of things and finding that 'some natures are
better than others', and we can avoid an infinite regress of such
comparatives only by coming to rest in one 'so pre-eminent that no
other nature is superior to it'.[79] It is the 'fact' of hierarchy which
prompts Anselm to look behind it to a general 'medium' of
evaluation, and upwards to a being who is the inexhaustible source
of all its distributed substance.

The sense in which dependent being is subordinated to greater
being is likened by Anselm to a craftsman fashioning an artefact
and, in an even more telling analogy, to the social honour gained
by association with a superior person. The subordinate is *created*

through dependence; it owes its existence, and not just its form, to a superior. In the first instance the universe owes its existence to God, who created it *ex nihilo*. This does not mean, Anselm points out, that He created it *from* nothing, as if nothing were some non-material substance upon which He worked, but rather that He made something from a thing which, before He worked upon it, was nothing. He explicitly likens the process to the forming of a social relation between two men, in which, 'the first man, who was formerly regarded as nothing, is now esteemed as truly something because of the making of the second man'.[80]

The same feudal imagery is at the root of Anselm's justly celebrated ontological argument.[81] The greatness of God is here expressed as a being 'than which none greater can be thought', from which Anselm claims to deduce His logical *and* actual necessity. The force of his argument once again rests with its social assumptions. Degrees of 'greatness' is an undisputed 'fact' of nature within the feudal world. In the ontological argument, Anselm exploits the equally 'obvious' assumption that materially observable degrees of being exist primarily as symbols of non-material essences contained within them. The created world is the plenitude of God's 'thought', and, as part of this order, man possesses his own image of Creation in the form of a rational intellect. It is given to man, that is to say, to exist in more than one way and herein lies his own particular 'greatness'. To exist in actuality and in the mind is 'greater' than to exist in either alone. And as God is 'than which none greater can be thought', the implication of his existence in actuality (as well as in thought) is irresistible.

To argue from necessary concept to necessary being is not the special discovery of Anselm. It is the common feature of feudal-Christian thought. The empirical world is a symbol; human experience, likewise, makes accessible only a small portion of reality. Each object and event discloses a portion of itself to the human gaze which, gifted with its own inner image of God's plenitude, can 'understand' the necessary connection between these symbols and their essential being in relation to which they are a kind of covering. Human understanding is, however, imperfect, limited both by man's place in the scale of being and by avoidable human foolishness. In order to understand it is essential first to believe. The *Proslogion* was first titled *Faith Seeking*

Understanding just to emphasize this dependence: 'I believe in order to understand . . . that I shall not understand unless I believe.'[82] Human freedom, particularly pride, obscures the symbolic relations of which the being of man is part; relations which the intellect can recover only when man places himself unconditionally and helplessly before Creation. Only in faith can the intellect pierce the symbolic veil of empirical reality and rest in ultimate, incorruptible, and universal truth.

More specialized theological problems connected with the notion of 'supreme greatness' were taken up by Anselm in his *Cur Deus Homo*, but once again his argument is best understood as a daring essay in social theory. Unlike his earlier philosophical works, it was occasioned by a specific controversy. The doctrine of redemption had received a vigorous new challenge from Judaism. How could Redemption be grasped rationally? How could God be saved from the indignity of contact with the world of evil? How could we accept the idea that He existed as a man; existed, that is, as something less than Himself?

This was only a more urgent form of the contradiction faced by every medieval theologian. Why did God exist in *imperfect* forms?[83] By definition, created being is 'less great' 'than which none greater can be thought', so how could God be responsible for all its limiting particularities?

Anselm had already tackled the more general point. To exist in more than one way is 'greater' than to exist in one way only, even where one of these ways is in itself perfect. The more specific problem was viewed as a logical implication of man's disobedience. Sin had frustrated the purpose of Creation, and must therefore be redeemed. Anselm views redemption in the light of vassalage. Services must be offered if the subordinate being is to enjoy the security and comfort of the superior's protection. But man's indebtedness to God is unlimited and could be redeemed only by an offer of services greater in value than the whole of Creation. Since only God is greater than Creation, only He can redeem man. The payment, to be effective, must none the less originate with man. Hence the necessity of God becoming a man. The Incarnation is, from this point of view, the supreme ritual of investiture: a token of the means whereby man can fulfil his obligations towards God.

In clarifying the central relations of feudalism, Anselm makes

91

them absolute. The symbolic hierarchy, terminating in the necessity of Supreme Being, is an abstraction drawn out of the experience of feudal relations themselves. It is the example of personal dependence in the social order which furnishes the central image of relative 'greatness'. The order of man and of nature is fixed because both are emanations from a hierarchy of essences of which they are the symbols. The physical world is thus a part, a small part, of the cosmos which, by synecdoche rather than metaphor, it represents.

Chapter Five

RELEASE

The social hierarchy of feudalism was just one imperfect symbol of the real distribution of divinely ordered essences; an enclosed and completed cosmic structure. Creation was a 'downward' movement in which everything which existed potentially came actually to occupy its specific place in the universe.[1] 'Mirroring' God's nature, man's soul (though not his body, which remained rooted in its earthly place) aspired to the reverse movement, ascending the ladder of being reunited with its Creator.[2] It is the very fixity of material things which allows human subjectivity – in overcoming the body, which is the resisting medium to its own symbolic actuality – to rise above itself. Transcendence is the real goal of life constrained by feudal relations.

The fixity of the empirical world is its greatest illusion; each frozen essence in fact is just another step towards an ultimate release. Within the framework of feudal-Christian categories such a possibility was entertained as something more than a distant vision of salvation. The mystical fervour of the Platonic tradition was resisted by many, but even the supreme rationalist, Thomas Aquinas, recognized its inspiring appeal and responded to it. 'In the hierarchy of being men cannot surpass the angels, who by nature are of a higher order,' he reasonably points out, but immediately adds, 'Yet man can pass beyond them in his knowing, as when he understands there is a being above them who can make him blessed and when quite possessed will give him complete bliss.'[3] It is this intellectual ascent which is man's true end and only source of authentic happiness. It is the 'perfection of the soul on the part of the mind which transcends the organs of the body'.[4] In the Angelic Doctor's view, however, corporeality cannot simply

93

be ignored as any such act of transcendence requires bodily perfection, 'so as not to encumber the mind's ascent'.[5]

The 'proofs' of God's existence favoured by Aquinas, therefore, took the form of just such an intellectual 'ascent' towards a point of conceptual dissolution. In the five 'ways to God', rational arguments released the mind's inherent tendency to reach beyond what was immediately present to it.[6] Each proof is a different path which the intellect can trace back to its source. The most direct route was the inverse of that sequence of 'movements' through which observable things came to be as they are. However 'fixed' empirical reality appears, it is in fact the outcome of a movement from 'potency' to 'act'. And as a thing can make this transition only by the agency of something already actual, the mind can fasten upon this agency as its 'cause'. The agency, as an actuality, must have itself made the transition from potency. The mind is irresistibly drawn 'backwards' from act to potency to agency to antecedent act, and so on. This sequence can terminate only in an 'unmoved mover'; a being which exists as pure actuality.[7]

The greater privilege afforded the intellect by Aquinas compared to earlier scholastic writers is more apparent than real. He has not really left the world of Anselm's *Monologion*. The only reason the intellect could surpass what was given to the senses lay in the mind's form as a 'mirror' of creation. Though expressing himself dissatisfied with Anselm's ontological proof, Aquinas's 'cosmological' proofs of God's existence share the same assumptions. Gilson puts it clearly: 'Our thought would never suffice to infer Him, unless the reality with which we are linked, constituted in its hierarchic and analogical structure, a sort of ladder leading us up to God.'[8]

The image of a ladder, or a tree, as a 'figure' linking the created world, human reason, and the hierarchy of essences was commonplace amongst scholastic writers. It was favoured not only by the orthodox but also by such diverse figures as Ramon Llull and Bernard Silvestris.[9] Indeed, this 'spiritualisation of the world went far beyond the Church and the strictly religious sphere'.[10] Even expressed in very general terms, however, the rationalist's way to God tended to exclusiveness. A certain kind of dialectical facility, and a rigorous training in the *trivium* and *quadrivium* was its precondition.[11] Yet the promise of salvation was general, possibly universal, and however the final 'release' from the constraint of actuality was conceived its possibility could not be made dependent upon a 'correct' argument alone. The authoritative *Summae* of Albert the Great, Aquinas, and Bonaventure were part of, and responses to, the enormous impact of Aristotelian philosophy during the late twelfth and throughout the thirteenth centuries.[12] Within the Church, however, and outside it, the *telos* of human happiness was integrated with the immediate experience of feudal relations in a variety of other ways.

LOVE

New monastic institutions, reform movements, and experiments, beginning with Cluny in AD 909 and culminating with the foundation of the Franciscan and Dominican Orders in the early thirteenth century, were intent above all upon the discovery of happiness. The longing for transcendence was the fundamental spiritual value pursued in a prodigious variety of practical, organizational, and liturgical innovations.[13] And, as with most radical departures, such movements appealed directly to the source of inspiration claimed by the corrupted institutions they sought to transform. Each fresh wave of monastic renewal conceived itself as more thoroughly imbued than the last with the original genius of St Benedict's *Rule*.[14] Cluny's exceptional freedom from local jurisdiction was secured specifically to ensure that its monks might serve 'God without interference, according to the rule of St Benedict'[15] – a claim to independence emulated by all its successors.

'This little rule for beginners' had in fact been framed in a spirit of generosity and compromise. It made provision for a sheltered way of life in a self-contained and self-sufficient community whose members adhered to a common life prescribed in a specific order of the day varied according to the season.[16] Periods of liturgical prayer, spiritual reading, and manual work regulated the outward forms of the collective life in such a way that each might aspire inwardly to 'that perfect love of God which casts out all fear'.[17] It appealed directly to those with serious rather than fanatical spiritual ambitions. Its deliberate avoidance of rigour, its reliance on the wisdom of abbatial guidance, its discretion, all conspired to allow, if it did not encourage, the growth of physical ease and moral indiscipline.

Cluny, in re-establishing the *Rule*, like many of the foundations that were to follow, exaggerated its precision and altered the balance of its practical guidance. It sought new standards of magnificence and splendour in its observance of the daily offices. All the monk's activities should be carried out in a spirit of worship, so at Cluny everything was subordinated to the demands of an exhausting liturgical cycle. In this it followed but soon surpassed a general trend.[18] Its extraordinary success was made visible in the richness of its ornamentation, in the continuous enlargements and refinement of its buildings, in the conspicuous display of its furnishings and decoration. The very sumptuousness of the setting seemed to demand continuous devotion. Hugh's great basilica, its proportion and elevation controlled by a precise harmonics, echoed to a continuous round of chanting and song, and sustained more completely than any of its rivals the aura of sanctity.[19]

As a result, the day (and most of the night) became clogged with ritual. Anselm, indeed, deliberately chose Bec in preference to Cluny so that he might be allowed sufficient time for study and reading.[20] Manual labour, not surprisingly, figured even less prominently in the daily round.[21]

The devotion of the monk remained a service to the community; he existed for others. But at Cluny, earlier and more clearly than elsewhere, the monk's way of life took on a purely personal value. To be utterly consumed by worship, his entire being taken up and ordered according to the relentless liturgical timetable, was a certain means of overcoming the resistance of pride or worldly interests to the acceptance of God's love. The privilege of the monastery was nothing less than a more intimate experience than could possibly be the case for those beyond its sanctuary of God's transcending reality.

The monastic ideal sanctified obedience. Strict subordination under a rule, whatever its content, was the preliminary act of renunciation essential to a proper deepening of religious experience. It was disobedience, therefore, that was the prime target of the reform movements. Beginning in north Italy and spreading rapidly throughout Europe, more ascetic and more authoritarian versions of the *Rule* were progressively embraced, less as an instrument of communal organization than as a proven technique of self-annihilation.[22] Thus, a century after its foundation, Cluny itself became an object of reforming zeal. Its very success became a handicap to the realization of its deeper purpose. Its openness, particularly its willingness to accept new adult members, allowed its rapid expansion.[23] New members brought with them grants of land encumbered by feudal obligations, and the habit of accepting gifts in advance so that landowners might end their days in the monastery drew Cluny into the same worldly network as the unreformed Benedictines.[24]

The foundation of Cîteaux at the end of the eleventh century provided a somewhat different model of monastic reform. Making a more decisive break with Benedictine tradition, the new order of Cistercians established houses in 'wild' and hitherto marginal or even uninhabited places. Grants of land were accepted only if they were of an uncultivated and inhospitable nature.[25] This physical isolation was part of the simplicity and modesty espoused by their Order. The rigour of their discipline, rather than the arduousness of their adopted rule, attracted many and quickly established their ascendancy among the reformed Orders. The internal organization of each house was not simply placed in the hands of an appointed abbot. A complex organizational structure of chapters and visitations ensured the centralized control of every aspect of communal life and maintained a uniformity of practice that was unmatched by any other social organization (ecclesiastical or secular) of the period.[26]

The success of the Cistercians, as measured by their growth in numbers at least, was phenomenal,[27] and once again brought its own difficulties. To support themselves in poverty the Cistercians had to clear large areas of land which, even then, were of little use other than for grazing sheep. Large tracts of such 'worthless' land were developed into *granges*, worked by lay brethren (*conversi*). They became, in other words, the most 'progressive' and innovative of agriculturists. It was difficult to sustain a genuinely primitive vocation in the face of growing wealth from the sheep trade. Their reluctance to use their riches in conspicuous liturgical display led them inadvertently to 'invest' in further land improvement that subsequently increased their wealth to embarrassing levels.[28]

Economic security encouraged a growing emphasis on study

and learning. Cistercian leaders in the century following the foundation of Cîteaux produced the finest psychological works of the feudal period. Where the developing scholastic logic of the schools confronted the task of integrating Aristotelian naturalism with religious orthodoxy, the Cistercian psychology sought a more perfect synthesis of Christian and Platonic traditions. 'Reason' for St Bernard or William of St Thierry could never move itself beyond the purely intellectual sphere. Logic and dialectic might 'prove' God's existence but in doing so would not bring Him any closer. Intellectual understanding is insufficient; 'there is nothing human in its operation'.[29] The real task is not to determine an abstract agreement between faith and reason, but to unite the human and divine soul in *caritas*.

Love more certainly than knowledge lifts man upward in the scale of being. Knowledge is more properly viewed as one aspect of the relation termed 'love'. Knowledge is founded upon a 'connaturalness' of subject and object. In knowing something our mind shapes itself to the object it apprehends. Now, as 'the soul's sense is love', our love of God, in 'contemplating' its proper object, transforms man's inner nature and lifts him towards perfection. The 'affections' conform the soul to the objects of its love.[30] We must therefore learn to love God. Such a love cannot originate in the human heart. The ascending ladder of being stops infinitely short of God's self-sufficient nature. Man's human self-improvement cannot by itself reach Him, and more than unaided reason can fully expose the inner necessity of faith. Authentic *caritas*, which is really the human form of faith, must originate with God; 'you have given me desire for yourself', remarks William in a formulation reminiscent of Anselm's *Cur Deus Homo*.[31] For no discernible 'reason', God in his infinite goodness has provided man with a means by which He might be approached. Love, when it is unconditionally offered to its proper object, bridges the abyss of transcendence. The problem of love is to arouse its passion in such a way as to avoid consuming itself upon anything less than God. Since 'it is God himself who loves himself in us',[32] the human subject need only learn to abandon itself to this inner movement in order to be saved.

The Cistercian's stricter asceticism as a prelude to personal ecstasy was far more demanding than external observance. In its initial stage the monk had to learn to put away 'all stirrings after worldly honours and delights and pleasures, and everything else that can and often does, arouse in me the lust of the flesh, or of the eyes, or that stirs up in me a wrong ambition'.[33] It is less the resisting flesh, however, than the more subtle and deceptive spiritual obstacle of pride that traps many into superficial obedience. It is useless to love God only for ourselves. Yet in loving God for Himself, we must also love ourselves. *Caritas* links the two and transforms us inwardly.[34] Its proper development requires the constant vigilance of a 'vehement and well-ordered will'.[35] It is the practice of authentic self-denial, in which the 'soul' is emptied of every secular tendency, that allows the spontaneous growth of spiritual love. God expands within the soul and lifts the truly devoted 'beyond themselves'.

Love's ascending movement is rarely consummated. The terrifying experience of infinite love is not simple enjoyment (which distracts the soul into its own particularities), but a willed helplessness in the face of God's *agape*. As 'everyone possesses you just insofar as he loves you', the path of spiritual love becomes a torment of frustration and disappointment.[36] The least inclination to a genuinely spiritual life reveals the recalcitrance of a will holding back from the extremity of a genuinely unconditional acceptance of love's transcending reality. Its *telos* is known and accepted intellectually. The knowledge that 'only those who love you truly are truly and uniquely and singularly happy, and that they are perfectly happy, who love you truly and perfectly'[37] did not make the unreserved acceptance of God's 'wounding affection' any easier. Even among the sincerely devoted, progress in spiritual love was painfully slow and uncertain: 'You do indeed send me at times as it were mouthfuls of consolation', admits William, but cannot restrain the impertinent reproach, 'but what is that for hunger such as mine?'[38]

The Cistercian path back to God, in principle open to all sincerely professing the faith, in practice became just as exclusive as the way of reason. Bernard of Clairvaux, its most illustrious exponent, makes clear in his *Steps in Humility* that rigorous subordination to the *Rule* is only a preparation for the arduous spiritual exercise the devoted voluntarily take upon themselves.[39] 'Contempt of your own excellence', is Bernard's paradoxical commandment, and he never tires of invoking it against disguised and perverted forms of pride.[40] Time and again he exposes the dialectical subtlety of self-abandonment. In willing the destruction of the will, those advanced in spiritual love could find their pride taking root again in acts of self-mortification. Ascetic zeal is as dangerous as secular distraction to the pursuit of humility. Learning and good works could similarly trap the unwary and offer a hiding place to the hard-pressed ego.[41]

To gain knowledge of God was to see him 'face to face', a possibility that demanded an absolute surrender in the face of his overwhelming reality.[42] Resignation, paradoxically, was the only effective means of spiritual advancement. And resignation, to be sincere, could not become a 'rational' means of such advancement but must itself flow spontaneously from inner faith.

The Cistercian retreat from the world and their subsequent economic and administrative innovations might give the impression of an Order that had in fact succeeded in extricating itself from feudal relations. This however would be far from the truth. Their doctrine of *caritas* was a profound realization and purification of the central principle of feudalism. Humility as a form of love was, like honour, an essence shared between superior and subordinate; an essence redistributed ritually to create a new person as 'the man of another man'. The monk in seeking to become 'a man of God' plunged into the absolute limit of feudal inequality and reciprocity.[43] He had nothing to 'offer' God but obedience, and as God's being was infinite, obedience was all that was required. And his only duty, consequently, was to accept his Lord's protection.[44]

It is in this context also that the Franciscan profession of poverty should be viewed. To be entirely dependent on alms was a form of ascetic trial which tested the faith of the believer. More than that, it was an open avowal of worthlessness. There is no sense here of a sentimental identification with 'the poor', or any desire for the performance of 'good works'. It was just because feudal society contained within its hierarchy degraded and worthless elements that the Franciscan could adopt poverty as a religious vocation.[45] Their experiment, however, for a time unsettling to ecclesiastical authorities, was rapidly regularized into a mendicant order.

The wholehearted devotion to typical monastic values, humility, surrender to the terrifying ordeal of God's love, transfigured the monk from being a spiritual functionary into a religious virtuoso on his own behalf. It was a potentially subversive individualism more than counteracted, however, by the peculiarly negative form in which it expressed itself. The monk sought an ultimate release – an ascent which, dispensing with the measured intellectual path of the cosmological argument, mounted beyond the ladder of existence to the direct apprehension of its first cause.

VALOUR

The questing knight is no less a spiritual figure than the monk. Allotted his fixed place in Creation, he none the less sought, in realizing his God-given nature, to transcend its immediately given boundaries. His aim, just as the monk's, was the happiness that came in the moment of release from the order to which he had voluntarily submitted himself. His quest was for justice, honour, and, above all, death.

The knight's route to eternity remained, in spite of a vigorous 'Christianizing' movement throughout the twelfth century, something very different from the asceticism of the *Rule*.[46] He did not seek the liberation of his soul through the disciplined neglect of his body so much as, in testing its physical powers, he allowed himself to become a secular instrument of God's will. The knight's body expresses, as a direct and natural symbol, a perfection as pure in its own way as the monk's soul. He is not merely powerful and active; he is beautiful. Perceval, brought up in the 'wild' because his mother was afraid of losing him to some adventure, instantly recognized the physical superiority of the first knights he happened to see. His hyperbole, 'they are more beautiful, I think, than God and all his angels', is in a sense quite justified.[47] Their physical being, incorporating and expressing the divine commandment of justice, *is* more beautiful than the intellectual idea, or the purely inner and soulful likeness, of God. Cligés, the most secular of Chrétien de Troyes' heroic characters, is possessed of a physical perfection quite beyond the range of accidental variation, 'In him is nothing that can be mended'.[48] In addition to stature, strength, proportion, there is a positively 'feminine' refinement: 'his hair seemed like fine gold, and his face a fresh-blown rose'.[49] The principle of plenitude, the cosmic fullness of being, has been poured into him: 'In framing him Nature was so lavish that she put everything into him at once, and gave whatever she could'[50] – a fullness that could not be due simply to nature, but rather was a token of a purpose that lay beyond it.

The true knight was 'fair' of feature; or, if false and corrupted, then 'coarse'. The moral distinction implicit in different physiognomies was more subtle and ambiguous here than among the holy orders. The monk should be emaciated, his whole body turned inward upon itself, cast down and worthless; or, if corrupted, fat, sleek, and complacent.[51] Among the chivalrous elite, some slight physical blemish becomes an incontrovertible symbol of moral imperfection.

Physical appearance, in other words, is not an isolated or accidental quality. It is part of a tightly knit complex of values; Cligés, for example, is described as 'a brave knight, so handsome, so noble, and so loyal'.[52] Such a list might be extended but could just as well be condensed to a single term: a 'true' knight cannot be anything other than handsome, brave, loyal, and noble. These are the essentials of knighthood, as humility is the essence of the religious vocation.

That is not to say that all such characteristics spring spontaneously from the knight's being. He must, like the monk, undergo a rigorous training. His body must be developed and tested, its passions trained and controlled. He must learn the etiquette of the court as well as the conventions of the battleground. He must become as adept in the formalities of love as in the techniques of combat. Ramon Llull, among his many works, produced around 1280 a small book on chivalry in which the range of knightly characteristics is precisely documented. The knight is chosen as one among a thousand, 'the most suitable, the most courageous, the most strong to sustain exertion, the most able to serve man',[53] intrinsic qualities which fit him, when they are properly developed, to the specific task of maintaining the secular order. His bearing and character therefore must be commanding, so that 'by love he restore charity and instruction, and by fear he restore virtue and justice'.[54]

It is these special qualities of character that justify his position of privilege, 'so very high and so very noble in the order of chivalry ... that it behooves also that the common people labour in the lands in order to bring forth fruit and goods whereof the knight and his beasts have their living'.[55] The functionalist view of society is clearly articulated by Llull; as the clerics 'incline the people to devotion and a good life, in like manner the knights, by nobility of spirit and force of arms, maintain the order of chivalry'.[56]

The 'order of chivalry' is as rigorous in its own fashion as St Benedict's *Rule*. The youth who would become a knight must learn the arts of warfare and courtly behaviour: how 'to carve at table, to serve and dub a knight'.[57] Hence the necessity of Llull's treatise which raises the training for knighthood to the same level of systematic preparation as that offered in the schools or monasteries. Personal association with and devotion to a master remains the definitive path to knighthood, but additionally it is necessary that 'the science be written and the art be shown and read in such manner as other sciences are read'.[58]

The knight is no more sustained by pleasure than is the monk. He is not ascetic, but his sensuousness, controlled and directed, is only a means to the accomplishment of a purpose beyond himself. The knight's 'natural' physical superiority expressed itself therefore in the virtues of the soul, in justice, wisdom, charity, loyalty, truth, humility, and hope. [59] Llull in fact 'deduces' these qualities of knighthood as a thirteenth-century theologian might discourse upon the 'names' of God. They are part of his knightly nature, and his nature is part of an inescapable order realized through his actions. We are once again within the orbit of the ontological proof.

Llull was lending intellectual weight to an ideological movement that was at its strongest in northern and western France from the mid-twelfth century. It was encouraged by ruling princes and kings as a means of developing, as a specific style of life, a defensive solidarity among their vassals, thus making the ruling class more resistant to the subversive influences of the towns.[60] The association between knighthood and the clerical orders had already been made, with compelling eloquence, in Bernard of Clairvaux's fierce support of the Second Crusade. 'Surely', he claimed, 'it is an intrepid knight, protected on every side, who clothes his body with the armour of iron and his soul with the armour of faith.'[61] The Cistercian leader could not, however, restrain his ascetic temper and lent his spiritual authority also to the foundation of the Knights of the Temple; a self-conscious and unsuccessful attempt to combine nobility and valour with monastic discipline, which Llull ignored in preference to Chrétien's more appealing image of the solitary, questing knight.

The knight organized but did not live wholly within a community. His essential nature was revealed only when he ventured beyond his castle walls into the wild. He was supposed, immediately prior to the ceremony of dubbing, to spend at least one night in fasting and prayer during which he was not allowed to sit or lie down. This imitation of the monk's nocturnal vigil drew attention to the sacramental character of the quest, and begins a physical and moral process of 'testing'.[62] The following morning he would hear a special sermon and celebrate a special mass distinguishing him from the rest of the faithful. Then he received his arms in an elaborate ceremony of benediction modelled on the bishop's consecration of the king's sword. Llull dwells on all the appurtenances of knighthood, whose symbolic values are revealed by the ceremony. The sword, for example, is in the shape of a cross to show that 'Our Lord God vanquished in the cross the death of the human lineage'. The spear signifies truth, which is 'a right and even thing'. The hauberk, being strong and closed on all sides, is 'a castle and fortress against vices and faults'; his spurs are 'diligence and swiftness'.[63]

Thus equipped, the knight 'sets out without mission or office; he seeks adventure, that is, perilous encounters by which he can prove his mettle'.[64] In the forest he confronts nature in the wild, not so much to subdue as to test himself against it. But in 'saving' the innocent victims of chaos he imposes a temporary and localized order. Divine justice exudes from his every action, protecting him and those he takes into his care. His survival depends upon conformity with the prearranged harmony of courtly ethics. His only assistance comes in the form of an occasional hermit. The knight redeems by direct contact, as it were, portions of nature that man's freedom has spoiled.[65] He becomes God's distant sensing device, bringing into divine contemplation (and thus ordering) disordered parts of the world. Through him God can familiarize himself with sin, with the consequences of human wickedness, a process which itself, within the sphere of the knight's action, redeems the corrupted world. The knight's courage is not only the physical bravery of meeting and conquering the fear of his own death, but the deeper valour of allowing himself to become the meeting point of absolute good and evil. The knight does not follow the path of innocence, of retreat from wickedness. He must carry the divine creative spirit

104

into the wild, risking his own soul in continually renewed adventure.

As an instrument rather than a vessel of God's love, the knight's quest is never ending.[66] It becomes the very core of his being; 'trial through adventure', Auerbach insists, 'is the real meaning of the knight's ideal existence'.[67] Such a series of encounters has no outward political or historical sense of direction and achieves no particular secular ambition. It is, rather, an inward drama through which the knightly virtues are gradually perfected. The knight's individual identity (the particular arrangement of his virtues) is not only retained but, through adventure, is enhanced. Unlike the monk who, to be ravished by God's love must become nothing, the knight, in acting, retains all his particularity.

This dangerous individuality, so much at odds with the ideal order the knight represented, must be mastered. The knight, that is to say, must run the risk of love as well as face the danger of battle. [68] Erotic love can wreck the quest. As part of the wildness of unredeemed nature the knight must confront and conquer its disorderly challenge; 'Love is a thing that copies Nature', and is therefore apt to destroy the social hierarchy.[69] And just as he must embark upon adventures which cannot, even in principle, be brought to a conclusive dénouement so he must love the unattainable. Courtly love plays with eroticism as the joust and tournament play with death. And it is no less in the game of love that the knight 'learned to control his violence, to reduce it to order'.[70] His particularity must be disciplined and confined, his love alights upon his lord's wife and becomes slavish devotion. It is all the more intense for remaining unfulfilled. The helplessness of the knight in the face of his overwhelming love charges it with mystical power.[71] Even the greatest knight, as the most zealous ascetic, may fail the test. Tristan and Lancelot were both close to perfection yet did not succeed in finally mastering themselves. True valour, as exclusive and demanding as genuine humility, offered no easier path to the sublime.

The knight's death is both the culmination of the quest and the readjustment of the secular world to the disturbing shock of his individuality. His unruly passion, if it is eroticized beyond the spiritual *affectus* of service, loyalty, and feudal subordination, must ultimately be extinguished. His death preserves his perfection; 'The death he inflicts is to the benefit of Christ; the death he receives is to his own benefit,' declares St Bernard with typical vigour[72] – a view echoed in Llull's conviction that no act can 'contribute to chivalry more than death'.[73]

Cervantes, looking back on a tradition of romantic literature which had become anachronistic, began by parodying its empty mannerisms. But Don Quixote, as a true knight (his honour is more to him than reality itself), triumphs over his Creator.[74] His passion, which is the knightly devotion to chivalric order, creates its own world in opposition to any mundane experience that might deny it. Yet there is more than a little of the Quixotic in his noble predecessors. The imitation is comic only because the world has ceased to believe in the transcending value of the quest. The knight, as the monk, had conquered death and lived uncaring of his own comfort because comfort was to accept the empirical world as the true reality. To go beyond this reality, to realize through a fixed but partial symbolic order the essential inner nature of being was his quest.

The privileged orders, as it were, confirmed the ultimate values of feudalism only by going beyond the everyday world's copy of its perfect harmony. The fundamental justification of the social order was not so much that God had willed any particular, mundane relationships (He had not since man, fallen from grace, used his freedom on his own behalf), but that, imperfect as it was, it could not help but be constructed 'in the image and likeness' of divine order. Its gradations provided a stairway to the greater reality it sought to express. Converted into subjective 'steps' or projected into the world as virtuous 'deeds', the monk and the knight were carried beyond the social world. On the one hand, 'contemplation' and on the other 'action' transcended the constraints of time and place to reach the realm of pure being; the 'love' and 'valour' which were identified as *happiness*.

SENSUOUSNESS

What of the much greater number of unfree peasants absorbed without prior conceptualization into the feudal hierarchy; the sublunary species, degraded, uncivilized and incapable of the refinement that existed as an aspect of the lord's benevolence? Can we find among them anything approaching a cosmology, a coherent view of their own world of experience? Nineteenth-century historians and sociologists assumed not. Frazer had no hesitation in including the European peasantry of his own day, let alone of the feudal period, in the same category of undeveloped culture as the primitive. 'Superstition' was the dull undertow of history from which only an educated elite had ever succeeded in shaking itself free.

In trying to describe the 'comportment' of the monk and the knight, however, a different way of reading the symbolic value of experience is being invoked. A cosmology does not exist exclusively, even among the educated, as a theory or set of theories about the structure of the universe. Including such abstract reasoning, it refers much more generally to the organizing principles implicit in everyday life. The European peasantry, existing in a condition of involuntary submission within the feudal hierarchy, developed an inarticulate cosmology profoundly at odds with either the Aristotelian scholastic, or the Platonic monastic versions of orthodox Christianity. Mikhail Bakhtin's important work on Rabelais has uncovered this neglected cosmology.[75] No longer immediately comprehensible, Bakhtin succeeds in demonstrating that Rabelais' imagery, his comic playfulness, his use of language, makes sense as the culmination of 'a thousand year development of popular culture',[76] a development which, as much as Christianity or Hellenism, adapted itself to the reality of feudalism. The fiction of Rabelais illuminates, as if from the inside, a world of popular festive forms: ritual spectacles created from comic verbal composition and uncouth forms of speech in which were preserved something of the powerful syncretism of the carnival.

Every feudal ceremony – liturgical, civil, legal, initiations, oaths – everything solemnized by public rite and sacrament had its comic counterpart. Parodies of prayers and psalms, uncrowning ceremonies, mock treatises (many themselves learned works), made outrageous fun of orthodox and authoritative opinion. All these counter-festivities 'built a second world and a second life outside officialdom'.[77] A world neither religious nor magical which, existing on 'the borderline between art and life' is a 'region of liberating ambiguity . . . which in reality is life itself, but shaped according to a certain pattern of play'.[78] This festive life, the carnival tradition, which is the people's 'second life', is subject to no regulation other than 'the laws of its own freedom'.[79] And, as in the case of the child's play world, 'while Carnival lasts, there is no other life outside it'.[80]

Carnival appears, from the outside, to have a certain inner logic, a meaning contained in its 'turnabout' and 'inside-out' gestural language. It seems to be a world constructed by simple inversion, a direct turning upon its head of the official feudal hierarchy: ceremonies of uncrowning, processions in which the clown takes the place of the bishop or king, feasts in which gluttony and indecency replace courtly etiquette, mock sermons in which the speaker imitates the sounds of an ass.[81]

Such reversals, however, were only the prelude to a more radical negation of officialdom. The established order was not merely parodied; its very existence was denied. There was a general suspension 'of all hierarchical rank, privileges, norms and prohibitions'. The carnival was existence 'hostile to all that was immortalised and completed'.[82]

Inversion is the carnival's own special symbol, which only later shrinks to express a secular political ideal. In the feudal period its meaning is unrestrained by any practical desire for a more perfect social order or any covert appeal to a subversive concept of social justice.[83] The carnival is complete in itself and need seek nothing 'beyond' its own sensuous fullness. A more profound reversal of conventional relations lay in its negation of the process of symbolic expression as such. Once established, Bakhtin insists, it is a world closed in upon itself: an exhaustive reality which criticizes the more profoundly by its indifference to, than by its caricaturing of the official world.

The carnival, in its inexhaustible aspect, can be represented to us who remain outside it (those within have no need of representations), as *grotesque realism.* This is not intended by Bakhtin to denote a particular 'aesthetic' standpoint but rather to evoke the recollection of a 'primordial' mode of experience. 'The cosmic, social and bodily elements are given here as an indivisible whole.'[84] It is a 'grandiose and exaggerated' somatic process; a collective, cosmic body existing in a continuous state of flux. It is the 'body of the people', fecund and degenerate, playing host to the popular culture of the 'primary process'. The entire world of human experience is rendered into bodily forms. Every somatic function is exposed and celebrated, every internal organ torn from its place of official concealment. Its 'brimming-over abundance' is continually emphasized in feasting, polymorphously, to excess.

Grotesque realism has an immediately 'degrading' significance. All that is 'high, spiritual, ideal, abstract' is given fresh existence as part of the sensuous cycle of death and rebirth. 'Laughter degrades and materialises', it is a 'perverse' movement of sensuousness, in opposition to all forms of asceticism and its ethereal creatures, mystical love or theological abstraction.[85] It literally 'brings down to earth' all human pretensions to spiritual transcendence. It realizes in its unrestrained sensuousness a quite different release from the petrified structure of the feudal hierarchy.

These bodily degradations arouse laughter, and it is as part of the history of laughter that Bakhtin views the popular culture of feudal society. Within the official world, 'that which is important and essential cannot be comical'.[86] And it was the developing 'seriousness' of officialdom, what might be called the feudal state, that forced laughter into its own, unofficial sphere. It was just because of this enforced separation and exclusiveness that laughter became 'marked by exceptional radicalism, freedom and restlessness'.[87] Subsequently, during a brief hiatus in the transformation of official society, it once again became generally available, and it was in that period that Rabelais (and Cervantes, and Shakespeare), were able to exploit its protean imagery. Their works are filled with carnival gaiety that mocks the 'tone of icy petrified seriousness' proper to 'serious' writing.

If the carnival was not simply an 'irrational' form of social protest originating among the lower orders, neither was it a cunningly devised 'safety-valve' through which, from time to time, the accumulated frustrations of the downtrodden could be harmlessly released.[88] They were genuinely popular forms in which the privileged also participated. The more significant towns devoted three months of each year to such ceremonials, feasts, theatrical shows, and public spectacles. They involved all the people and temporarily made them one. The 'feast of fools' was held at least once (more commonly, three or four times) a year in most places. During this festival, 'grotesque degradations of various church rituals and symbols' were sanctioned. A justification of these rites was published by the Paris School of Theology in 1444, defending such practices, 'so that foolishness, which is our second nature and seems to be inherent in man, might freely spend itself at least once a year'.[89] Easter laughter (*risus paschalis*) and Christmas laughter similarly drew into their festive excess clerics, schoolmasters, lawyers, and artisans. The theologian and the philosopher were equally impatient to celebrate the feast days so that they might be temporarily released from 'the oppression of such gloomy categories as the "eternal", "absolute", "unchangeable"'.[90]

The system of grotesque bodily images ensured 'the victory of laughter over fear'.[91] Everything abstract and spiritual found itself somatically duplicated. The soul descended to the 'material bodily lower stratum'. The carnival body, unlike the monk's (skeletal) or the knight's (athletic) body, was dominated by the belly, bowel, and genitals. Neither a vessel nor an instrument, it dissolved into the unbounded metamorphoses of digestion, elimination, and procreation. Through its somatic functions, the body continually over-reached its apparent boundary and devoured the earth. It died and was reborn. It remained perpetually unfinished, incomplete. The carnival face is reduced to a huge, gaping mouth, 'the wide open bodily abyss', through which the world passes. It is a common, collective body, 'a point of transition in a life eternally renewed'. [92] Through it, hierarchy is transformed and denied. The carnival body, which 'can fill the entire universe', engulfs the cosmos in 'sublunary' sensuousness, overwhelming God's fixed and abstract order with man's own generative powers.

The carnival, then, does not oppose one ideology with another, one unofficial picture of the cosmos to the authoritative tradition of theological and philosophical reflection. It reduces intellect to substance. In Bakhtin's wonderfully penetrating analysis, carnival is revealed as the continuation of the 'primary process', the persistence of *fun* in the midst of the cosmology of *happiness.*

Yet it cannot be the fun of the primary process; not quite. The paradisaical innocence of primordial experience can never be repeated. Fun cannot be planned; it is simply 'given'. The carnival, however 'absorbing' to its participants, could not remain wholly unconscious of its relation to the official world of fixed and stable categories. Indeed, the carnival can be viewed as a species of the comic only because it is the *recollection* of fun rather than the primary process itself. Laughter, as Freud clearly demonstrates, is a particular kind of nostalgia.[93] It is not fun, but fun's incommensurability with anything 'serious' that rouses laughter. Thus, the carnival is a huge joke at the expense of the official medieval world view. And like any joke it can only partially succeed; in arousing laughter it lifts a veil upon the vanished world, only to let it fall again.

Just as in childhood play is gradually subverted by its dependence upon the adult world, so the carnival cannot altogether forget its relation (albeit one of negation) to orthodox cosmology. A process of rationalization corrupts its liberated forms by assigning to them a purpose. In Switzerland, as early as the fourteenth century, the carnival had acquired a much more restricted politico-religious significance.[94] And by the sixteenth century – quite generally, it seems – it had become a focus of, rather than a release from, a whole range of social conflicts. At Romans, for example, Mardi Gras had become a 'secular' event with 'a twin purpose; pure and simple enjoyment, first of all, but social, political, municipal protest as well'.[95] Mock battles could and did become genuinely violent. As subversive street theatre it gained a completely new value, but, more importantly, in making 'enjoyment' its *purpose*, the carnival ceased to profess the wholly absorbing ontological givenness that Bakhtin claims to be its original genius.

The body, transformed into intellect, or love, or valour could grasp, beyond its particular and definite order, the transcending reality of the cosmic hierarchy. The asceticism of disciplined thought, humility, or valour allowed the soul, in its longing for the ultimate release, to rise above all material symbols and discover *happiness*. The ways to God, however, were unending quests. Reason opened into the abyss of the cosmological argument, love struggled to escape from self-absorption, the chivalrous knight was tested to destruction. The *telos* of human existence, the reality which gave form to experience, was reached only in death. The carnival's immediate and undemanding recollection of the play world offered a different kind of release from the chain of being. It slid downwards, liberating the human into pure sensuousness. Human reality, however, could no more be wholly materialized than it could be infinitely spiritualized; it moved between the two, stretched into an order that remained distressingly incomplete and insurmountable.

A passionate longing for the infinite thus expressed itself

differently among the different orders of feudal society.[96] These relations, of course, are for the most part imaginary. More often than not the knight was barbaric, the monk corrupt, and the carnival a tawdry display of bad manners. Ideals of loyalty and friendship could scarcely conceal the exploitation and brutality of everyday life.[97] It is ideology we are dealing with, not history. It is above all bourgeois ideology that informs, by oblique comparison, the subject matter of so many discussions of 'feudalism'. The symbolic hierarchy is a 'model' of order which bourgeois culture has discarded. The more 'ideally' it is described the more absurd and contradictory it appears. In its insistence that the empirical world is no more than a 'figura' of reality, and in setting for itself goals which, in its own terms, were unattainable,[98] the present-day commentator, by assigning them to the past, rids his contemporary culture of uncomfortable beliefs.

Chapter Six

NATURE

The images of the mirror and the ladder characterize the symbolic reality of happiness, and serve to distinguish it from the primordial givenness of fun. The cosmic body contains all possibilities within itself; cosmos and psyche are one and the same process of metamorphosis. The world of happiness, however, is divided between macrocosm and microcosm. And, once divided, it can be endlessly subdivided into an order of qualities. The world of inner experience (psyche) and the order of the universe (cosmos) are no longer identical. Neither are they 'empirical' categories describing different segments of the world as it can be directly experienced. In their simultaneous dependence upon God's unconditional and unlimited beneficence they are the separate 'mirrors' of divine being. As such they are related analogically, so that man's unique privilege is, by virtue of qualities inherent in his soul, to have the potential for knowledge both of the world and of divine things.

Such a view would seem to allow to nature no more than a superficial coherence. The various orders of the empirical world are related directly to the invisible essence which informs them rather than to one another. The longing for happiness, which is central to the bourgeois reconstruction of feudalism, gave to nature an insubstantial reality. It lacked the inner necessity of its own purely physical laws.

Nature existed, for the most part, as an intermediary between divine hierarchy (cosmos) and inner release (psyche). The questions it posed demanded interpretive rather than explanatory answers. The revival of the *quadrivium* (geometry, arithmetic, astronomy, music) as part of the liberal arts curriculum in thirteenth-century Paris was defended, therefore, on the grounds of its utility as a preparation for the reception of revealed truth which was the subject of more advanced study, the *trivium* (rhetoric, dialectic, theology), and not for the light it might cast upon the natural world.[1] Within a universe of analogies, these 'mathematical arts' were essential intellectual tools, supremely 'useful' in a sense quite different from that espoused by earlier and later periods. Since the time of Boethius, the most practical of such arts was held to be music.[2] It was, like arithmetic which it resembled in dealing with discontinuous quantities, a system of pure relations. Detached from the constraints of corporeality, musical harmonies were expressive of the deeper order of creation.[3] And as this was an order which penetrated (imperfectly) the human form, music, like reflective thought and love, had therapeutic and 'elevating' powers. Every description of the mundane, physical world was bound up with the moral and aesthetic qualities of its local 'harmonic' order.

The meaning rather than the actuality of nature was the essential issue. The maintenance of ancient learning was, with some exceptions, the principal source of 'observation'.[4] Compilations, abstracts with occasional glosses, and encyclopedic collections of traditional authorities were the basic materials of study, in theology and philosophy as much as in 'natural history'. Isidore of Seville's *Etymologies,* Pliny's *Natural History,* and increasingly the works of Aristotle were reproduced, rather than discussed, as intellectual models. But such collections did not in themselves constitute 'knowledge' of the natural world. Rather, they formed the materials with which the mind could synchronize its own inner harmonies, 'losing' itself in contemplative 'participation' in the world.

Nature, an aporetic entity, was none the less formed from its own set of secondary, physical relations. Physical things were bound together by a network of causes which, formally analogous to the essential structure of the cosmos, could also be grasped separately and superficially as a material phenomenon. Beginning in the twelfth century, scholars influenced by Arabic commentaries on Greek philosophical and scientific works, as well as those speculating within the Platonic and Pythagorean tradition, approached nature in a new light.[5] This constitutes a 'renaissance' only in the most general sense of being a rebirth of scholarship and intellectual self-confidence; it did not anticipate in its values and precepts the Italian Renaissance, far less the scientific revolution.[6] It was a development that lay wholly within the cosmology of happiness; its discoveries aroused rather than satisfied the longing for the 'real' knowledge that would come only from surpassing the sensible realm to claim the gifts of reason and love.

Historians, responding to the same dissolution of certainties we have noted in modern treatments of the irrational, were not slow to invert an earlier orthodoxy. The medieval world, from being stigmatized as a confusion of symbols, steeped in ignorance and prejudice, was suddenly rediscovered as the intellectual source of all that was valuable in classical bourgeois thought.[7] Medieval humanism and naturalism were, in consequence, thrown into a new prominence. More recent and detailed scholarly attention has revealed, however, the extent to which nature remained only a small part of the medieval cosmos embedded in the larger structure of the feudal hierarchy.

FORM

Nature may be only part of the cosmos, yet its existence betrays a material *form* distinct from, but replicative of, the order of other intelligible essences. Matter, properly speaking, could not exist other than formed into the coherence of nature. It is both a part and an emblem of the whole.

The *Cosmographia* of Bernard Silvestris offers a particularly vivid account of nature's original formation.[8] Inspired by the *Timaeus* and the works of Erigena, the *Cosmographia* is 'a landmark of twelfth century humanism',[9] which presents the process of creation through the interaction of a number of 'theophanies' or figures representing elemental cosmic powers.

Silva (also called Hyle), 'still a formless chaotic mass', holding 'the first beginning of things in their ancient state of confusion',[10] approaches God complaining of the failure of Noys ('the consummate and profound reason of God'[11]), to shape her in 'the image of a nobler form'.[12] Silva represents, that is to say, the chaotic substance from which all matter will be drawn. It is 'intractable, a formless chaos, a hostile coalescence, the motley appearance of being, a mass discordant with itself'.[13] It contains 'the original natures of things diffused through her vast womb'.[14] As 'pure' substance, Silva is without extension, or location, or coherence born of inner necessity. As created being, however, Silva is possessed of an unquenchable longing to 'return' to God; thus, 'yearning to emerge from her ancient confusion, she demands the shaping influence of number and bonds of harmony'.[15] Creation is the materialization of the order potential in its limitless turbulence: 'The elements come before you, demanding forms, qualities, and functions appropriate to their causal roles, and seek those stations to which they are almost spontaneously borne, drawn by a common sympathy.'[16]

Silva's demand is timely, coinciding with the divine impulse to order: 'I will produce a form for Silva.'[17] Not an arbitrary arrangement but the realization of a uniquely necessary order, Noys 'effected a balance of properties among her indisciplined and recalcitrant materials, joined them with means, and so bound them together in arithmetical proportion'.[18]

Hyle, 'once given definition by visible images of the ideal',[19] abandons her ceaseless transformations. Under the continuing guidance of Endelechia (the cosmic soul which exists 'by a sort of emanation'), the elements are separated and located in fixed relation to one another. Thus the 'totality of created life unfolds in ordered progression from the nurturing womb of Silva'.[20] Nature is the product of this second Creation; it was 'from the intellectual universe the sensible universe was born', fashioned into a 'continuum, a chain in which nothing is out of order or broken off'.[21] The cosmos extends itself spatially in an unbroken chain of causality. The plenitude of intelligible essences, Noys, fashions itself into a hierarchy, thus bringing into existence the physical manifestation of nature. It is a hierarchy of 'real' powers and influences. The more perfect, quintessential matter is placed 'above' degenerate forms whose cycles of generation and corruption it controls. Everything that can exist finds its appropriate place in the cosmos, including all future and past events: 'For that sequence of events which ages to come and the measure of time will wholly unfold has a prior existence in the stars.'[22]

The substance of nature, however, retains something of its original intractability; its 'wild and perverse quality cannot be perfectly refined away or transformed'.[23] It is only the continuous action of Endelechia that counteracts the tendency of created matter to return to a condition of incoherent sensuousness. The very form of nature is the outcome of opposing tendencies; of the longing for 'perfect' form, on the one hand, and the undertow of original substance, on the other. These tendencies are the two fundamental images that have previously been isolated as fun. If nature fully satisfied its longing for perfection it would become identical with Noys, and its materiality would be volatilized back into the thought of God. Equally, if it were to give way completely to the tug of Hyle, it would lose its material form in the chaos of pure sensuousness. Nature, like man (its microcosm), seeks release from the feudal order imposed upon it by the act of creation. Unable, however, to rediscover the absolute freedom of reason, or the infinite transformative power of the primary process, nature fills out the cosmos with a fixed structure stretched taut between the two.

The creation of a cosmos turns the primordial experience of fun into an unattainable idea. Happiness is the concept or memory of fun, the unreachable *telos* of nature, and therefore of man. Empirical existence is an imperfect compromise, tending simultaneously to both the pure abstraction of reason (Noys) and the absolute concreteness of sensuousness (Silva). If either, and therefore both, extremities were actually attainable, the cosmos would collapse upon itself and resume the untroubled playfulness of the cosmic body.

Silvestris, as if anticipating Freud (or rather Bachelard), psychoanalyses nature and discovers that its reality is the outcome of a process of 'inhibition' upon the primary process.[24] Nature is, first of all, a wish. It is filled with the desire to transcend itself; it is the spontaneous inner movement which is the source also of the microcosmic order. Noys determines 'to complete the success and glory of my creation with man',[25] a being marked by 'the distinctive attribute of dignity'.[26] In man is concentrated the opposing tendencies of the cosmos: 'His body will issue from the depths of chaos and his spirit from the powers above.'[27] His nature is therefore nature itself, the unique link between substance and intellect.

Alan of Lille, whose *Plaint of Nature* is the direct successor to the *Cosmographia*, expresses the same idea with particular clarity: man's nature was formed 'according to the exemplar and likeness of the structure of the universe so that in him, as in a mirror of the universe itself, Nature's lineaments might be there to see . . . so in man there is found to be continued hostility between sensuousness and reason'.[28]

The link between megacosmos, as Silvestris terms the super-lunary universe, and microcosmos is more than analogical: 'The whole appearance of things in the subordinate universe conforms to the heavens . . . and it is shaped to whatever image the motion of the heavens imparts.'[29] The unity of nature is sustained in a double aspect, as intelligible spirit and as matter formed and shaped by Noys. The stars are quintessential but still natural beings, endowed with the capacity directly to apprehend God's will, which they execute by means of nature's own necessity.[30] Physical forces, expressive of the divine idea of nature, flow inwards from the more exalted spheres of Creation. As the central focus of celestial forces, the sublunary world becomes a much more significant place than was allowed within the traditional Christian cosmography.

Nature, however, was not to be understood exclusively in terms of such forces. The intelligibility of nature remained a form of divine reason operating through secondary causes. The humanism of the twelfth and thirteenth centuries was, in this respect, strictly limited in its ambitions. It was less radical, for example, than the Arabic science which, in part, served as its inspiration. The heterodox astrology of al-Kindi, for example, was founded upon a belief in the direct power of the heavens upon the 'radiation of occult influences from the stars'.[31] The binding forces of nature, that is to say, were held to be inherent in the substance of the universe itself. But for Silvestris, as much as for Anselm or William of St Thierry, substance was not even material before it had received the imprint of divine wisdom. Nature's necessity remained for him the order of perfectly chosen symbols.

An apparently more 'naturalistic' account of the origins and structure of the universe was advanced by Robert Grosseteste, Bishop of Lincoln. He identified the formative power by which 'first matter' was created with the physical reality of light:[32] 'For light of its very nature diffuses itself in every direction in such a way that a point of light will produce instantaneously a sphere of light of any size whatsoever, unless some opaque object stands in the way.'[33] This is just the principle of self-generation required to account for the physical cosmos. The infinite expansion of *lux* acts upon the absolute simplicity of 'first matter', 'drawing it out along with itself into a mass the size of the material universe'.[34] In the very act of creation nature, as it were, oversteps itself, the pure actuality of the firmament contains nothing but 'first matter' and pure form. It is incapable of the internal differentiation of motion and cannot therefore affect our senses. This marks the boundary of the physical universe, its outermost, infinitely expanded limit.

Not all matter is expanded to the limit of its potentiality. *Lux* internally reflected as *lumen* is focused back upon the centre of creation. On its return journey it gathers matter into concentric spheres of uniform 'density'.[35] The matter within these spheres still contains the potentiality for movement, but constrained by the absolutely expanded limit of the firmament all motion within them returns to itself upon an endlessly circular path. Matter confined to the central 'sublunary' region 'is more corporeal and multiplied', and, being compressed and mixed, one element with another, beneath the lowest point of perfection, its movement is rectilinear. Heavy objects tend even more to take up a central position, while lighter ones seek a separate and more elevated position.[36]

The hierarchy of nature, still ordered by qualitative differences, is here united, formally and physically, through the action of light, which is the unique connection or influence between heaven and earth, and the explanation of all earthly phenomena is found in its presence and action.[37] It is the absolute simplicity of light as a form-giving power that encourages a 'geometrical' approach to nature. Point, line, angle, proportion; the grammar of light is reducible to a few essential relations. The generality of such relations can be rediscovered over and over again at any point in the natural hierarchy of Creation. And just as the physical universe expanded from a point, so we can rediscover its fullness in the merest speck of dust. The smallest particle 'is a complete and inexhaustible treasury of all the primary mathematical constituents from which the whole universe is constructed'.[38]

This view of light as a formative principle was certainly influential in the development of a sophisticated science of optics. But it was not, for Grosseteste, a matter of physical theory alone. *Lux* existed in ways quite different from the physicality of light. His views are rather the culmination of a medieval tradition of the 'metaphysics of light' than the beginning of a modern science of optics.[39] Physical light is a symbol of spiritual light, the formal as well as the efficient cause of nature. Actuality was drawn out, held together, and formed into a natural hierarchy by the integument of a divine light.

It was only one of the qualities of light to act as a force, and even in this respect was never the bearer of a purely mechanical power. The influences and sympathies conducted from the centre of the universe to its circumference and back again were laden with purpose and meaning. In identifying a 'natural mechanism' for the expression of the divine will, Grosseteste, rather than establishing a domain of independent necessity, sought to implicate God more intimately in the operation of Creation. It was therefore the identity of formal and efficient causality in light that placed the science of astrology on a sound basis. The relation between the heavens and the sublunary world is primarily, but not exclusively, analogical. Certainly it is not, as Arabic and Jewish writers had assumed, a matter of inherent necessity.[40] John of Salisbury, among others, objected to naturalistic astrological hypotheses: 'For they impose on things a certain fatal necessity under the guise of humility and reverence to God, fearing lest his intent should perchance alter, if the outcome of things were not made necessary.'[41]

It was just because astrology was not a 'natural science' but part of the interpretive schema of the medieval world view that it became so well developed. It thus played a central role in medical theory and practice. The microcosm of bodily humours were 'mixed' in much the same way as the earthly elements, and both were subject to an internal harmonics controlled by the distribution of the heavenly bodies.[42] The balance of humours within individuals depended, therefore, on the state of the heavens at their birth, and no general diagnostic or therapeutic axioms could be applied without careful adaptation to the cosmic situation of the patient.[43] Health depends upon maintaining the original harmony between microcosm and macrocosm. Man falls ill because in exercising his inner freedom he offends against nature. Nature by itself would never suffer such irregularities. Albertus Magnus, the most learned naturalist of the thirteenth century, points out that 'there is in man a double spring of action, namely, nature and the will; and nature for its part is ruled by the stars, while the will is free'.[44]

Albertus, who was also one of the most respected and authoritative of theologians,[45] held that such control could only be exercised by the descent of a real power through the hierarchy of being. Light and other celestial 'virtues' impressed upon the sublunary world the pattern of divine things. Nature strove to conform to this pattern, just as man strove to return to God. The modelling forces of the cosmos could be enhanced and 'focused' by careful manipulation. At an appropriate time, a particular planetary conjunction, a particular house in the ascendant, an astronomical 'seal' or 'signature' could be traced on the smooth surface of a stone, or gem, or piece of metal. We are assured that 'Marvels are worked by such images';[46] a force flowing into them is gathered and concentrated by the image. This 'natural magic' is not unlike William of St Thierry's theory of love; the image is a part of corrupted nature so formed as to be receptive to the transforming power of superior celestial virtues. Nature, as well as man, could be redeemed if it were made ready to accept God's love. If ordered, its elements distinguished and formed into separate spheres, there would be no obstacle to the penetration of the most sublime cosmic forces. As it is, the materiality of nature only accidentally exists in a receptive mode, and divine illumination appears as an unpredictable 'marvel' of nature rather than as its inherent tendency.

The 'formative powers' of the heavens none the less operate continuously and to some effect. In *The Book of Minerals* Albertus describes the 'mineralizing power' through which stones are formed.[47] Earth and water, when confined together in an appropriate place, are condensed into stone by the action of celestial virtues. It is the earth giving birth to itself, a relatively simple process because stones, like metals, are homeomerous substances – that is to say, perfectly simple, uniform, and identical in 'occult' or inner qualities as in external accidents. And, being internally undifferentiated, stones and metals are less demanding of specific conditions of place or astral disposition.

Albertus thus admits the possibility of alchemical transmutation. Though an art practised by pagan philosophers, it exists also in a legitimate form: 'For whatever the elemental and celestial powers produce in natural vessels they also produce in artificial vessels, provided the artificial are formed just like the natural.'[48] The generation of metals can therefore be stimulated, albeit with difficulty, in the workshop. The practical difficulties, however, are immense. All metals are formed from stable mixtures of two fundamental substances: sulphur and quicksilver. Each resulting 'specific substance' is a metal with its own peculiar qualities, including 'affinities' with particular celestial virtues. Metals act as conductors, so to speak, of celestial virtues. Hence their importance as medicines, or as ligatures to be worn as protective devices. In association with favourable astrological circumstances, 'the elemental and celestial powers of the material', captured in the process of its formation, are released, 'healing' affected organs in its vicinity.[49]

Alchemist and doctor alike 'purify' the unhealthy mixtures of corrupted nature. In manufacturing and transmuting metals the alchemist is acting as a physician to the natural world, separating its elements and bringing them together in permitted combinations. The alchemist seeks an elixir which is a medicine that cleanses the womb of nature. Gold, as the purest of metals, is the form in which the metallic principle 'ought' to exist. There is nothing illicit in the artificial arrangements of sublimation, calcination, and distillation, because 'nature itself performs the work, and not art, except as the instrument, aiding and hastening the process'.[50]

Celestial virtues are contained also in particular plants and animals. The herbal and bestiary, therefore, as well as the lapidary, formed the corpus of popular scientific knowledge in medieval society.[51] For many later commentators they have been interesting as examples of credulity and observational inaccuracy. From them it appears that medieval writers had no real interest in the natural world. Their purpose, however, was classificatory rather than naturalistic description. What mattered was the isolation of significant or salient 'features' of an animal or plant; in other words, their distinguishing features, not as members of a species so much as items in a symbolic register. A particular animal is of interest not in terms of its 'horizontal' relations with other similar species, but with reference to the 'vertical' relations with the higher powers that shape it. If the blood of a camel, for example, 'be put into the skin of the beast called Stellio, which is like a lizard, having on his back spots like stars, and then set on any man's head, it shall seem that he is a giant, and that his head is in heaven'.[52] The relation between camel and stellio, or both with man, is not interpreted then as part of a natural 'ecology' but as part of a system of cosmic symbolism.

If the created sublunary world could be disassembled and recombined in terms of essential proportions only, and each nature confined to its uniquely appropriate place within the overall design of the cosmos, then 'matter', in becoming perfectly formed, would cease to exist. It is its imperfections which give to the material world all its characteristic 'substance'. But nature, as much as man, constantly strives to go beyond itself; to realize its ideal form as order and thus escape the perpetual inner movement of generation and decay. It strives for a pure ideality, to conform itself perfectly to the macrocosm, and to allow itself to be shaped by divine celestial powers. Nature, no longer free to revel in the freedom of first matter, also longs for happiness. But it is a longing that cannot be satisfied; nature is fixed in its imperfections, stranded in a material realm between formless substance (primary process) and insubstantial form (reason).

FLUX

Nature, as the unfortunate realm of material existence, is in a state of perpetual movement. It continually strives to overcome the violence inherent in its disorderly appearance, to actualize its form against the tendency of matter to revert to the carnival of pure substance. All movement is a quest to actualize the paradise of appropriately ordered places.

Substance and place are defined in terms of each other. The actuality of any object implies a particular spatial relation to the cosmic order as a whole. Space and time, in other words, can never be abstracted from the material, sublunary world and treated as 'empty' extension.[53] If everything were 'in its right place' there would be no motion, no generation or corruption, no 'material' life whatever. All forms of movement, as transitions from potency to act, are either 'violent' and disorderly, as when a heavy object is lifted clear of the ground, or 'natural', as when the same object is released and falls.[54] The latter movement is inherent in the order of nature itself. Thus, similarly, fire tends by its essence to rise above the air and assume its 'correct' position directly beneath the sphere of the moon; or air released under water will rise 'naturally' to the surface. The 'lightness' of air is an inherent quality and as such defines a specific place as well as particular 'physical characteristics' of a substance. The two are only conceptually distinct for us because, no longer living in a feudal society, we cannot immediately apprehend the world as a figurative order. Forced or 'violent' motion, on the other hand, is not an inherent tendency but requires the continuous application of a force external to the moving object.

As change of place implies the alteration of substance and the transformation of substance implies change of place, all 'movement' is simultaneously an 'intension' or 'remission' of sensible qualities. 'Mechanics', the science of physical change, deals with far more than simple physical systems or the 'laws of motion'. Changes in colour, and especially in temperature, were often regarded as paradigmatic of the physical transitions to be treated by a mechanics of qualities.[55]

The special case of 'local' motion none the less raised some of the most intriguing and difficult problems. The development of kinematic ideas throughout the medieval period demonstrates a more precise, but no less feudal, conception of a natural order.

All 'mixed bodies' (containing some combination of the four sublunary elements) could be moved from one place to another as a consequence of a motive force exceeding the resisting or restraining force offered by the inherent heaviness of the object and the *plenum* through which it moved.[56] This covered such obvious everyday cases as the turning of a waterwheel or windmill. An external force, the flow of water or air, was the motive force responsible for the rotation of the wheel, which continued just so long as this motive force was effective. This was no different in principle for sentient actions, where the motive force took the form of an intention or purpose within the soul of the acting creature, and its material reality offered its resistance to the immediate realization of this *conatus*.[57]

The interesting and more difficult cases involved projectiles. A ball could be thrown, or an arrow fired from a bow, resulting in an appreciable motion that was apparently independent of the initial, external, propelling force. Though creating obvious difficulties, the most popular accounts of such cases utilized the *plenum* as an intermediary motive force. In the most common instance, the air displaced by the forward movement of the projectile was thought to be re-formed immediately behind it (thus preventing the formation of a vacuum), and by its agitation conveyed a continuing thrust to the moving object.[58] Rather obscurely, the initial violent force is gradually absorbed by the resisting medium, resulting, as the natural downward motion of the heavy body asserts itself, in a typical decaying trajectory.

During the 'first feudal age', that is to say, local motion, including violent projectile motion, was held to depend upon the continuous application of an external force. Local social action, similarly, required the continuous application of a superior power. The activities of the manor depended entirely upon the presence, actual or symbolic, of its feudal superior.[59]

During the later medieval period, however, in the new 'secular' and urban atmosphere of the schools that were developing somewhat apart from the traditional centres of feudal culture, a rather different analysis was proposed. These innovations, particularly those associated with Merton College and the University of Paris during the first half of the fourteenth century, so impressed Pierre Duhem that he effectively dated the scientific revolution from them.[60] Important as they were, however, they fit more easily the context of an 'advanced' feudal than an 'early' capitalist culture. The greater attention paid by these authors to secondary causes and the mathematical analysis of local motion did not yet take precedence over the transcendental goal of all knowledge.

The Oxford 'calculators', Thomas Bradwardine, William Heytesbury, Richard Swineshead, and John Dumbleton, developed a new quantitative approach to the analysis of local motion.[61] This arose, none the less, from a traditional scholastic distinction between the intensity of a quality (for example, temperature or speed) and the quantity or magnitude of the same quality (correspondingly, heat or distance). There seemed an obvious and important distinction between the 'heat' in a given quantity of water at a particular temperature and that in a double or treble quantity at the same temperature. Similarly, 'speed' could be simplified to a measure of the 'instantaneous velocity' of a body at a given moment, and should be distinguished from the 'magnitude' of the velocity sustained over a given period.[62] The idea of 'speed', in its simplest form, was therefore a completely abstract notion and could only be defined as a potentiality; as the distance that would be traversed in a given duration at a given intensity. The technical significance of the notion of 'instantaneous velocity' was in allowing a much clearer description of accelerated motion. By geometric methods, simple cases of uniform acceleration – that is, of a uniformly changing instantaneous velocity – could be resolved into equivalent statements of non-accelerated motion.[63] Time, space, and velocity in all this remain qualitatively distinct and possessed of their own inherent qualities. Mathematical manipulation has no implication here for physical theory. There is no suggestion of a real interchangeability of such terms.[64]

After 1530 the Merton analysis was taken up and elaborated in Paris by, among others, Nicole Oresme. He aimed at a qualitative and symbolic 'geometry of qualities' at the most general level. The intensity and magnitude of any quality could be represented graphically, and the resulting figures used to suggest the distinctive characteristics of the phenomena under study.[65] Tempting as it might be to suppose, his method in fact did not amount to an anticipation of co-ordinate geometry or the technique of covariance. It was the dimensions of a *single* quality that were 'mapped' in relation to each other. And the resulting figures were not read as abstractions from physical dimensions but interpreted symbolically as the 'configurational' essence of the phenomenon itself. Thus, 'not any quality can be imagined by any figure'.[66] Colour or sound, for example, could be represented figuratively, and Oresme claims that the aesthetic appeal of either depends not upon simple arithmetical harmonies (such as pitch) but on the simplicity and purity of their overall configuration.[67]

Some 'astrological' and 'magical' effects, not attributable to the defect of the interior sense organs, could similarly be accounted for on the basis of configurational 'sympathies'.[68] Hence the medicinal properties of particular gems and plants,[69] and the 'fitting accord' between the 'configuration of qualities' in magnetized metals which accounted for their mutual attraction. Such a configurational conformity, in a more complex fashion, accounted also for the mutual attractiveness of human friendship.[70] Likewise, discordant configurations are held to account for the phenomena of repulsion, enmity, or the 'fact' that 'hot goat's blood fractures the diamond'.[71] And, in a more general sense

Oresme claims that 'change in this harmony or deformity can be one of the causes why heavenly bodies emit below more benign, and at other times less benign, influences'.[72]

Oresme's 'method' is simply a refinement of the analogical model. The 'harmonies' within the macrocosm shape and form the material microcosm as their image. The most significant of the causes of physical change in the sublunary world is discovered in the form of things. It was in fact the inability of scholasticism to conceptualize natural relations in abstract fashion, rather than its excessively 'metaphysical' disdain of observation or experimentation, that acted as a brake upon its 'scientific' development.[73] The description of any part of the world had to be made in terms of some quality inherent in an object. The world was a plenitude of such objects. The 'dimensions' of existence were themselves the qualities of nature. Space and time could not be treated as the separate and 'empty' containers of the physical world; they were an ordered part of its content. Nature could not therefore appear universal, manifesting at best a series of localized uniformities 'accidentally' generated by the interaction of such qualities.

The continued discussion of problems of projectile motion, culminating in Jean Buridan's sophisticated impetus theory, is, once again, a development within the general feudal picture of the world rather than a breakthrough into fundamentally new dynamical concepts.[74] Buridan raised a number of empirical objections to the traditional Aristotelian treatment of projectile motion: that, for example, a lance pointed at both ends does not fly more slowly than one blunted, and offering greater resistance to the 'push' of the ambient air, at its trailing end. And on a barge, immediately after hauling has ceased, everyone feels the air resisting rather than assisting the continuing forward motion. 'Therefore', he concludes, 'it seems to me that it ought to be said that the motor in moving a moving body impresses in it a certain impetus (*impetus*) or a certain motive force (*vis motiva*) of the moving body'.[75] An external force, on contact, is taken up and becomes an inherent quality of the moving object. The ambient air, or other medium, gradually overcomes this impetus. There is, additionally, a positive connection between the quantity of matter and the effective force required to impart an impetus to it. Once impressed, impetus does not decay except by the action of an opposing force. Heavenly bodies moving in a 'pure' medium offering no resistance, thus maintain for ever their perfect circular orbits with undiminished velocity. And a freely falling body accelerates, not from some occult 'attraction' it feels for its proper place in the universe, but from the continuous addition to its impetus of the constantly acting force of its own weight (*gravity*).[76]

Projectiles depend for their motion upon an 'impressed force'; that is, upon a power 'borrowed' from some exterior and superior being. Motion is not itself a condition or state of being; it remains a transition within a partially ordered hierarchy. Local motion as impetus was no more independent of the cosmic hierarchy than it had been in the traditional Aristotelian conception of natural and violent motion. Its origin and its *telos* transcend the empirical domain. Each local motion is the outcome of an impressed force which is itself the consequence of a 'higher' force or an act of intelligence. All motion is ultimately traceable to God.

In the context of the more highly developed 'second feudal age', the projectile is conceptualized as a vassal rather than as a serf. Nature is composed of relations of mutual obligation which constrain the projectile to act 'as if' continually directed and controlled by an extraneous power. It is self-propelled only because nature has been sufficiently well formed, in that domain, to guarantee a 'correct' outcome. It can be trusted because it has absorbed into its very being not a blindly acting force but the transcendental *telos* of the cosmic hierarchy. Buridan considers particular examples only. He does not propose an image of nature as the outcome of undirected forces impressed upon helpless objects. Each motion can be referred upwards, towards a directing intelligence, so that nature's flux is merely a further development of that ideal form which already exists in the mind of God. It is a continuous sorting and resorting of recalcitrant matter. The *dramatis personae* have changed, from the theophanic Silva and Noys, to the apparently more realistic 'gravity' and 'impetus', without however altering the figurative status of nature.

That 'nature' had come into greater prominence by the fourteenth century is hardly to be doubted. The implicit model of vassalage, rather than of serfdom, is expressive of a genuine social development. The connection between these different aspects of social relations can be pursued in yet another metaphor. The money economy, which played only a minor part in the organization of social life in the earlier phase of feudalism, came to play, especially in the developing urban areas, an increasingly significant role.[77] This was commented upon by scholastic writers themselves, notably by Oresme, in a short treatise, *De Moneta*, written about 1355.[78]

Oresme begins by pointing out the rational origin of money; as 'exchange and transport of commodities gave rise to many inconveniences, men were subtle enough to devise the use of money to be the instrument for exchanging the natural riches which of themselves minister to human needs'.[79] Money, which cannot be directly consumed, materially, aesthetically, or psychically, is 'artificial riches'. It does contain, none the less, a worth of its own, otherwise it would be useless in facilitating exchange. The fact that money could be exchanged for commodities demonstrated the existence within it of some inherently valuable quality. Precious metals are thus ideally suited to minting since they contain a high value within a conveniently small weight. Oresme supposes indeed that coins originated as a means of overcoming the inconvenience of weighing out specific quantities of precious metals each time a transaction took place. The prince's head, a symbol of truthfulness, replaces the laborious process; stamped upon each coin it differentiates and guarantees its value.

Money, to be effective, must be 'genuine'; it must possess an almost 'physical' power. Its inherent value must precisely match that of the commodity for which it is exchanged. But as the value of a commodity depended upon its intrinsic qualities, each was, by definition, incommensurable with another. Money was used therefore to sustain the fiction of equivalence between two inherently dissimilar objects. Like the symbols of vassalage, money presumed and therefore created a relationship of equality.[80]

It was inconceivable on this basis that money could become a general 'medium' of exchange. Its use was limited and controlled by the 'transcendental' needs to which it was put. It operated only as an intermediary between the arousal of a need and its eventual satisfaction. The 'reality' to which money responded did not reside in the commodity itself, but in the 'spiritual' realm which lay beyond it. Money departed from and returned to the *telos* of a guiding intelligence. It could not of itself generate a 'system' of exchange, because each act of exchange was circumscribed by some specific social relation itself bound to its own place within the feudal hierarchy. As the efficient cause of exchange money must, so to speak, draw breath and renew itself after each expenditure of energy. It cannot spontaneously regenerate itself as the 'multiplication of species' typical of the superior forms of being such as light or loyalty.

Within the social world, that is to say, money was a kind of 'impetus', and commodities were the imperfectly formed 'material nature' strung out between the primordial chaos of carnival and the realization of society's inner rational values. Money was the immediate, efficient cause of the movement of commodities from one place to another. It imparted to the commodity a directed tendency, sufficiently powerful to overcome the resisting gravity (inherent value) which rooted it to the spot. In doing so it 'spends' itself and becomes exhausted – only to be renewed by the inextinguishable need regenerating itself within a directing soul.

Money is the impetus of commodities; impetus is the money of nature. They are the means by which things change place, and in doing so actualize their inner qualities. As intermediaries, money and impetus can be construed as both cause and effect of the flux of the temporal world. Money effects a redistribution of material things in accordance with the tendency of an implicit order. [81] Money, in the social sphere, is no more a solvent of the feudal hierarchy than is impetus of the traditional conception of nature. Both man and nature, degraded and corrupted but capable of redemption, are unstable. The superficially hectic activity of both betray a continuing effort to receive more completely the forming spirit whose presence counteracts the tempting slide into chaos. Both seek the ultimate release into pure, non-material being.

Nature and man are united in their longing for happiness. Their common exigency is the want of God; a longing to discover the permanence of being, as opposed to the transitoriness of existence. To perfect themselves they need only resign themselves, unconditionally and absolutely, to the supremacy of a Rule. But both nature and man are confined within the limbo of the material world because of their inability or unwillingness to be obedient. Both therefore suffer the deceptive flux of experience, the impermanence of material forms into which they can never pour their true being, the subordination and dependency of low status within the hylomorphic order of the cosmos.

THE MIRROR OF GOD

Feudalism is an invention of bourgeois society. It is the product of a particular kind of historical consciousness, a way of describing the past which cleanses bourgeois society itself of relations it deems anachronistic. It cannot be described, however, as another form of the 'irrational'. It is not another instance of fun. What is attributed to feudalism (and thus rejected by capitalism) is a particular species of reason. The categories of medieval culture are constructed with meticulous regard to the law of non-contradiction, but its unity and coherence depends entirely upon an order of symbols: an order which reveals itself at every turn as a hierarchy of being.

The cosmology of happiness must therefore be reconstructed through a double relation. It is, first of all, an historical phenomenon. This presents us now, as it did not for the originators of the notion of 'feudalism', with certain ambiguities. The core of feudalism can be seen in terms of the slow unfolding of a reality whose fullness we have ourselves come to occupy (the urban values of the second feudal age, the twelfth-century renaissance, the growth of the state). But it can also be seen as a system of relations irreconcilably opposed to those of bourgeois rationalism (hierarchy, personal dependence, the symbolic interpretation of nature).[1] This ambiguity is itself merely a sign that contemporary scholars no longer share the larger world of assumptions and values common to the historical sciences during the period of high capitalism.

The cosmology of happiness, secondly, can be examined as an internal relation – internal, that is, but not self-sufficient. Where the cosmology of fun (mistakenly) believes itself to be an absolutely exhaustive reality, the world of happiness is sustained by the continuous application of the force it exercises upon a primordial, unformed reality. In some respects, feudalism barely mastered the restless and subversive spirit of fun. Nature, in which man includes himself, is in a permanent state of flux. The intoxication of the carnival is never far off; certainly, it is too close to the surface of life to have been forgotten. Fun has therefore to be controlled, a necessity met by transforming it into a self-conscious goal. This immediately opens an enormous 'gap' between the world of experience and the 'ideality' of pure being,[2] a gap which feudal culture devotes itself to attempting to cross. Fun made absolute remains aloof. The object of unquenchable longing, its presence transcends and at the same time organizes the world devoted to its helpless pursuit.

A series of separations accompanies this transformation. The empirical world becomes a small segment of a larger cosmos. Every object and event are tokens of the meaning that summons their existence. Substance and form, the sensible and intelligible, body and soul, are held on the point of inner dissolution. These relations fall into a unique hierarchical order which both defines existence and provides the means of salvation from it. Happiness is the summation of this order, the self-transcending movement which is the *telos* of all human and natural motion. The infinite regress of the 'ways to God' turns earthly activities into a rational tool for the attainment of an ultimate release from striving and change.[3] The mind seeks to raise itself to the apprehension of its own prototype.

The ultimate tendency of happiness is repose, tranquillity, and rest; to dwell in the fixity of its own essence.[4] It seeks to overcome ceaseless metamorphoses and in doing so establishes the degraded world of nature as its symbol.

The cosmic body, confined within the sublunary world of imperfect being, therefore, cannot fill the universe. The order of the world cannot be immediately sensed or felt, but it can be known analogically through the development of proper self-knowledge. Man, as the image and likeness of his Creator, is the mirror of the cosmic order. His mind, limited and weak as it is, none the less participates directly in a hierarchy charged with the divine presence.

Occupying a strategic position within the great chain of being, man is also the focus of universal secondary causes; a point through which higher forces act upon objects in the lower world. He thus comes to know, through reflection on his own experience, as well as by the revelation of his inner being, the order of things.

Unlike the cosmology of fun, the irresponsibility of which in this regard constitutes its central social meaning, the cosmology of happiness grudgingly recognizes the necessity of labour. It is, however, kept in its 'place'. The recurrence of material needs, a blemish upon human nature, is dignified only to the extent that it remains embedded in social relations which derive their authority from a higher sphere. The direct satisfaction of material needs can never be the basis of social activities within a cosmology of happiness.[5]

We view the cosmology of happiness now from a somewhat different position. It was its closeness to the bourgeois world that excited the historicizing imagination of an older generation of scholars. Now it seems an oddly remote world.[6] The primitive or the lunatic are more disturbing and more subversive; the repressed becomes more dangerous than the simply ignored. We find it harder to accept the inner meaning of symbols than we do the arbitrariness of signs.

In allowing the domain of the symbolic to fall into neglect, the bourgeois world has apparently succeeded, much more completely than might have been expected, in ridding itself of its undischarged longings for redemption. Yet we would understand nothing of feudalism had such longings fallen completely into abeyance. In using them to furnish a psychological portrait of an age we have 'transcended', we allow them at least a vicarious satisfaction.

Part Three

PLEASURE

The only rational passion is self-love.

Jean-Jacques Rousseau

We are certainly not to relinquish the evidence of experiment
for the sake of dreams and vain fictions of our own devising: nor
are we to recede from the analogy of Nature, which is wont to be
simple, and always consonant to itself.

Sir Isaac Newton

Chapter Seven

HARMONY

The problem of the transition from feudalism to capitalism has excited considerable scholarly attention, both as part of a general reassessment of the writings of 'classical' sociologists and, more recently among economic and social historians, as a perspective that breaks free of confining disciplinary boundaries.[1] It has figured much less prominently within the conventional history of art and culture (including the history of science), whose distinctions, fixed by Burckhardt, assign a positive value to this intermediary zone as 'humanism' and the 'Renaissance'.[2]

The transformation of both cosmos and psyche implied in the categorization of 'capitalism' and 'feudalism' is not a straightforward historical issue. *Both* terms belong to the cultural world of capitalism. Thus the effort to define in a more precise way the boundary between the two, and to specify the historical process connecting them was bound to become an ultimately solipsistic exercise.[3] The real history of such a transformation remains unreachable. Attempts to write such histories, however, reveal the bourgeois world view at its most vulnerable. Through them we can observe the formation of the bourgeois world's central values and indispensable metaphors. The symbolic realm of happiness is thrust back into the incoherence of the past, the anarchism of fun is finally suppressed, and reason takes its place at the centre of world history.

The foundation myth of capitalist society thus involves two rather different transitions: the movement from *happiness* to *pleasure*, which can readily be conceived as the 'subjective' aspect of the transition from feudalism to capitalism, and the movement from *fun* to *pleasure*, which has a much more diffuse range of historical reference as the domestication of the primary process. The order of symbols, on the one hand, and the multiplicity of systems of signs, on the other, gave way to the mechanistic relations of cause and effect. They gave way, however, from within an already constituted bourgeois society. The cosmologies constructed by 'mirroring' (symbols/happiness), and 'somatically' (signs/fun) are the creations of and belong still, if negatively, to the world mapped out by bourgeois reason.

It was in a conception of reason self-consciously purged of happiness and fun that 'man became a spiritual individual', a being 'not afraid of singularity'.[4] At the origin of the bourgeois world, 'self-realisation and self-enjoyment of the personality became a goal'.[5] It was only much later that pleasure became somewhat detached from reason and it was possible to imagine the development of the latter as a process dependent upon the repression of the former.[6] In fact it depended on the repression of fun and through this the establishment of pleasure as a legitimate value for, and organizing principle of, personal and social life. It describes, in other words, a different kind of ideal society, one founded upon non-hierarchical relations.

UTOPIA

The movement from happiness to pleasure is first of all the discovery of Utopia; the replacement of the transcending *telos* of human action by a secular vision of perfection. Utopianism, of course, is endemic to western society.[7] A society continually stretched between the promise of salvation and the degraded spectacle of everyday life is prodigal of secular, as well as purely religious, forms of the ideal. The Renaissance stands out none the less as the period of classical Utopias and the period during which Utopian thought comes closest to realizing its systematic ambitions. The classicism of the Renaissance was a fairly self-conscious attempt to reach back beyond the fixed symbolism of the orthodox religious world view and recover a more potent image of human perfectability. Direct borrowing from classical culture was, however, slight and constituted no more than a point of departure from which Utopian visions, architectural, literary, and artistic, could take flight.[8]

Informing the varied expressions of secular longing was a

rejection of hierarchy as the necessary condition of all existence.[9] At the outset of the Renaissance, anticipating in a thorough and comprehensive fashion its fundamental social logic, stands Nicholas Cusanus, whose *On Learned Ignorance*, aiming to make no original theological advance, in fact radically transformed the medieval Platonic tradition.[10]

Cusanus begins with the absolute unknowability of God. The simple maximum, as he terms Him, cannot be reached by any reflective science. All knowledge is founded, he claims, upon comparison. Where there is 'comparatively little distance' separating the 'object of enquiry' from the 'object regarded as certain', it is a relatively simple matter to make correct judgements leading to valid knowledge.[11] But God is infinitely remote from living man; this is the real meaning of his divine perfection. No hierarchical sequence can ever terminate in authentic transcendence. However extended and exalted, it must always cling to earthly knowledge and material forms. Thus, 'the infinite as infinite is unknown ... since it is beyond all comparison'.[12] Although human knowledge and human will is endlessly self-expanding, it cannot become identical with the simple maximum, which remains indivisible and without any distinction whatever. 'The Good cannot be reached by any series of inferences that begins with an empirically given,' he points out briefly, undermining the entire scholastic tradition of the 'ways to God'.[13] All valid knowledge must begin with the recognition of this radical ignorance. The entire edifice of hierarchical logic is consequently subverted. In direct opposition to the scholastic principle of 'relative greatness', Cusanus calmly asserts that 'from the self-evident fact that there is no gradation from infinite to finite, it is clear that the simple maximum is not to be found where we meet degrees of more and less'.[14] If we begin with the created world, we must remain enclosed within it. The cosmology of happiness had viewed reality as a hierarchy of emanations from a transcendental order just to avoid this logical problem. It was God who, in creating the world, passed into the forms of things and thus offered a miraculous 'way up' for man. The decisive transformation effected by Cusanus lies in his beginning from the created world, and therefore having to deny the possibility of arriving at God by a process of abstraction, confining himself within its limits.

The created universe itself constitutes a maximum, but a

'restricted maximum' which, while existing in a form which maximized its own inner potentialities, could not be other than limited in an absolute sense. The original dependence of the universe upon God was not called into question, what could no longer be supported was the view of existence as a gradation of His absolute and necessary being. The empirical world, formed by God, existed in itself, and persisted on the basis of its own inner laws. God was removed to an infinitely distant point, and could be found once again only by discovering the minimum which lay as incommensurably 'beneath' all possible comparison as the maximum lay 'above' it. 'God's being which is unity', he tells us in a typical passage, 'is not abstracted by the mind from things, nor is it united to or immersed in things; it is therefore, beyond anyone to understand how the plurality of things is a development of the unity which is God.'[15]

This leads Cusanus to many apparently paradoxical expressions of the nature of creation. The cosmos is limitless but bounded, a place in which 'the centre and circumference are identical'.[16] Not only are we condemned to utter ignorance of God, but all our relative judgements of created being are threatened by our inability to 'fix' the end points to the chain of being. All judgements of 'place' are similarly suspect in a universe without centre or circumference. Any arbitrarily chosen point might, equally, serve as its centre. In some sense God undoubtedly is the centre of Creation, but in a more restricted physical sense 'the earth is not the true centre of the world' and 'the circumference of the world is not the sphere of the fixed stars'.[17] And what applies to the earth is true of all other places: 'in the heavens there are no fixed, immovable places'.[18]

The entire feudal cosmic structure is dissolved and reformed into a multitude of self-centred points. Man, roused from his lowly position of centrality, inhabits a world as dignified as any other celestial object; possessing its own light and heat and as free to move as any other. The categorical distinction between superlunary and sublunary has been shattered, and with it the possibility of a qualitative physics of place.

The full implication of this radical break with the symbolic-hierarchical conception of the cosmos was not immediately evident. For the most part, 'Renaissance cosmography' was a mixture of what now appear to be incompatible elements; some belonging still to the scholastic religious tradition, while others, often ill-articulated and confused, belonged to a world not yet properly formed. What did become clear at once was a new philosophical independence. Natural reason was no longer harnessed to the task of clarifying and elaborating revealed truths. As an instrument for the exploration of nature it reflected the dignity and power of man as a privileged being, conscious of his own actions and master of his own instincts.[19]

Man's knowledge of the world was no longer contemplative. His mind was not a mirror so much as an active and expanding principle of mastery. The 'image and likeness' of God was manifest actively in his creative powers, rather than passively in his form. And as the mind could no longer be guided by an abandonment of the will, its restless self-expansion found enjoyment in its own existence. Of divine beatitude, that happiness promised but vainly sought within the heart of the old society, Lorenzo Valla rhetorically asks, 'who can call it anything better than pleasure?'[20] It is pleasure and not virtue that is to be 'sought on account of itself'.[21] Valla, a 'Christian Epicurean', doubted, in the absence of an authoritative order of symbolic valuation, the wisdom of men's actions, but he was confident of their motives: 'men act for the sake of pleasure'.[22] Reason aspires to coincide with the amplitude of our experience of the world, but cannot go beyond it. Reason thus becomes a form of pleasure, an intelligent appreciation of the world of created things.[23] Pleasure receives a fresh value as 'the conserving principle of life'.[24] Human aims must be realizable, and realized, within the constraints of a secular world no longer camouflaged as degraded matter. God is not renounced, but the passion that had sought its release in the happiness of seeing him 'face to face' spent itself in the erection of secular Utopian visions.

Throughout the Renaissance, which acts as a sort of cultural buffer between the cold rationality of capitalism and the symbolic order of feudalism, Utopian excitement continually renewed itself in all the human arts. Petrarch's revival of classicism in letters not only inaugurates a collective search for civic virtues which did not reach its zenith until the eighteenth century, but also it has as its prelude the first of those solitary walks which became characteristic of bourgeois self-reflection. The ascent of Mont Ventoux, a strenuous and unusual journey prototypical of the wilful detachment from social life of Montaigne, or Rousseau, or Kierkegaard, is a new moment of personal inwardness as well as a new evaluation of 'nature' as a beautiful object. The ego, just as celestial points in the 'restricted maximum', crystallizes into an independent unity and discovers itself in isolation.[25]

The artist, no less than the philosopher, confronts the challenge of comprehending and representing the natural world without the aid of symbolic interpretation. If the cosmos is no longer an order of fixed, dependent relations, how is it to be conceived? How is it constructed? What guarantees its coherence and persistence?

The artist's answer was as clear as the philosopher's, and considerably in advance of the scientist's.[26] The theory of harmony, taken from Greek sources rather than from Boethius, and developed outside of any direct religious preoccupation, provided the fundamental starting point.[27]

The first and most striking application of secular harmonics was in architecture. Compared particularly to the Gothic cathedral's crushing realization of divine superiority over the human frame, the Renaissance church was scaled to the intimate proportionality of the human body. Panofsky highlights the difference succinctly: 'medieval architecture preaches Christian humility; classical and Renaissance proclaims the dignity of man'.[28] And in painting it meant that 'the painter is no longer to work "from the ideal image in his soul" . . . but from the optical image in his eye'.[29] The space thus defined was continuous and infinite. Exact geometrical perspective and proportionality was developed from the notion of a visual image produced by straight lines projected on to a two-dimensional plane. Classical attempts at perspective drawing had constructed the visual field as a virtual sphere, centred on the eye and composed of straight lines of varying angle but equal, finite length. In Brunelleschi and Alberti's new theory the lines of sight were infinitely extended and the picture produced by introducing an orthogonal plane of projection artificially cutting them off.[30] The single 'vanishing point' of the projection plane mirrored the unique position of both the artist and the subsequent spectator of his work whose individual view of the scene it recreates. Even the largest frescoes could thus become an intimate communication.

Now the artist 'strives to represent only the things that can be seen'.[31] And by adopting the visual perspective of a single privileged observer there is created upon the surface of the canvas an infinitely receding space as if framed by 'an open window through which I view whatever is to be depicted there'.[32] Perspective painting is therefore much more than a purely technical innovation; it 'opened not merely a new phase in the practice and theory of the visual arts but a new age in which reality came to be viewed and understood in mathematical terms'.[33] The Renaissance canvas is filled with objects detached from one another, not simply physically but metaphysically. They form a unity only by virtue of the subjective reconstruction of the spectator whose position is the focal point of all the formal harmonic relations exploited by the artist.

The use of individuated perspective and proportional drawing implies a separation of space from place. The 'objective' world is decomposed into a series of superimposed visual fields, the space of any one of which is identical with any other. Space, that is to say, is abstracted from any particular observation and is represented as the undifferentiated medium of any possible arrangement of objects. An object, indeed, is defined by its location in space, by its separation from and relation to other objects. Space itself lends nothing to representation because it is the 'container' and not the substance of objects. And all the physical qualities of an object can be rendered by the conventions of drawing just because they are independent of 'place'. The nature of things depends upon the proportions of its own parts, its interior harmonic arrangement, and not upon an invariant and all-encompassing structural order. The cosmos is therefore a structure of relations that can be described only from the point of view of a single, privileged, human observer.

The revival of Pythagorean conceptions of order provided a secular model of such an indwelling order.[34] It was a vision of the cosmos as an artistic work; the product of an artificer. The fundamental proportionality of nature could be rediscovered in the human body. Vitruvian figures, popular throughout the Renaissance, have therefore a quite different significance from the medieval 'microcosmic' theory of the body. Now, rather than belonging to different orders of being related analogically, body and cosmos are harmonized because they are both 'natural'.[35] What is at issue here is not a piecemeal criticism of the scholastic tradition or its religious assumptions, but a systematically different conception of the cosmos, a difference which makes itself felt in every particular.

It is from this perspective, the humanistic rejection of hierarchy and the revival of Pythagoreanism, that Copernicus's astronomical innovations are best understood. In the atmosphere of north Italian humanism, indeed, much of *De revolutionibus* could appear conservative.[36] The universe is held by Copernicus to be a physically finite sphere, 'the most perfect shape'.[37] And within it, 'the motion of the heavenly bodies is uniform, circular, perpetual, or compound of circular motions'.[38] Many of the traditional scholastic conceptions were retained, or modified only implicitly. His radical Pythagoreanism was moderated, more than was general among artists, by a Christian Platonism. In the *Narratio Prima*, for example, in which Rheticus first reports his master's discoveries, a conventional 'metaphysics' justifies observational astronomy. The outermost cosmic sphere thus 'was studded by God for our sake with a large number of twinkling stars, in order that by comparison with them surely fixed in place, we might observe the positions and motions of the enclosed spheres and planets'.[39]

The technical advances claimed by the Copernican system were two-fold: first, by a much more accurate determination of the phenomena of 'trepidation'[40] – that is to say, the very slight, long-term 'third movement' of the earth in relation to the backdrop of 'fixed' stars, more precise knowledge of which was essential to a rational reform of the calendar. Second, and more significantly, it proposed a considerable simplification in the geometrical constructions required to 'save the phenomena'. It reduced, in other words, the apparent motion of celestial bodies to an ordered and interconnected system of relations.[41] The real fault of Ptolemaic astronomy was not its prejudice in favour of uniform circular motion (a Pythagorean principle shared by Copernicus), but its incoherence as a unity: 'It is as though an artist was to gather hands, feet, head and other members for his image from divers models, each part excellently drawn, but not related to a single body, and since they in no way match each other, the result would be monster rather than man.'[42]

When Copernicus claims, therefore, that 'there is no one centre of all the celestial circles or spheres', he is being much less daring, in terms of cosmological speculation, than Cusanus. But, more significantly, when he argues that 'the centre of the earth is not the centre of the universe, but only of gravity and of the lunar sphere',[43] unlike Cusanus, his conclusion is not simply a philosophical assumption, and is consistent with the most complete and precise mathematical analysis of astronomical data. These were arguments which were bound, sooner or later, to carry weight amongst those who could comprehend them.[44]

The technical demonstration of the earth's movement was far in advance of any other astronomical work of the period, but its technical superiority alone does not fully account for the Copernican achievement. It is as part of a general cultural movement that the meaning of his work becomes clear. It is as the most sophisticated demonstration of man's inherent dignity that *De revolutionibus*, without ever being widely read, remained a vital work and aroused controversy for well over a century. In spite of the odd mixture of prescient mathematization and scholastic archaism,[45] it marked a decisive shift, by no means confined to 'advanced' and 'radical' cultural movements, in the relation of man to his cosmos.[46]

Rather than participating, by virtue of being its natural symbol, in the plenitude of creation, man becomes a real part of nature, proportioned and harmonized according to simple mathematical relations replicated throughout the universe. Man could understand the universe by observing the order of which he was part, rather than by interpreting himself as its symbol. The cosmos took on a purely objective aspect, first and foremost as an object of beauty. It was not sufficient to 'save the phenomena' by arbitrary geometrical manipulation; the diagram of the world had to express the 'real' harmonic relations sustained at the heart of nature.

Copernicanism, therefore, was not surprisingly embraced by the movement of humanism long before it became a 'scientific' model of the universe. Cast adrift in virtually boundless, if not infinite, space, remote from God, man's perfectability lay in his own hands. The longing for happiness, outliving its 'natural' medium of religious symbolism, enjoyed a period of intensification in innumerable Utopian visions. This entire movement was supported by institutions that lay outside of the traditional feudal hierarchy. North Italy had never been a genuinely feudal society, and from the late medieval period flourished as a distinctively urban, secular culture. Northern humanism similarly grew up outside traditional scholastic institutions.

This entire movement spread throughout northern and western Europe with the growth of the book trade, the first internationally successful enterprise organized along capitalist lines.[47] The most renowned writers of the age worked, from time to time at least, as publishers' copy-editors. Erasmus, Budé, and Rabelais were all employed at different times by Sebastian Gryphus in Lyon.[48] The spread of the book was phenomenal. In 1470 there was just one press in Spain, at Seville; by 1500 there were thirty, many in towns with no previous tradition of intellectual activity. And in Germany towards the end of the sixteenth century, at a time when universities had suffered a marked decline, books were the principal commodity at fairs.[49]

The printer and bookseller was the most 'progressive' figure of the age, the first to establish, through the direct application of capital to the process of production, an ideal world of commodities. Highly sophisticated techniques of manufacture were allied to a refined division of labour supplying a large and growing market for these luxury goods. Through them 'ideas' appeared in the market-place; freed from all restraint and authority, they anticipated rather than realized the inner freedom they were to claim as the core of the bourgeois world view.

In northern Italy, scholasticism, and especially the Aristotelian tradition of natural philosophy, far from being an obstacle to the innovation was itself the vehicle for a new secular vision of man and nature. Randall goes so far as to claim that the 'critical Aristotelianism of Padua' led directly to Galileo's revolutionary science.[50] This was certainly not the only, or even the major, source of Galileo's inspiration. The commercially successful north Italian cities of Padua, Bologna, and Pavia nurtured a humanist version of Aristotelianism that was important in its own right and serves as an indicator of the growth of a secular civic culture.

Influenced particularly by Averroës, a number of writers tackled the traditionally difficult problem of a 'naturalistic' theory of the soul. Medieval authorities held the human soul to be indivisible and immortal; separate in function, that is to say, from the bodily parts with which it was 'associated'. Yet, if the soul were genuinely immortal, it could not participate in any way in organic life; and if it were 'personalized' in particular individuals, it must be corruptible. The solution proposed by St Thomas had been to hold that the soul was quite separate from the body and participated in organic life by virtue of a 'miraculous' dispensation. The Paduans, however, insisted upon a more rigorously naturalistic approach. Pompanozzi argued that 'a natural bodily function can behold rational truth'.[51] Man's nature is thus 'a mean between mortal and immortal'.[52] The human duality comprised both a functioning organism and a power of thought and will which, because it was not differentiated and specific to a particular organ, could in some sense be called immortal. The intellect is this means which is neither 'totally free from the body or totally immersed in it'.[53] The intellect requires material conditions but is not reducible to such conditions; the soul grasps the universal.

The loss of the traditional doctrine of immortality and, therefore, the withdrawal of immediate divine supervision of individual human actions, makes way for a genuine moral philosophy. Just as God's retreat from the physical cosmos opened up a space into which human reason at once expanded, so a new secular individualism was at once invaded by imperative social norms. Both sought conformity to 'nature' rather than God; to an inherent harmony rather than a transcending spirit. Human action, therefore, ought to be guided by the 'natural' tendencies of human nature.

The idea of the immortality of the soul is transformed into a reflective image of the structure of nature. In Zabarella can be found a genuinely naturalistic psychology, in which the soul becomes identified with wholly organic functions. The soul is the 'principle of the animate body'.[54] It must therefore, like the body, be extended and divisible. It becomes an 'active intellect' which 'illuminates' sense images, disclosing their rational structure. Man's special nature is his ability to look inward and, by discovering the harmonic pattern within his own mind and temper, find the key to unlock the secrets of the universe. In other words, humanity becomes the central object of curiosity and study, out of methodological necessity rather than for reasons merely of self-glorification.

The empiricism of these writers was founded upon fundamental human certainties rather than upon the direct observational science of nature which, to a limited extent, it encouraged. It was a desymbolizing of the medieval cosmos, and, along with 'the sober recognition of its finite conditions', it retained the 'lingering odour of immortality', which Randall ascribes to all forms of humanism.[55]

The individuation and secularization of happiness (pleasure) was the 'inwardness' of the scientific revolution. Begun but never consummated in northern Italy, it preceded the development of the classical scientific world view. In Telesio and Patrizzi we can see the fullest possible development of Aristotelian naturalism. Telesio redefined the Aristotelian *potentia* as a natural force and the efficient cause of all change. The physics of 'place' was thrown over in favour of an account of motion as the outcome of purely mechanical 'forces'. The artist's extension of space as a medium, qualitatively identical with itself, turned out also to be pre-condition of a new mechanics. Space itself is conceived as devoid of all qualities and powers, and cannot be the cause of any physical change; 'therefore place (*locus*) must be made the container for all beings whatsoever' and 'remains perpetually the same'.[56]

Patrizzi, drawing directly upon Cusanus, also recognized the significance of the formal emptiness of space. Space 'is the accident of no earthly things',[57] and must be conceptualized as a 'hypostatic extension subsisting by itself and inhering in nothing else'.[58] Geometry, therefore, as the analysis of space, is the most fundamental of the sciences.

In this tradition force replaces form as the fundamental organizing principle of matter. Nature can then be seen as a system of mechanical forces expressive of deep harmonic relations which reappear in all its aspects. There is no need any longer for an analogical discourse; man is not like nature, man *is* nature. Man's own dignity therefore implied the rehabilitation of fallen nature as a whole; its discovery as an object of beauty.

ARCADIA

A second transition is possible. While primitive societies cannot leap into capitalism other than by the external force of colonization, a direct intellectual transition from fun to pleasure is suggestive of a real possibility[59]: an Arcadian fantasy, springing to life as a development within the Platonic tradition, in contrast to the Utopian movements which were for the most part an aspect of progressive Aristotelianism.

148

The refounding of the Platonic Academy by Marsilio Ficino, under the patronage of Cosimo Medici, reached back beyond the Christianized medieval conception of form as a divine emanation to the apocryphal sources of a more recondite wisdom. Works attributed to Pythagoras, Orpheus, Zoroaster, and Hermes Trismegistus figure in his writings alongside those of 'official' Neoplatonism.[60] Ficino indeed was as devoted a Pythgorean as Alberti, and at times seemed to have sought to revive a mystical religious cult in its name. It is a mysticism, however, that completely rejects the ascetic discipline of self-annihilation. Pleasure is the core of Ficino's spirituality. First and foremost, man's superiority and dignity are exemplified in the delight of the senses, particularly in his instinctive love of music. As both 'the human spirit' and the 'material medium' for the transmission of music is constituted by 'air', harmonic sounds can exercise 'a peculiar power' over the human soul.[61] Both are 'a living kind of air', and it is the inner harmonic movement of music (its playfulness) that makes it more effective as a spiritual nostrum than any painting or building.

Music had of course been long recognized as a therapy for the soul, and played an important part in medical astrology throughout the feudal period. Ficino's theory of 'astral music', however, goes much further. Music becomes a means of 'elevating' the soul of the healthy rather than restoring the bodily functions of the ill. Its psychological effects are universal. Music 'captures' planetary and astral influences. Celestial movements, conforming to deep harmonic patterns, are (as Copernicus and Kepler also believed) the ideal 'form' of music. These 'celestial figures act by their movement irradiating the soul with harmonic rays and motions', penetrate everything and 'constantly effect the spirit secretly, just as music does openly, in the most powerful way'.[62] The spiritual elevation which Ficino sought in special theurgic practices is an expansion of the ego; an identification of the subjective microcosm of fun with the macrocosm in such a way that man realizes his divine nature without loss of his temporal existence. The soul is led through the enjoyment of Orphic music 'to the deep and silent memory of the harmony which it previously enjoyed'.[63]

This sustained inner relation to divine harmony is even more marked in Ficino's celebrated student, Pico della Mirandola, whose *Oration* proclaims with a new confidence the status of man as a privileged creature. It is his separateness from nature that marks man's decisive advantage over any other created being. Man had been allowed the privilege of an unlimited freedom, 'the Creator gave him the germs of every sort of life'.[64] Man 'possesses all possibilities within himself',[65] but unlike the child unconscious of a world other than his own subjectivity, man through the inexhaustibly self-expanding forms of knowledge and love relates this inner infinity to an outer world of determinate being. Pico does not regard this as a tragic self-alienation but rather as an ideal of pleasure; the recognition of the infinite self in the world, rather than the longing for the infinite beyond any practical selfhood.

The preservation of inner freedom, therefore, becomes a moral imperative. And it is to preserve human autonomy that Pico attacks traditional astrological practices. He is not opposed to Ficino's recondite Pythagoreanism, which is an advanced psychological technique of self-discovery, so much as to the routine ascription of personal fate to 'occult causes'. If man is governed by stellar influences, rather than being merely conditioned by natural forces, his inner freedom is hopelessly compromised.

Ficino's philosophy is centred upon a revival of Eros; an aspiring upward and insatiable desire that is now, however, 'a human passion and not a divine goal'.[66] It does not terminate in a moment of ecstatic release but maintains a continuous reciprocal relation with its dialectical image. Man's implication in corporeality is not a punishment for his original sin but the unique configuration that allowed the inner plenitude of human nature to express itself. Ficino and Pico, by reaching back to pre-Christian spiritual sources, indeed avoided the entire narrative of human history as penance. Nature was to be the secular ideal of beauty: 'nothing in the world is misshapen or to be despised'.[67]

The individual soul is infinite; how otherwise could it conceive itself? Its insatiable desire can never be satisfied and it remains conscious of a sense of loss, 'wistfully reminiscent of another world ... thin and disembodied and ever trembling on the verge of the Christian mystery'.[68] But it holds back, suddenly suspicious of religious salvation, and remains bound up in an Arcadian nostalgia for the departed world of fun.

Neither Ficino nor Pico were Kierkegaard, any more than Zabarella or Telesio were Newton; yet an anticipation of bourgeois psychology underlies their revival of Platonism as the intuition of natural science colours the latter's progressive Aristotelianism. What is equally significant is the general splitting of the hierarchical and integrated cosmology of the feudal period into the separate domains of subject and object.[69] This division, which is of course only a tendency in the Renaissance, is not a distinction between, on the one hand, an ordered and coherent view of the world as a system of interrelated forces (objective nature), and a disordered chaos of sentiments (subjective humanity), on the other. The Florentines were just as precise and coherent about their psychology as the Paduans were about their natural science. Both subject and object were ordered by immanent harmonic relations which tended in time to be viewed as relations sustained by mechanical 'forces'.

Florence and Padua epitomize these tendencies. Neither was complete. The movement towards empiricism fell short of systematic observation, and as Lucien Febvre reminds us, it remained extraordinarily difficult throughout the sixteenth century to be anything other than Christian.[70] The most radical became unorthodox Christians, and even the Florentines shared with the departed world of happiness its metaphysical quest for being. In both cases it was less a secular apprehension of existence than a growing, inward sense of selfhood. And the world beyond the self, only in consequence of its capacity to satisfy its desires, became likewise dignified.

More generally, if Cusanus expressed with particular clarity and coherence the intellectual transformation of happiness into pleasure, anticipating the central structural features of the new cosmology, then Giordano Bruno spanned the emotional gulf between fun and pleasure, bringing that phase of the transition to a close.

Bruno, the most extravagant individualist of the Renaissance, began his adult life, like Rabelais and Luther, as a monk. It is hard to imagine anyone less suited to a cloistered existence. He soon took up a life of itinerant and reckless disputation. Travelling widely throughout Europe he challenged every Aristotelian orthodoxy in the very places where conventional truth was most powerful. He quickly gained the immense and scandalous reputation that we might suppose he most desired.[71]

He not only eagerly embraced Copernicanism, and intuitively

grasped its most radical implications, but he also ventured far beyond Cusanus in redrawing the cosmographical map. He followed the Florentine Platonists, back beyond Christian and Greek sources of rationalism to the 'original' wisdom of the Egyptians, and in particular to the mythical figure of Hermes Trismegistus.[72] Where Ficino in translating the *Corpus Hermeticum* had done so in the belief that it would provide a key to the understanding of his beloved Plato, Bruno took the Hermetic writings as a self-sufficient view of the world and found in them the kind of conceptual language he craved. It was a renaissance of a rather different sort. It was less the humanization of the cosmos through a relocation and revaluation of man within its physical structure than a diffuse but powerful spiritual eruption of man throughout the entire universe. It was very similar in fact to the ecstatic dispersal of human subjectivity that is so characteristic of the cosmology of fun. Locked within the historical and conventional constraints of mere fun, however, such a vision remained a mute cosmology. The continuous repression of fun throughout the development of both feudal and capitalist societies has created a powerful prejudice against the idea that any primordial, 'unreflective' experience could express a sophisticated and coherent view of the world.[73] The psychiatric and anthropological evidence noted in an earlier chapter challenges such a view; Bruno's writings, and the revival of the Hermetic tradition to which they were central, is a yet more eloquent denial.

For Bruno, even more markedly than for Ficino, the cosmos is a living organism. Its coherence resides in the single spiritual entity which animates its endlessly diverse appearances. The points of light we discern in the night sky are (as they are for children of five), living creatures endowed with life and movement.[74] This is not an obscure 'anagogic' mode of thought. The universe is alive with souls, it is 'a world crowded with souls, with masses of souls, of souls which join together – which irradiate each other'.[75] The peculiar quality of individual humanity, the subjective structures of desire, will, love, power, are replicated throughout the cosmos. And replicated in an endless variety of forms. Bruno pushes the new conception of space to an extreme. Infinity, and therefore divinity, is a predicate of matter as well as of absolute being. The infinite becomes an 'empirical' category, so that the actual physical cosmos need not be a limited 'copy' of the limitless divine idea, but an endless extension within which an inconceivable plurality of worlds jostle together. The fact of infinity unifies and integrates the cosmos in a new way. It is not a fixed totality so much as a structure which can be thought of as a musical harmony, an endless series of internal chords sustaining one another, each being a resonance within an infinitely extendable harmonic order.[76]

The human being is the real centre of such resonances, a node or concentration of celestial vibrations. The Hermetic writings provided Bruno with an initial clue. Nature, as the Paduans had guessed, was a system of forces inherent in matter, but these forces should be understood on the basis of a 'psychic' model, as emotional rather than mechanical 'tensions'.

The *Asclepius* had described the manner in which cosmic powers could be drawn into the statues of the Egyptian gods, filling them with life. These texts were believed to be Egyptian in origin and to have been composed by a contemporary of Moses. In them are expressed astrological and magical conceptions, and a vision of human knowledge as a self-expanding participation of man in the cosmos. Knowledge, a practical, rather than a contemplative or analytic art, was identical to summoning and controlling the cosmic forces of which man was the natural focus. In fact, the *Asclepius* and other Hermetic texts were written no earlier than the second century AD, but for the Renaissance Hermes was not only a real person but for many the fount of wisdom from which both Greek philosophy and Christianity had been corrupted. As Frances Yates notes, 'this huge historical error was to have amazing results'.[77] The Florentines were at times wary of Hermetic heterodoxy, but Bruno, unperturbed, claimed it as original religion as well as original philosophy.

Ficino, extolling at least the philosophic superiority of Hermes, claims the *Pimander* to be a clearer guide to creation than the *Timaeus*. The Egyptian genesis paralleled the Christian myth with uncanny accuracy; but the Fall is not a consequence of wilful disobedience so much as of narcissistic curiosity. Seeing himself reflected in nature, original man, a star-demon, himself drew aside the heavenly sphere and descended to earth, stepping as it were into his own shadow. In doing so he fell under the power of other stars and remained confined by natural forces. Hermes passes on to his son Tat the secrets of regeneration so that he becomes the first in the apostolic succession of Magi, cosmic 'operators', able to reconstitute themselves as divine beings through the manipulation of the celestial forces they gather within themselves.[78]

The figure of the magus was rehabilitated, though not without risk. Ficino's Orphic music seems modest compared to the cosmic powers sought by Cornelius Agrippa, who had 'cohabited with the elements, vanquished nature, mounted higher than the heavens, elevating himself above the angels to the archetype itself, with whom he then becomes co-operator and can do all things'.[79] This immediacy, typical of the cosmology of fun, is charged however with a powerful egoism foreign to primordial experience. The liberty of signs, the infinitude of metamorphoses, in becoming systematically elaborated into a theory of the universe loses its innocence; and in its Arcadian form tends either to fantasy (Fludd) or the cult of the personality (Bruno).

The magus was a reformer as well as a wise man. He sought a type of knowledge which, abstruse and recondite, was none the less primarily practical and unphilosophical. He aimed at the regeneration of life in accordance with a purer inner harmony. He held out the hope therefore of an earthly salvation. Secular individualism was taken up and magnified into a new cosmic design. From the theoretically unlimited powers at man's command, a new society could be created, an order realizing the harmonic code to which the cosmos clings for its coherence.

Much has been made of the practical, utilitarian aspect of the Hermetic search for original wisdom.[80] The ideological tone of Hermeticism, however, stressed the participation in, rather than control of, nature. Human aims could be realized through magical techniques but, in the process, these aims were themselves transformed. The Hermetic quest is a rediscovery of the natural 'harmony' of which man is a part. It will ultimately overcome the necessity of employing any particular technique to wrest from nature its inexhaustible store of energy. The Hermeticist does not place between himself and nature a mechanical contrivance as a means. This again betrays a mode of thought typical of the cosmology of fun. The magus strives, through symbolic/magical participation, for psychic mastery of the universe.[81]

Bruno aimed to construct a wholly internalized map of the cosmos, in the form of a mnemonic code, that would allow him to participate in the whole of creation. The archetypal images and 'natural signs' alluded to by Albertus Magnus seem a crude and stumbling science in comparison. Bruno grasped the core of the magical tradition. As the subjectivity of the cosmos, human memory can be ordered and subdivided according to its own harmonic code. The cosmos is a kind of psychic system.[82] By engraving upon his own memory the astrological 'seals' preformed to the pattern of stellar forces, Bruno could call down into himself unlimited cosmic power.[83] He could himself become the divine being who stepped for the first time into nature from his heavenly abode.

The distinction, so important for us, between signs and symbols is dissolved by Bruno. His epistemology appears at times to be founded on a notion of immanence, at other times to be a version of transcendentalism. Categories merge and reform in his thought. His fascination to modern writers lies less in his martyrdom than in his uncanny facility in the construction of images: the mediators between the world of nature and human subjectivity;[84] constructions which interpenetrate in ways which the scientific tradition was soon to discard.

He did not simply reject orthodoxy. His threat to authority was the more profound insistence on the unlimited freedom to create personal cosmic images. His pursuit of inner freedom was directly linked to the organization of nature unmediated by official religious, philosophical, artistic, or political institutions. The pursuit of personal pleasure (and pleasure is now firmly a qualification of the ego) and the development of knowledge of the world became identical. The distance between the sensible and the intelligible, which was only just being established in a new way, was completely abandoned. The longing predicated upon the absolute distinction between matter and spirit was finally quenched. No wonder Bruno seemed passionate! He proposed a cosmology centred on the individuated ego but retaining all the transformative charm of play.

Bruno's 'frenzy' is uncategorizable as a cultural event. His is the first 'absolute ego' of the bourgeois age, as he is its first schizophrenic; the paradigmatic heroic martyr to the madness at the core of classical rationality.

Hermeticism did not die immediately upon Casaubon's definitive dating of its 'sacred texts'. Robert Fludd produced a typically extravagant programmatic cosmology three years after the Egyptian myth had been exploded.[85] And, as an underground intellectual and spiritual movement, it persisted a good deal longer, exerting an obscure and gradually diminishing influence on the developing 'official' scientific world view.[86]

Cusanus and Bruno exemplify the complexity of Renaissance cosmography. Their thought moves in opposite directions. Cusanus cuts off all possible access to the transcendental and thus leaves man master of the created universe. Bruno projects man into the infinite and allows nothing to be placed beyond his reach. In the former, the secular spirit is confined within a finite universe and realizes its longing for the absolute by apprehending nature's concealed harmony. The world becomes an object of beauty, an object which presupposes an individuated as well as a secularized 'observer'. The latter takes individual subjectivity as a new absolute and identifies it with the empirical universe. Instead of describing the ego as a detached point from which to observe creation, it is creation itself. These movements are geographically and politically specialized. The first belongs to Venice and Padua, the second to Florence. One is associated with a capitalist transformation of trade, and the other with manufacture.[87]

Burckhardt insists that the Italian Renaissance was a culture expressive of the individualism not only of the artist or scientist, but also of the Prince. The individual patronage of scholarship and the arts was part of a disintegrative process of competition among ruling families who scarcely bothered to claim legitimacy for their power.[88] Their brutal realism was in fact one of the most conspicuous aspects of the new secularism. Neither capitalism as a civilization, nor an integrated and systematic world view expressive of its fundamental principle could thus develop. That step required the additional one of centralization of authority and political organization soon to become familiar as the nation state.[89]

The transition was discovered when the necessity of protecting the origins of classical rationality became important; and that was only when suspicions of rationalism developing within bourgeois society could be identified. The Renaissance played an important role as a buffer between the cold logic of science and its calculative rationality on the one hand, and the symbolic hierarchy of feudalism on the other. The fundamental aesthetic categories of the Renaissance embody the political serenity that the Prince failed to discover in the everyday social world. In either form, as patrician detachment or as frenzied participation, it could never become a democratic vision of the world. As Utopia or Arcadia, pleasure was a oneness with creation which was limited to the enjoyment of a few. It was in that sense a politically expensive cosmology, one whose intoxicating freedom had first to be strenuously combated before, much later, it could be allowed to re-emerge.

Chapter Eight

MECHANISM

How can the classical bourgeois 'scientific' world view be grasped as a social relation? Efforts to establish a sociological account of the scientific revolution of the seventeenth century have met with only limited success. Merton's influential monograph provoked an extensive literature devoted to elucidating the social and cultural 'context' of such a revolution but did little to encourage a sociological analysis of the transformation in the conceptualization of natural processes.[1] Different sociological traditions have offered other accounts of the historiographical background but have similarly taken the theoretical transition as given.[2] It is difficult not to sympathize with Koyré's bewilderment, 'I do not see what the *scientia activa* has ever had to do with the development of the calculus, nor the rise of the bourgeoisie with that of the Copernican, or the Keplerian, astronomy.'[3]

At the same time, it has become increasingly obvious that the rise of science does not simply explain itself. Historians of science have become sensitive to the 'extrascientific' influences that shaped the central ideas of any of the major figures of the scientific revolution. These influences are most often viewed as the 'philosophical' or 'aesthetic' predispositions of particular personalities and as such require, apparently, no further explication. The most illuminating historical studies have been those which have attempted to trace such personal intellectual 'prejudices' to a variety of neglected intellectual traditions.[4] Science, however revolutionary its implications, is thus understood as part of a continuous history of 'ideas'.

Science, once granted the privilege of its own inner 'rationality', can never be wholly reabsorbed into the totality of social life. Yet if

158

this privilege is denied it, no amount of contextual investigation can make it emerge as a distinctive cultural phenomenon. This methodological difficulty is just one aspect of the general intractability of historical understanding to which allusion has been made at various points. It is a difficulty which will be avoided (but not solved) by viewing science within the context of the four related cosmologies outlined here. In a general sense, the *content* of the classical scientific world view can be grasped in relation to the fundamental structure of capitalism in the same way that the medieval world view can be understood in relation to feudalism. It should be borne in mind, however, that this is by no means the complete picture. Science, as well as superseding the symbolic interpretation of nature, established itself by repressing in its own way the prodigal disorder of fun.

The language of hierarchy and release provided a common context within which feudalism as a 'social structure' and as a 'cosmology' could be described. In fairly obvious ways the medieval world view encourages a 'sociological' reading of its own content. The notion that nothing stands on its own, that all appearances are connected symbolically with an ordered hierarchy of essences, discourages the tendency (central to the classical scientific tradition) to 'explain' isolated phenomena in terms of specific chains of cause and effect. This is just the approach which, initially, can be used to grasp the nature of the scientific cosmology. An 'interpretive' rather than a 'scientific' understanding of its specific form is a prerequisite of any possible 'explanation' of its emergence within western society.

Fun is a society without exchange, without labour, and without money, while happiness is a specific order of such necessities, an order defined within the limits of personal relations. The 'society' of fun is an hermetical, self-sufficient body, that of happiness a hierarchy of dependence. Capitalism (and its scientific world view, pleasure) is primarily a system of relations among *commodities*. And it is just in a proper understanding of commodity relations that we can see the fundamental features of classical science.

For the sake of clarity the general characteristics of the scientific revolution can be summarized in advance. Koyré describes the transition from the medieval to the classical scientific view of the universe as ' the destruction of the cosmos' and its replacement by the 'geometrisation of space'.[5] There is, that is to say, a 'substitution of the homogenous and abstract – however now considered as real – dimension space of the Euclidean geometry for the concrete and differentiated place-continuum of pre-Galilean physics and astronomy'.[6] Rather than a 'qualitatively and ontologically differentiated whole', there is an 'open, indefinite, and even infinite universe, united not by its immanent structure but only by the identity of its fundamental contents and laws.'[7] It is furthermore '*only* in this abstract-real (Archimedean) world, where abstract bodies move in an abstract space, that the laws of being and of motion of the new – the classical – sciences are valid and true'.[8] The new science seeks to formulate laws of nature which are everywhere the same, to absorb qualitative distinctions into an absolutely general theory of matter in motion, and give expression to the emergent totality of the universe as a system of forces.

This conception of a universal order is fundamental also to an understanding of capitalism as a system of commodity relations. Marx describes the capitalist mode of production first of all as an 'immense collection of commodities',[9] each one of which is an 'external object' whose 'use value is wholly independent of its exchange value'.[10] All the sensuous qualities which distinguish one object from another and in relation to which the human subject satisfies a multiplicity of needs are suppressed in favour of a purely quantitative measure of value; 'As use-values, commodities differ above all in quality, while as exchange-values they can only differ in quantity.'[11] Commodities are marked, therefore, to a peculiar degree, by abstraction. 'Not an atom of matter enters into the objectivity of commodities as values,' argues Marx, conceptualizing, rather as Galileo had the dynamic relations among bodies, exchange relations, as both abstract and real.[12]

If commodities are pure quantity then they vary only in the relative proportion of their 'common substance'. In terms of their exchange-values, 'all commodities are merely definite quantities of congealed labour time'.[13] Thus, although human labour is infinitely varied in terms of its intrinsic qualities, for the world of commodities, it is 'reduced to human labour pure and simple'.[14] Labour power ceases to be distinguished by particular human attributes, every commodity contains 'the same kind of labour, human labour in the abstract'.[15] Thus, as soon as an object emerges as a commodity, 'it changes into a thing which transcends sensuousness'.[16] Labour power, in itself being transformed into a commodity, creates only 'alienated' objects which, lacking specifically 'human' attributes, appear possessed of the necessity of nature itself. Or rather, nature, from the perspective of the bourgeois world, is endowed with the necessity, universality, and abstractness of the commodity.[17]

Of the commodities filling the bourgeois world, labour power has a special significance as the source and measure of all others. Labour power cannot be used, however, in actual exchange, as a 'general equivalent' of any commodity. As a real process, exchange requires an 'objectification' of the process of labour – that is, as money. Labour power and money belong together as the essential commodities of the capitalist world, the commodities in the absence of which no others could exist.[18] They form, so to speak, the framework for the whole process of circulation.

Equivalence, however, is not identity. The universality of labour power and money, the fact that it is everywhere the same and provides the essential framework for exchange, should not obscure their separate tendencies. They refer to two quite separate dimensions of the social world, and from them are created the two fundamental axes of the classical scientific world view. Money, we might say, is the 'space' of the capitalist social world. In place of the limited 'personal' space of exchange in feudal society money as a commodity generalizes the circulation of all commodities. Money 'infinitizes' space. It allows all existing commodities to enter into relations of exchange. It is indifferent to the physical limitations of space. In the place of the qualitatively differentiated human 'space' it constructs an ideal, empty extension through which can pass every possible commodity. Money, therefore, 'is the absolutely alienable commodity, because it is all other commodities divested of their shape'.[19] Labour power, on the other hand, has a special relation to time. It similarly divests human time of all its 'irrational' qualities and substitutes for it an absolutely uniform duration, an infinite extension containing all possible interactions.

In terms of the classical scientific world view, the connection between space and money is of particular importance and has been alluded to by a number of writers.[20] It would be misleading, however, to suggest that money in some way 'led to' the scientific revolution. We have already seen that money in feudal societies, just because it was part of a system of personal relations, could be seen as embodying fundamental aspects of *its* cosmology. It is only when money becomes a general commodity that we can see in it the formal aspects of the 'geometrization of space'. Marx, viewing the development of society as the movement of a totality, points out therefore that 'money does not create the antitheses and contradictions' of capitalist society, but rather it is 'these contradictions and antitheses which create the seemingly transcendental power of money'.[21] It is only therefore within capitalist society that 'money is the imperishable commodity'.[22]

The natural world, the universe conceptualized by the scientific revolution, is best understood sociologically as a universe 'filled with commodities'. Individuated objects are distributed within the infinite extension of space and time and ordered by simple underlying and universal laws of nature. It is a world 'liberated' from the arbitrary or the subjective; reducible in the final analysis to matter in motion. Neither matter nor motion, however, can be taken as conceptual givens, and to understand the social logic of the classical world view its fundamental scientific terms must be more fully reconstructed.

MATTER

It is altogether odd to discover in the greatest of modern Pythagoreans, Johanne Kepler, the spokesman of a new material- ism. At first sight Kepler appears to belong completely to the world of the Renaissance. There is in him a conviction as powerful as we find in Ficino, or Alberti, or Bruno; a conviction that the cosmos is an immanent order of harmonic relations. So strong was that conviction indeed that Kepler, a more gifted mathematician than any of his immediate predecessors, began his 'scientific' career by explaining the Copernican system exclusively in terms of such relations.

Kepler's intellectual ambition was nothing less than to reveal 'the constructional laws which, in the mind of the Creator, directed the creation of the universe'.[23] The problem of planetary motion, for the first time treated systematically by Copernicus, left in Kepler's mind a number of unsolved problems. The most conspicuous of these were the number and spatial distribution of the planets. Why should there be just six, and why should their orbits about the sun fall in precisely the paths that they did? Clearly Kepler 'considered it perfectly reasonable to seek the architectonic principles which determined the structure and composition of the Cosmos'.[24] His initial solution, given in his *Mysterium Cosmographicum* which was published with his teacher Maestlin's help in 1596, was to produce the most beguiling of all cosmographical maps. Kepler was quite convinced that 'the almighty and infinitely merciful God, when he created our moving world and determined the order of the celestial bodies, took as the basis for his construction the five regular bodies which have enjoyed such great distinction from the time of Pythagoras and Plato down to our own day'.[25] It was in fact a three-dimensional 'model' of the solar system. There were six and only six planets because there were, and could only be, five regular solids. The relative distances of the orbits of the planets could be represented as lying within the spheres consecutively inscribed within, and circumscribed about, a unique arrangement of these solids.[26]

The significance of this work, however, was somewhat different from those inspired by Alberti or Ficino. As the world is 'the corporeal image of God',[27] Kepler was not content with the revelation of a formal harmony. His 'model' had to replicate the real physical relations existing among the planets. It was the first genuinely systematic cosmology of the scientific epoch. To demonstrate the true system of the world, Kepler was forced to pass 'from astronomy to physics or cosmography'.[28] Kepler in fact rejected the numerological tradition, at least in its more florid Renaissance examples. In an important polemic directed against Robert Fludd, Kepler defended the use of mathematics constrained by realistic physics. Mathematics ought to be purely instrumental and logical, it could not itself 'contain' reality so that the truth of the world could not be grasped directly through mere arithmetical manipulation.[29] For Kepler, mathematics was a rational means to clarify and present a physical conception.[30]

In attempting to outline a physics of the cosmos, Kepler also rejected the Aristotelian tradition which subsumed physical properties under an ontology of place.[31] His search for an architectonic cosmic principle united heaven and earth within the same uniform space. It was the space within which terrestrial mechanics was valid.[32] In extending the physicality of the earth throughout the visible universe, Kepler was heir to the Arcadian

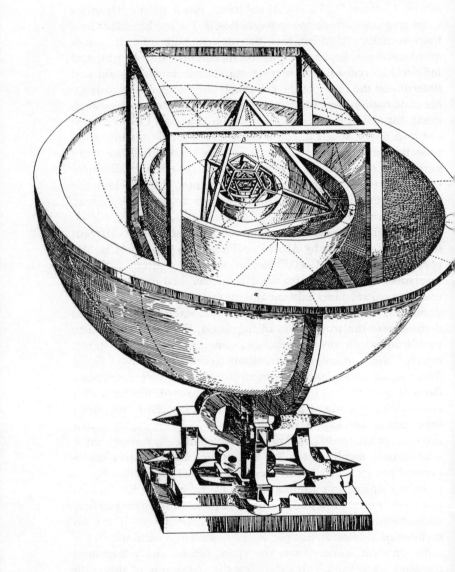

Kepler's Mysterium Cosmographicum

fantasy. Celestial bodies were no different, qualitatively, from the earth, and therefore could be understood in terms of the laws regulating familiar matter. Kepler's physical realism was responsible for the completion of the first phase of the Copernican revolution. It was inconceivable to him that the motive force ordering the complex interplanetary relations should emanate from a geometrical point. In the original Copernican scheme, the centre of the earth's orbit, rather than the physical body of the sun, occupied the centre of the universe. Kepler objected that 'a mathematical point, whether it be the centre of the world or not, cannot move and attract a heavy object'.[33]

It was the physical magnificence of the sun which, bathing the heavens in light and heat, held the celestial bodies in their proper paths. The physical centrality of the sun, however, was not fully explicated in the first version of Kepler's cosmology. In fact his initial exercise in physical harmonics was only a limited success. Impressed by the advances in observational astronomy made by Tycho Brahe, Kepler could not ignore the imperfect 'fit' between his three-dimensional model and the observed paths of the planets, especially that of Mars.[34] He became briefly, and somewhat uneasily, Tycho's assistant, appropriating for his own use an enormous mass of observational data on the latter's death.

Mars became the test case for his new celestial physics and the centrepiece of his *Astronomia Nova*. Referring all planetary motions to the actual position of the sun necessitated at first the reintroduction of all the unwelcome geometrical complications, the elimination of which had been the driving force behind the Copernican revolution. To understand the motion of the planets as instances of local motion, however (as the *result* of the application of a continuous force which had its origins in the body of the sun), left him no option. The new point of view had the advantage at least of a plausible, intuitive explanation for the apparent variation in the speed of the planets. The variation, in fact, was quite real and depended simply upon the varying proximity to the sun.[35] The historic assumption of uniform motion was abandoned, and with it the necessity of 'artificial' mathematical constructions.

Kepler conceived of the force propelling the planets as having a physical aspect, rather than being exclusively material in nature. He could still talk of a 'soul' or 'spirit' animating its body and being the cause of its light as well as the source of its physical power. More generally he likened its force to the power of a magnet.[36] Yet Kepler's revolutionary astronomy was allied to conventional physics. The sun's power is a continuously acting *raptus*, a solar wind which 'pushes' the planets around in their orbits. It is not an 'attractive' force. Kepler pictured the sun as the central hub of a large rotating wheel. The trailing spokes, spiralling outwards, swept the planets around its centre. To account for the variations in the speed of planetary motion, Kepler had then to introduce both a resisting inertia inherent in each body and particular 'local' motive forces belonging to each planet.[37] The complications multiplied to the point where it appeared likely that the new astronomy would be stillborn.

Having abandoned uniform motion in one respect, Kepler finally saw that the simple solution to both the cosmographical and the mathematical problems lay in abandoning it in a more radical sense. The doctrine that celestial motion was of necessity circular had been central to Copernican as well as Ptolemaic astronomy. But if astronomy was to be an extension of *terrestrial* physics, this need no longer be the case. Upon the earth elliptical and parabolic motion was commonplace; if this were the case also for celestial motion, then the orbits of the planets could be simply understood as ellipses within which the sun formed one focus. The simplicity of a uniform and universal force replaced the simplicity and universality of form. The world system was the consequence of a force acting everywhere in the same way. The speed of the planets depended simply upon the quantity of force acting upon them. All celestial bodies could be considered as qualitatively identical, varying only in terms of their inherent power to resist the force of the sun's *raptus*.

Kepler's notion of inertia was therefore of a resisting power. The ontological distinction between rest and motion was still fundamental to his physics. Matter 'preferred' rest, and its inertia was related to its gravity. All bodies, by virtue of their materiality alone, possessed a power of 'mutual affection': 'If two stones were removed to some place in the universe, in propinquity to each other, but outside the sphere of force of a third cognate body, the two stones, like magnetic bodies, would come together at some intermediate place, each approaching the other through a distance proportional to the mass (*moles*) of the other.'[38] This is not yet a theory of universal gravitation. The nature of the force retains a spiritual quality and, more significantly, is not itself responsible for the movement of the planets, which depends upon the unique force of the centralized sun.

Kepler's great 'scientific' success, his three laws of planetary motion, were buried amidst the many false starts and blind alleys of his major works.[39] They were for him no more than indicators of the validity of his underlying cosmic vision. Buoyed up by his success in resolving the long-standing enigma of the orbit of Mars, Kepler returned to the 'deeper' problem of the harmonic structure of the solar system. His fresh insight into elliptical orbits, and the more accurate observational material, prompted a fresh approach. The harmonic structure of the cosmos could be captured most completely in musical notation. A 'note' determined by its angular velocity could be assigned to each planet. Then, by comparing its lowest value (when its velocity was least, at aphelion) with its highest value (at perihelion), a musical 'interval' describing the eccentricity of the orbit could be computed. Kepler found indeed that the ratios of such intervals were almost entirely consonant.[40] The *Harmonice Mundi* completed the task begun in the *Mysterium Cosmographicum* and was quickly forgotten. In spite of his own deepest instincts, Kepler had turned astronomy into a physical science.

MOTION

Galileo's mode of universalizing nature moved in the opposite direction to Kepler's. Rather than project terrestrial phenomena into the heavens, he united them by absorbing the celestial into the earthly and gave a scientific twist to the Utopian vision.

Galileo began his professional career as a fairly orthodox Aristotelian.[41] But he was already a convinced Copernican when, in 1609, he heard of a Dutchman's success in fabricating a telescope.[42] Galileo immediately constructed one for himself, improving it several times before turning it upon the heavens. The immediate results of his observations, published in his *Siderius Nuncius*, were to confirm dramatically the physical assumptions of Kepler's astronomy. The general and apparently inescapable conclusion to which Galileo was drawn was that celestial bodies were no more 'perfect' and incorruptible than the earth. He could thus clearly distinguish the shadows cast by the light from the rising sun falling across the moon's irregular surface.[43] Not only were there 'protruberances and gaps' on what should have been a perfectly smooth surface, but they were sufficiently large for him to be able to estimate their dimensions.[44] Even more damaging to the traditional cosmography was the discovery of Jupiter's four moons. The great aesthetic appeal of the pre-Copernican system was its unification about a single centre. To discover not only that the sun and earth were centres of rotation, but that other planets were as well, seemed to introduce unacceptable asymmetries. Many of course denied the evidence, but they found it difficult to persist in their objections and even more difficult to resist the implication of the spectacular increase in the number of visible stars. The Milky Way, resolved into countless separate points of light, challenged the old theological assumption that God had placed them there as a cosmic ornament that man might enjoy and contemplate.[45] Would God have been so prodigal as to fill the universe with unobservable objects? The argument became strained.

Three years later, his *Letters on Sunspots* demonstrated in a series of careful observations that blemishes also occurred regularly on the sun's surface.[46] But in the longer term the observations through which Galileo became celebrated were not his most important contribution to the formation of the new world view. It was in mechanics rather than astronomy that Galileo's daring showed itself to its fullest effect. His arguments served to make the Copernican cosmos physically plausible. In making the rotation of the earth a physical as well as an optical necessity, Galileo destroyed the 'common-sense' assumptions which had sustained scholastic science.

In the fourteenth century Oresme and Buridan had considered the diurnal rotation of the earth as an interesting logical possibility. But while the *optical* evidence could be interpreted equally on the assumption of a stationary or a rotating earth, they did not for a moment suppose such a rotation to be an actuality.[47] Copernicus, on the other hand, had certainly taken the earth's rotation to be a real phenomenon, and Galileo recognized this.[48] Yet Copernicus had made no attempt to provide *physical* arguments to support his belief, and it was this deficiency that Galileo set out to make good, particularly in the brilliantly argued

'Day Three' of his masterpiece, *Dialogue on the Two Great World Systems*.

The physical objections to accepting the earth's rotation were of two fundamental kinds. The first raises the question of the simultaneous rotation of everything upon the earth's surface, and the second points to difficulties associated with the notion of centrifugal force. These were both arguments which had weighed heavily with Oresme.

If the earth revolves, moving from west to east, and completes one revolution every twenty-four hours, then why does a perpetual wind not blow from east to west? How do birds succeed in flying equally well in any direction? Why does a heavy object dropped from a tower, or the mast of a sailing ship fall parallel to the tower or mast rather than 'behind' it? And, a more recent example discussed by Tycho Brahe, if two similar cannon are fired, one due west, the other due east, why should the shot be carried a comparable distance in each case?[49] Kepler had answered these objections by supposing the earth to be possessed, like the sun, of its own dynamic force, its *raptus*, which holds everything close to its surface 'as if by chains'.[50]

Galileo's solution was more radical. There is no need of a crude 'mechanical' explanation of motion in these cases. Without recourse to any special 'scientific' demonstration or specialist knowledge, Galileo shows that our everyday understanding of motion, when fully analysed, is perfectly compatible with such phenomenon. Traditional 'physics' had not dealt with rotation at all, and Galileo points out that it does not fit into the Aristotelian categories of 'natural' or 'violent' motion. It is movement without change of place and therefore could not be understood as if it were a special case of rectilinear motion. Instead, Galileo offers an account based upon a simple 'thought experiment'.[51] A perfectly spherical hard ball will 'naturally' roll down a smooth inclined surface. It will, furthermore, accelerate as it does so.[52] And, if already in motion, it will gradually slow down if the plane is inclined, even very slightly, upward. Suppose, rather than any actual inclined plane, an absolutely horizontal, frictionless surface. Clearly, upon such a surface, discounting any extraneous forces, a ball would continue in its state of motion indefinitely; 'if such a space were interminate, the motion upon it would likewise have no termination, that is, would be perpetual'.[53] Equally, a ball at rest would remain at rest. In contrast to Kepler, Galileo does not assume that matter has a preference for the state of rest. Indeed, by defining motion as a continuous *state* rather than a condition of *change*, Galileo assumes an absolute indifference of matter to the distinction between movement and rest. Matter will naturally preserve its *state* of motion, including rest, unless acted upon by an external force.

This anticipates but does not accurately present the classical law of inertia.[54] Rectilinear motion is not preserved. To see this we have to understand the meaning of an 'absolutely horizontal plane'. One possible definition would be a tangent to the earth's surface at any particular point. Now, as movement in any direction upon such a plane would in fact be away from the earth's centre, motion upon it would naturally be diminished. The only motion which could conceivably be conserved would lie along a plane upon which every point was equidistant from an unmoving centre; that is, upon the surface or 'parallel' to the earth's surface. It is in fact *circular* motion which is conserved, and it is for this reason that the traditional objections to the earth's rotation fail. No force need be applied to a projectile, or anything not permanently anchored to the surface of the earth to ensure that it will 'participate' in the earth's motion. As the truly 'natural' motion of rotation, all share equally and effortlessly in its perpetuation.[55] What Galileo does come very close to comprehending is the purely 'secular' significance of circularity. However important empirically, circularity should not be viewed as an inherently privileged form; it is simply the consequence of two forces interacting upon each other: the tendency of motion to conserve itself, and the tendency of matter to cohere about its own centre.[56]

Nor did Galileo formulate the classical conception of gravity. The tendency of matter to cohere about its own centre was not generalized into a new cosmic principle. Indeed, he did not use it in a systematic way to counteract the other major argument against the earth's rotation. Small objects placed on the rim of a spinning top are thrown off at a tangent by its rapid rotation. Rotation seems in some way to 'counteract' gravity. Why are we not similarly shrugged off the surface of the rapidly spinning earth?[57] The answer requires the same leap in scale as the Copernican insistence that the 'fixed stars' lie at an inconceivably vast distance from the earth (and therefore show no parallax). The earth's rotation does indeed make us 'lighter' than we would be if it were stationary, but such is the immensity of the earth's gravity that its speed of rotation cannot fling us into space.

Resisting Bruno's radical infinitization of space, Galileo could not grasp the full significance of his own mechanics. It was only by enclosing the universe that, ultimately, its order was guaranteed. And in spite of his mechanical account of circular motion, it retained in his work the aura of perfection.[58] It is, however, an earthly perfection, a feature of the empirical world itself.

Perpetual recurrence, the conservation of quantity in a specific ideal form, describes an important aspect of the new society as clearly as it laid the foundations of a new mechanics. The social logic of Galileo's physics becomes apparent when it is contrasted with the impetus theory to which, at the outset of his career, he was committed.[59] In the feudal scheme, there is no 'conservation' principle and the entire process of 'exchange' (change of place) occurs only under the continuous application of an external force. With Galileo, nature becomes genuinely self-moving. Once set in motion, the entire dynamic system of the world continues without the necessity of any additional 'motive force'. This, of course, is just the character of exchange in a capitalist society.[60] To reach his fundamental insight, Galileo had to abstract from the world of real experience 'ideal' conditions of perfectly hard surfaces and frictionless planes. In the social world, as we have seen, the commodity form is a similarly 'ideal' reality within which all particular qualities are swallowed up into pure quantity. Money, in effecting this abstraction for us, has a special affinity with thought. It behaves with a certain philosophical cunning to reveal the necessity (and conceal the origin) of our more immediate social 'world system'.

FORCE

Kepler and Galileo introduced into the analysis of nature, incompletely and imperfectly, a new social logic. Secular realism (matter/labour) and a principle of conservation (motion/money) became the indispensable preconditions of a scientific cosmology. Newton, developing the view of nature as a system of interacting forces, produced such a cosmology.[61]

Once again, the purely historical problem can be avoided. Newton as a personality remains hidden. His 'science' was clearly only part of an intellectual endeavour as baffling to our notion of 'reason' as the work of Bruno or Kepler. The 'how' of the Newtonian revolution however is, for the moment, less significant than the 'what'.[62] The shock of discovering that he was not a cool-headed rationalist should not distract us from the central significance of the genuine rationality of his systematic science. In arguing that this rationality is itself the expression of something 'extra-scientific' need not lead us into biographical reconstruction or an endless search for the specific 'sources' of his various arguments. It is not what was 'in his mind' so much as the 'form' into which it was shaped as a 'finished product' that is of immediate interest.

Newton's 'mathematical way' introduced quantitative analysis into the scientific description of the 'system of the world'. The reasoning involved did not aim at reproducing the force of geometric demonstration alone, but sought to establish the limits of variation of actual physical phenomena in such a way that the empirical world could be *deduced* from its fundamental theorems.[63] This involved an initial simplification, in which physical intuition played a part as vital as that of mathematical acuity. The complexity of the material world was at first reduced to a 'one-body-system' in order to clarify the elementary forces implied in our most primitive conception of matter. Such a body is distinguished by 'extension, hardness, impenetrability, mobility and forces of inertia', but not of necessity by gravity.[64] Gravity is a universal but not essential characteristic of matter as such. It can be detected only relationally and has no meaning in reference to a universe containing only a single body.

His attribution of inertia to an isolated body is a corollary of his notion of 'absolute space' which, 'in its own nature, without relation to anything external, remains always similar and immobile'.[65] It is possible, in other words, to conceptualize an isolated body as in motion. In relation to such a body in absolute space (rather than an ideal ball on a perfectly smooth inclined plane), Galileo's principle of inertia could be seen as a special case of a more general law which applied universally to rectilinear motion. By assuming that the relation among bodies was governed by a simple, inverse square law of gravity Newton was able to demonstrate the mathematical *necessity* of Kepler's laws of planetary motion.[66]

In spite of the dazzling success of the mathematization of the general dynamics of the solar system, Newton's *Principia* fell short of what many had already come to expect from a 'scientific explanation'. Few of his contemporaries felt able to challenge his mathematical reasoning and were consequently less critical of his scientific account of nature than they might have been.[67] Even so, his ascetic denial of causal explanation could hardly go unremarked. Even more rigorously than Galileo, he eschewed 'hypotheses',[68] and refused to be drawn as to the 'cause' of gravity. 'What I call attraction may be performed by impulse, or some other means unknown to me,' he suggests. And attraction is similarly just 'any Force by which Bodies tend towards one another, whatever be the Cause'.[69] For many this was hardly sufficient. What 'mechanism' was responsible for the effects Newton had so brilliantly analysed? In allowing for the possibility of 'action at a distance', was he not allowing disreputable 'occult powers' an unnecessary role in an otherwise exemplary science? A modern scholar goes so far as to describe Newton's masterpiece as 'less a battle-cry for the new science than a confession of failure'.[70]

Yet 'gravity' was no more or less occult than any of the familiar 'primary qualities', the underlying mechanism of any of which was unknowable and accepted without difficulty as the 'givens' of nature.[71] Newton's instincts on the matter were, later, fully justified. The 'rationality' of science, as of human action generally, depend not only upon a context of untestable assumptions but also upon the 'irrationality' of values.[72]

The immediate impact of Newtonian science, however, did not depend only upon his general solution to the problem of motion. In addition to uniting heaven and earth in a coherently ordered system governed by a simple, quantifiable law, he extended this order 'downwards' as a general theory of the structure of matter. The primary qualities which distinguished the isolated single body in space could be used to designate the nature of any elementary particle of matter whatever. The 'stuff' from which nature was composed, that is to say, was ideally like space itself: simple and 'conformable to itself'. Primary matter was everywhere the same and the bodies constructed from it differed one from another by virtue of the quantity and internal organization of the matter they contained. Newton's corpuscular philosophy was expressed with particular clarity and freedom in the *Queries* appended to his *Opticks*. In the justly famous *Query 31*, he writes: 'it seems probable to me, that God in the beginning formed matter in solid, massy, hard, impenetrable, moveable particles, of such size and figure, and with such other properties, and in such proportion to space, as most conduced to the end for which he formed them'.[73] Matter was ideally reducible to 'inertially equivalent' particles, each, whatever its 'figure', the possessor of precisely the same degree of 'materiality' as any other.[74]

The differentiation and cohesion of bodies depended upon interparticulate forces of the kind (inertia and gravity) that ordered the motion of celestial bodies. Newton was much less successful in specifying quantifiable laws in relation to such forces. The *Opticks* was experimental where the *Principia* had been mathematical and inspired a scientific tradition in many ways at odds with the philosophical intentions of its founder. The 'mechanical philosophy' borrowed the authority of Newton but tended to a cruder form of 'corpuscularism'.[75] Newton's 'third law' seemed to provide a simple mechanical principle through which to explain almost all events. If quantity of force was conserved, impact only redirecting and never destroying motion, then a physics of contact might be presumed to explain the continuity and coherence of nature.[76] The perpetual interaction of particles was conceived as a 'market' where forces were exchanged and every outcome was governed by a simple law of equivalence.

The idea of mechanism is not to be confused with that of a 'machine'. Mechanism, as a characteristic of the classical scientific world view, should be understood as the view that the empirical world is the 'resultant' of a number of continuously interacting 'forces'.[77] It is by no means limited to the theory of the 'transmission' of such forces through physical contact. The machine, especially sophisticated clockwork, had a privileged position during the seventeeth and eighteenth centuries as an illustration of the providential *design* of nature, rather than as itself an architectonic model of the cosmos.[78] The machine is not therefore, in itself, a construct from which genuine knowledge of the world can be derived. 'Forces' remain unknown; they constitute the irreducible relationships among particles of matter.

As well as the 'passive' force of inertia Newton also sought to elucidate the operation of the 'active' forces of gravitation, chemical 'attraction', and fermentation.[79] The most important technique of investigating small-scale interparticulate forces he considered to be the observation of the interaction between light and matter.[80]

The cosmos of capitalism is a mechanism not because it is constructed upon the analogy of a 'machine' (no matter how sophisticated) as an instrument of manufacture, but because its order is the outcome of the continuous interaction among indwelling forces. Nature can be conceptualized as an autonomous system of 'exchange relations' whose fundamental ordering principles must simply be accepted as 'given'.

The cosmology of happiness is focused upon use-value; upon the 'irrationality' of the human subject. Feudal relations are always personal relations. The triumph of exchange relations, their separation from any human 'value', is expressed therefore in a new vision of the cosmos as a realm of universality and necessity.

Chapter Nine

EGOISM

The correspondence between the macrocosm and microcosm was transformed, during the Renaissance, into a new metaphor of 'inside' and 'outside'. The cosmos became a unified but unbounded structure ordered through universal laws of nature. The intimate reality of human experience could no longer be conceptualized as a metaphor or symbol of an all-inclusive order. The human world must conform to the same laws as the rest of nature. Yet the dignity of man, which had been one of the central ideas of the Renaissance, seemed to imply a privileged status for the human.

This difficulty, the paradox of Renaissance humanism, lies at the intellectual heart of bourgeois psychology. No sooner had the claim to dignity been established, through the assimilation of man to nature (in Arcadian or Utopian forms), than it was lost. Man did not inhabit a centralized and degraded world; he assumed the dignity of celestial motion. But, in consequence, his soul was no longer the mirror of the world. The unique advantage of his position as an observer of the cosmos was lost. True, the cosmos was no longer held to be the means to proclaim an ultimate truth; but even as a mechanism it appeared baffling. Its regularities could be formalized through the quantitative language of mathematics, but the cause of its order remained mysterious. Was human nature similarly unknowable? Was it not equally contrary to presume that man was a mystery to himself?

These purely intellectual considerations were, additionally, ways of describing the new social reality of commodity exchange. The commodity embodied a dual reality. As exchange value it expressed the universality of necessary relations. It created a world of 'nature' within which man could live, a world which appeared to have sprung up magically around him and now sustained itself by the power of its own indwelling forces. As use-value, however, the commodity was ultimately relinquished to the 'irrationality' of the human subject.

The division between microcosm and macrocosm gave way, then, to the distinction between object and subject. 'Human nature' was, just as much as the physical cosmos, inexhaustible. As 'subjectivity', the human was an interior cosmos. But what was its specific mechanism? A series of attempts were made to 'save the phenomenon'; to 'explain' human reality in terms of universal laws and, at the same time, clearly distinguish between the peculiar quality of human experience and all that lay beyond it.

Cartesian dualism is the first step towards the self-consciousness of the bourgeois ego. The radical distinction between matter and mind, itself hinting at the momentous split between the view of the 'object world' exclusively in terms either of exchange or of use, cleared the way for the direct 'application' of physical concepts to human activities. This, of course, was not Descartes' intention, but as so often demonstrated in the history of ideas, innovators are powerless to exercise control over the effects of their ideas. Once the 'qualitative' character of human experience had been confined to a particular category, it could be fully explored by the methods developed within quite different fields of study.

These methods were to be Newton's, rather than Descartes', and the disciples of Newtonian 'corpuscularism' rather than those of their master. Although, especially in France, controversy surrounded the ultimate significance of Newtonianism for a philosophy of nature, his 'methods' were freely borrowed and uncontroversially adopted as the foundation for a new 'science of man'. Just as the cosmos had to be described in a form commensurate with the new social reality of commodity exchange, so the human 'qualities' hidden within such a reality could appear philosophically only as elements within a theory of the market.

Society was composed of commodities; individuated objects defined by the universal attribute of exchange-value (labour). The cosmos was composed of individuated objects, bodies, that differed only quantitatively (by mass) and thereby established invariant relations with other bodies. The 'internal' reality of human nature, in spite of its 'ultimate' irrationality as 'pure subjectivity', could be grasped as a similar 'internal market' upon which some universal quantities were exchanged. First, therefore, human nature became a universal definirg criteria, the 'species being' of man. This had not been the case, of course, in feudal society. There only particular human beings existed, specific qualities held fleetingly within a living subject. The bourgeois revolution, opposing all feudal restriction upon internal 'freedom' of the market, created the individuated 'ego' as a new historical actor.

The human was internal and subjective, but after that it was infinite, necessary, and universal. It was not a differentiated part of the cosmos, but a cosmos in its own right, a cosmos viewed from a different vantage point. The social world, which was the 'model' for both cosmologies, the order of nature and of the psyche, was self-generating and self-justifying. All that was required for its perfect order was that the individuated subjects that composed it should be free to act 'rationally'.[1] This rationality was in turn guaranteed by the universality of 'human' attributes.[2] The difficulty, as outlined in the first chapter, was primarily to account for the failure to realize in practice this universal nature. Society was imperfect, and human beings corrupted to the extent to which reason remained locked up within the constraints of a symbolic order the bourgeois world had already outgrown.

The intellectual problem confronted by the bourgeois psychologist was then to explain the peculiar attributes of human nature in such a way that reason could be deduced from it as a genuinely universal phenomenon. This was attempted in a number of different ways, with each succeeding school believing itself to be overthrowing the very foundations of its predecessors. But from the perspective of a greater distance, we can see the common assumptions which guided all these efforts. In broad outline, three different but related approaches can be distinguished, characterizing three different types of internal market. They can be designated as sensation, sympathy, and desire.

SENSATION

'Sensationalism' was the first and perhaps the most successful version of bourgeois psychology. Its initial appeal rested on its claim to offer a systematic and comprehensive account of subjectivity adequate to the new 'scientific' world view. Hobbes of course, even before Newton, had proposed a radical materialistic psychology. [3] It was, just for this reason, rejected. 'Mechanism', not 'materialism', was the key to the new intellectual order. Even nature could not be grasped as a purely 'material' phenomenon, and a properly scientific approach began with the recognition of the *limits* of our commonsense notions of physical causality. The world contained nothing but matter in motion, but that was not to say that every phenomenon was reducible to the effects of 'contact' and collision among its elemental particles. One of the leading 'corpuscular' philosophers, Robert Boyle, insisted upon the importance of 'emergent' and functional relations for any reasonably satisfactory account of even simple physical phenomena. The 'texture' of matter, its internal organization, gave rise to many of the 'sensible qualities' through which we recognized it. [4]

Sensationalist psychology, in its more sophisticated variants at least, was fully alive to these distinctions. Even where it appeared to be an attempt to reduce psychology to physics, or physiology, it was careful to preserve the 'dignity' of its subject matter by making the particular and unique quality of subjectivity its ultimate frame of reference.

David Hartley, for example, the most ambitious of such psychologists begins his treatise *Observations on Man* with a scientific declaration of intent, inspired directly by *Query 31* of the *Opticks*. He devotes his initial discussion to an analysis of sensation, defined as 'those internal feelings of the mind, which arise from the impressions made by external objects upon the several parts of our bodies'. [5] He defines sensations, in other words, in terms of their presumed causes rather than phenomenologically, and by this procedure hopes to avoid the arbitrariness of individual variation and internal judgement. It appears, indeed, that he is interested exclusively in the physical details which allow such causes to be operative. 'We are to conceive, that when external objects are impressed on the sensory nerves, they excite vibrations in the aether residing in the pores of the nerves (medullary) by means of the mutual actions interceding between the objects, nerves, and aether.' [6]

Hartley's major interest, however, soon emerges: it is ideas rather than sensations which form the real subject matter of his book. The simplest idea is an 'internal representation' of a sensation. Hartley insists upon sensation as the ultimate 'source' (that is, cause) of all subjective phenomena, and explicitly rejects Locke's division of the experiential sources of human knowledge into sensation and 'reflection'.[7] There are no superadded internal human 'faculties' which are not themselves traceable to the invariant operation of external sensory causes.

Simple ideas, therefore, through a process of 'coalescing of vibrations' which are their causes, give rise to more complex perceptions. Hartley is soon unable to sustain the cumbersome language. Following Locke more closely than he might have wished, he uses the purely subjective term 'association' to designate the psychological end result of such a mechanical process. 'Simple ideas will run into complex ones, by means of association,' he claims, referring both to the cause and effect of the manner in which the mind becomes filled with its specific content.[8] The general process of associating ideas becomes more complex and more characteristic of human subjectivity through the interaction of two particular types of 'vibrations'. A new 'level' of association is developed through language. The range of possible association is enormously increased by the use of words, which connect present sensations to the virtually infinite reservoir of past and projected future experience. It is the associative power of words which, in fact, creates the 'arts of logic and rational grammar'.[9]

Second, as 'sensible pleasures and pains' are 'the most vigorous of our sensations',[10] they are also the most effective in forming associative links in the human mind. Hartley indeed proposes a 'quantitative' view of pleasure and pain. Pleasure is conceived as an 'optimum' level of vibration of the nerve fibres, such 'that pain should be nothing more than pleasure itself, carried beyond a due limit'.[11]

The doctrine of association is used more 'mechanically' by Hartley than by Locke. Having established its centrality as force ordering the internal experiential cosmos, Hartley attempts to derive from it *all* the specific characteristics of human nature. As association is founded upon a 'natural' process of cause and effect, the connections in the mind, which are derived from them, ultimately form themselves into a 'rational' picture of the world. The human subject retains something of the occult quality of the 'mirror of the world'.[12] Sensory vibrations finally sort themselves out into a structure which corresponds to, though is qualitatively distinct from, the world of nature which is their source.[13]

If reason is a correspondence between internal perception and the order of nature, morality equally owes its structure and validity to the process of association. Those valuations and judgements formed at the earliest stage and continually reinforced by the agreement of others 'appear like instincts'.[14] He does not hesitate to claim that 'the moral sense is therefore generated necessarily and mechanically'.[15] And as the rational mind is nothing more than the internal perception of the inherent order of empirical reality, all human activity is consequently assimilated to nature: 'By the mechanism of human action I mean, that each action results from the previous circumstances of body and mind, in the same manner, and with the same certainty, as other effects do from their mechanical causes.'[16]

This makes Hartley's conventional moralizing all the less convincing. In seeking the absolutely universal and necessary foundation of human psychology, he had in fact obliterated the very quality of inner freedom through which subjectivity makes itself felt.[17] Association might be the organizing principle of the inner life, the principle from which all its operations could be deduced, but where intimate knowledge of nature could, reluctantly, be forsworn, it was much more difficult to deny to the human subject authentic knowledge of its own quality. Hartley's work proved popular, as had Locke's, establishing a tradition of reductionist psychology which remains unbroken. Although the most generally accepted tradition of bourgeois psychology, it is neither the most rigorous nor the most persuasive version of sensationalism, nor is sensationalism the most compelling school of thought within that tradition. In providing a 'model' of 'scientific' psychology, however, its unfulfilled promise continues to exercise an evident fascination.

La Mettrie, seeking a view of human nature consistent with both Locke and Descartes, produced a psychology that was more coherent and more scandalous than that envisaged by Hartley.[18] Much more clearly than the English empiricists, he recognized that 'the *inner* connection between physical cause and psychic effect remains unknowable'.[19] His aim, therefore, was not so much to produce a materialist metaphysics or a complete demonstrative science of man as to reveal the inherent consistency of a particular view of the *human* mechanism.

La Mettrie specifically rejects the Cartesian 'beast-machine' as an inadequate model of the human organism. Nor does he admit to a simple continuity between sensation and some putative law of association to account for human reason. The specific and universal feature of man does not depend upon his organism, it *is* his organism. 'Man's preeminent advantage is his organism,' he declares, finding the special privilege of the human in the *structure* of his body:[20] 'The human body is a machine,' he claims, but a machine of a special sort, one which 'winds its own spring'.[21] The Cartesian dualism was dissolved in his notion of the organic machine as 'a genuinely self-sustaining system':[22] Unlike any man-made machine, man himself is self-moving and purposive. Genuine materialism could not afford the simplicity of a theory of motion. The complexity of internal relations defined the human organism in a particular and characteristic fashion. 'But since all the faculties of the soul depend to such a degree on the proper organisation of the brain and of the whole body, that apparently they are but this organisation itself, the soul is clearly an enlightened machine.'[23]

Sensation ceases to be a direct mechanistic principle of human perception, and expresses an inner structure adapted to the forming of images of an external reality. Thought, he claims, 'is so little incompatible with organised matter, that it seems to be one of its properties'.[24] La Mettrie sought to '*Vitalise* the Cartesian "dead mechanism" approach to biology',[25] forcefully expressing the human character, as well as material form, of his man-machine. 'To be a machine' is not, as in Hartley, or for that matter in Baron d'Holbach, to be 'nothing more than a passive instrument in the hands of necessity', [26] but is rather 'to feel, to think, to know how to distinguish good from bad, as well as blue from yellow'.[27] In man necessity, so to speak, escapes from itself.

La Mettrie's psychology is a veritable aesthetics of mechanism. The whole tendency of French materialism in fact, far more than its British counterpart, was towards a systematic conception of 'personality'. This is the case particularly in its most distinguished eighteenth-century representative, Condillac, whose explicit indebtedness to Locke and Newton should not conceal the extent to which his sensationalism was very different from theirs.

The ultimate point of reference for Condillac's 'system' was the sense of self which, as inexplicably as the reality of the exterior world, presented itself to us as an accomplished fact.[28] But unlike Locke, or Descartes, its immediacy cannot be taken at face value. The fact that the external world appears to us as outside ourselves does not mean that we must simply accept the continuous presence within us of an observing self, any more than it unambiguously proves the reality of such an external reality itself. It too must be constructed out of sensations. Psychology therefore becomes a fundamental part of the study of reality; a study which should aspire to the systematic rationality of Newton's deductive system. Condillac therefore aims 'to reduce to a single principle everything concerning the human understanding'.[29] His attack upon metaphysical system building did not diminish the confidence with which he proposed the 'true' psychological 'system of the world'. The multiplicity of subjective qualifications of matter were interrelated in conformity with a simple underlying law such that 'man himself is a system'.[30] And in renouncing the arbitrary assumptions of the philosopher, we free ourselves to discover those systems 'which the author of nature has made'.[31]

In order to demonstrate the interactive reality of the sensory world, Condillac conducted a 'thought experiment'. He imagined, not a man-machine, but a 'statue-man' into which he introduced in turn the various senses of the external world. This can be seen as a systematization of the discussion of the perceptual world of the blind, or the newly sighted, which had aroused considerable interest since the work of Locke.[32] Particularly extending Berkeley's insight into the relationship between sight and movement, Condillac tried to develop a sensationist psychology that was at least phenomenologically adequate to the richness of human perceptions.

Sensationalism, in spite of Condillac's rigour, proves insufficient to the descriptive task. No account of the interior world of subjectivity can really begin until the sensory system, articulated and mutually interrelated, is awakened to the experience of pleasure and pain, and in so doing create from within its own interactions a *personality* dependent upon, but not reducible to, such experiences.[33] Condillac's sensationalism almost imperceptibly leads to the assertion of the 'self' as the conscious, moral, and rational principle of the 'human system'. Lockean empiricism shades into a new solipsistic metaphysics: 'whether we ascend, to speak metaphorically, into the heavens, or whether we go down into the abyss, we never leave ourselves; and we never perceive anything but our own thought'.[34]

The emergence of the self as the active principle of a materialist psychology was inevitable once the classificatory division between matter and spirit (exchange and use) had been established. It was expressed less systematically but more brilliantly by Diderot, for whom 'individuality is the subjectivisation of reality' rather than simply a 'feeling point of nature',[35] and less brilliantly but even more comprehensively by d'Holbach. The latter indeed goes so far as to propose the self as the *gravitational* principle of the subjective world.[36]

Rather than attempting to 'explain' subjectivity as a particular effect of matter in motion – as we find, for example, in Hobbes, and thereby assimilating man to a universal category of nature – sensationalism progressively accepted the separateness of human nature as a 'spiritual' phenomenon. As such it had to be described and understood in terms of its own internal relationships. The human subject then became an interior cosmos, a psyche, qualitatively distinct from, but ordered homologously to, the 'system of nature'.

SENTIMENTS

If the perceptual world of the human subject could not be reduced to a simple stream of sensation, then neither could its moral and aesthetic dimensions be explained solely upon the basis of utility. Yet just as sensationalism at first seems to be the 'natural' psychology of the bourgeois age, so utilitarianism seems to express its immediate ethical assumptions.

Sensationalism in psychology and utilitarianism in ethics were in practice linked, historically and systematically, through the powerful and somewhat diffuse ideological movement that developed around the new scientific world view. If the material world could be intellectually reconstructed with reference only to the essential qualities of matter and motion, as Descartes and Hobbes supposed, then the inner psychological world could be fully explicated in terms of the ego alone. The self was, from one point of view, the end result of a complex interactive sensory process; but from another point of view it constituted the elementary particle of social life. The essential character of the self, therefore, must 'contain' all that was necessary for the logical reconstruction of society as an ordered whole. Since the fundamental property of the self was an ego – that is, nothing other than an unavoidable tendency to act in conformity with itself – all human activity could be grasped as modifications of this fundamental principle.[37]

Hobbes proposed a consistent psychology of the ego as part of his general materialistic philosophy. As clearly as in d'Holbach, the ego emerges in his work as the active principle of the psychic world, the invariable relation through which the psyche became constructed.[38] Though systematic and compelling, Hobbesian psychology did not become central to the development of English utilitarianism. His uncompromising secularism was too easy to reject, and it is only in retrospect that we can appreciate the adequacy of his vision of the market as a subjective reality.[39]

It was Bernard Mandeville's *Fable of the Bees* that established a successful tradition of 'ego psychology' in bourgeois social thought. It is important to realize, however, that just as Condillac or La Mettrie were intent upon laying bare the sensory *foundation* of the perceptual world, so Mandeville's concern was the egoistic root of moral action. He did not argue, therefore, that terms such as 'wickedness' or 'vice' were without meaning, only that actions should be understood in terms of the private motives prompting them and judged, in part at least, from the consequences flowing from them. The usefulness of vice to the community, which he was at pains to demonstrate, did not abolish its ethical wickedness.[40] Nor was his aim to excuse the many vices which were manifestly useless. It is difficult to read his work now without separating his 'functionalism' (the usefulness of vice) from his utilitarianism proper (the self-regarding root of all moral judgement).[41]

Mandeville demonstrated at greater length and with more vigour, if not always with greater subtlety, than anyone else the self-deception of 'conventional' morality: 'The nearer we search into human nature, the more we shall be convinced, that the moral virtues are the political offspring which flattery begot upon pride.'[42] No man is 'proof against the witchcraft of flattery',[43] and it is the love of self, pride, which is the most powerful of human motives. The need to maintain a good opinion of ourselves is also the most civilizing of forces; 'a dextrous management of our selves'[44] depends not simply upon the satisfaction of appetites whose source lies within the individual, but requires that our actions be approved by others. To this end, we disguise our real feelings and control our spontaneous inclinations: 'a man need not conquer his passions, it is sufficient that he conceals them'.[45]

Mandeville therefore interpreted *all* altruistic actions as motivated by pride and selfishness: by love of praise and fear of blame. A compassionate act was simply incomprehensible other than as a relief of the conscience, and conscience was only a mechanism to preserve our pride under the scrutiny of public opinion.

Bentham's later axiomatization of utilitarianism adds little that is basically new, either psychologically or logically, to Mandeville's provocative text. If 'the principle of utility neither requires nor admits of any other regulator than itself', then everything follows as a matter of definition,[46] and the theory becomes as tautological as the action it explains. The internal world of the subject is organized as a system of exchange governed by the inherent tendency to conserve pleasure.

Objections to utilitarianism, particularly in the work of Shaftesbury, and the important Scottish writers on moral philosophy centred upon the identification of pleasure with the self. Human nature from this perspective also appeared to be wholly self-regarding, but a generous definition of the self made the point in a less cynical fashion.

Shaftesbury, indeed, more than Hobbes or Locke, deserves credit as 'the first moralist who distinctly takes psychological experience as the basis of ethics'.[47] Assuming philosophy to be 'the study of happiness', he recognizes self-knowledge and self-mastery as the central practical and intellectual problems of the age.[48] He criticizes Hobbes on the grounds of the limiting conception of man forced upon him by his dogmatic materialism. Hobbes, in failing to acknowledge anything 'which naturally drew us to the love of what was without or beyond ourselves', painted an unrealistic portrait of human selfishness.[49] But 'moral' action consisted just in this reaching beyond the immediacy of the sensuous self; if it did not, it is difficult to see how the language of morality could have developed. The utilitarians thus 'have made virtue so necessary a thing, and have talked so much of its reward, that one can hardly tell what there is in it, after all, which can be worth rewarding'.[50] Why make an appeal to ethical principles if a narrowly conceived natural propensity to utility were in fact exhaustive of the store of human motives and sufficient to regulate social behaviour? There would be no need to deceive and flatter if *all* ethical action were a disguise. It is only the authenticity of our moral sense which makes hypocrisy possible.

The self, that is to say, is no mere reflex of a pleasure principle; or, to be more precise, pleasure is not mere sensuousness. An ethical and aesthetic sensibility, as much as natural appetite, makes demands upon the human subject, the satisfaction of which draws him into a series of psychic exchanges. The self is an internal cosmos and not an 'atom' of society. It is therefore continually possessed of an inner dynamic, movements through which it seeks to complete itself as a system, and 'thus is everyone convinced of the reality of a better self'.[51] The special task of philosophy is to facilitate such an inward development, 'to teach us ourselves, keep us the self-same persons, and so regulate our governing fancies, passions and humours, as to make us comprehensible to ourselves, and knowable by other features than those of bare countenance'.[52]

Francis Hutcheson insisted just as vigorously upon the variety of the internal senses. Beauty, as well as virtue, is an irreducible aspect of selfhood and one without which our intellectual capacity for reason would hardly be roused. The new scientific world view is itself the finest expression of the ordered expression of this internal sense.[53] The discovery of such arduous and impractical truths of reason would never result 'were we not conscious that mankind are pleased with them immediately by this *internal sense* of their *beauty*'.[54] Science cannot be explained, that is to say, by reference to its practical advantage, any more than ethical conduct can be grasped as enlightened sensuousness.

Hutcheson's moral theory is more fully developed than that of Shaftesbury. Accepting the reality of an internal moral sense, he goes on to show the social mechanism developed from it. Virtue can be conceptualized as a 'good' possessed by an individual to the degree to which he receives the 'approbation' of others. But unlike other goods, its accumulation, existing in the admiration of others, generates emulation rather than envy.

Hutcheson's writing, particularly his critical attack on Mandeville,[55] was widely admired, and influenced both Hume and Smith. But it is in Smith's *Theory of Moral Sentiments* that bourgeois psychology could be said to have come of age. Smith admits that 'every man is no doubt, by nature, first and principally recommended to his own care',[56] and that a 'regard to our own private happiness and interest'[57] is a necessary element in virtue. Yet utility cannot wholly account for any of our actions: 'How selfish soever man may be supposed, there are evidently some principles in his nature, which interest him in the future of others, and render their happiness necessary to him, though he derives nothing from it except the pleasure of seeing it.'[58] The core of moral life is *sympathy*; that is, a spontaneous 'fellow-feeling' in the spectacle of life. 'Whatever is the passion which arises from any object in the person principally concerned, an analogous emotion springs up, at the thought of his situation, in the breast of every attentive spectator.'[59] This clearly is a principle which could play little part in the moral life of feudalism and contains an implicit assumption of individual freedom upon a market. Sympathy is founded upon the formal equality and the qualitative indifference of the personality. The individual 'imagines' the situation in which he perceives another and experiences, as a result, the sentiments appropriate to such circumstances. It is not a direct imitation of another's passions; the sight of a lunatic, for example, may arouse intense sorrow in the observer, irrespective of the manifest feeling of the madman.[60]

Apart, therefore, from sensuous satisfaction, 'nothing pleases us more than to observe in other men a fellow-feeling with all the emotions of our own breast'.[61] The formation of the self in its broadest sense is an intersubjective process and is itself a source of pleasure independent of the emotional content of the feelings it sets in motion. The mutuality of sympathy is thus the ground upon which we urge our sorrows upon others since, in arousing their fellow-feeling, we reduce our suffering by the addition of a fresh source of pleasure.[62] Furthermore, as another cannot fully experience the intensity of our own emotions, we will, in enlisting the consolations of fellow-feeling, moderate the violence of our own passions.[63] Sympathy, the mutual interdependence of selves, provides Smith with a regulatory motive more powerful than any individualistic calculus of utility. If pleasure depends upon the feelings of others, then the 'civilized' restraint of conduct through moral codes becomes immediately comprehensible. There is an inherent 'economy' of pleasure which tends towards the expression of selfhood as personal identification with such codes. The person is thus conceived as the moral consistency running through innumerable forms of subjectivity, guiding the 'choices' enacted in individual behaviour.

Smith makes sympathy the organizing principle of his psychology, deducing from it the naturalistic detail of personal life that is conspicuously absent from the sensationalist tradition.[64] The psyche becomes an internal market in sentiments. Selfhood is a tendency to pleasure, to the accumulation of approbation and virtue, all of which depends upon the exchange of sentiments. Sentiments, we might say, are the commodities of the psychic world, whose circulation is governed by an internal law of gravity that we experience as personal identity:[65] the internal 'mass', or seriousness of the personality.

This psychology is not simply fitted to a general ideological view of modern society as the liberation of the universal subject from the constraints of feudal particularism; it accounts for the particular social features of capitalism. Sympathy is an asymmetric principle, since 'mankind are disposed to sympathise more entirely with our joy than with our sorrow', and because of this 'we make parade of our riches, and conceal our poverty'.[66] The Scottish school, with their original historical perspective, had shown that there is nothing 'natural' in man's desire for riches, and Smith concurred; 'what is the end of avarice and ambition, of the pursuit of wealth, of power, and preeminence?' he asks.[67] Our love of commodities is stimulated by something other than the material benefits they might bestow: 'to be observed, to be attended to, to be taken notice of with sympathy, complacency, and approbation, are all the advantages we can propose to derive from it'.[68] It is in order to distinguish himself that the rich man displays his wealth; he 'glories in his riches, because he feels that they naturally draw upon him the attention of the world'.[69] The individual seeks to complete his personality by becoming distinguished from all others. This is a process more easily and generally accomplished indirectly, not, that is to say, by arousing the sentiments of others to the personality itself so much as to its halo of commodities.[70] And as our appetite for distinction is in principle unlimited, there is no possibility of satisfying the demand for commodities.

DESIRE

The various traditions of bourgeois psychology, sensationalist materialism, utilitarianism, the theory of moral sentiments, all had as their common focus the emergence of the 'self' as a central philosophical and practical issue. The self, however, whatever the manner of the approach, resisted the blandishments of reason. Shaftesbury indeed had taken this inspired coyness to be its hallmark.[71] Personal identity lay not merely beyond the reach of accurate conceptualization, but also aggravatingly evaded its own existence. More often than not it refused, as it were, to crystallize within the flux of subjectivity. We could hardly doubt its reality, since our experience was apparently founded upon its continuous presence,[72] and yet it was never present to us in its completeness.

The romantic urge to grasp the self in its entirety exposed the impossible depth of universal subjectivity. As the individual was an inner cosmos, each was as unbounded as the universe of objective and 'absolute' space and time. The self, as one possible arrangement of the infinite variety of subjectivity, could be understood (as the universe was understood) as the continuous operation of a specific mechanism rather than as a particular arrangement of elements or parts. Associationism, utility, and moral sensibility were therefore attempts to outline the 'laws of the psyche' rather than to provide a 'phenomenology' of the subject.

The perpetual incompleteness of the self, which is the hidden motive of bourgeois psychology, is best described as desire. If pleasure is the pursuit of the ego, desire is the endless quest for the self in its 'finished' form. The transcendental infinite in which the subjectivity of happiness rested is transformed into the perpetual torment of the self's search for concrete 'authenticity'. The discovery of our own personality, its emergence as a wholly determinate and necessary structure, becomes the 'religious' duty of the bourgeois individual.[73] This is a discovery of 'inwardness' and inevitably so as the 'double freedom' of capitalism extends to each individual an identical legal and political identity. The act of distinguishing one person from another, in a condition of ideal equality, must therefore be the work of an interior personality. The endless variety of such personalities is the corollary of the complete inner freedom of subjectivity. In reaching beyond itself, hopelessly, to complete itself, the incipient 'self' progressively discounts all the personalities which it is not.[74]

The self as a relation of desire, a segment of the infinite interior freedom of the subject, is an assumption of bourgeois psychology unexamined by the majority of its conventional practitioners. It is hinted at in Hume's scepticism and agonized over by Rousseau. Perhaps only in Hegel does it become the foundation for a systematic reconstruction of reality.

In a famous passage in the *Phenomenology*, Hegel analyses the growth of self-consciousness as a dialectic of desire. Simple undifferentiated consciousness, the 'I', which is 'absolute mediation,'[75] conscious of itself only as the immediacy of the given world, as the 'dizziness of a perpetually self-engendered disorder',[76] postulates itself as a coherent structure by desire. The more it desires, the more concrete and specific it becomes. Desire is the specific mode of existence of the self.

In desiring commodities the self seeks to confirm itself as distinct from any other self, but what it seeks in any object is the spiritual reality which lies at its core. The original disordered state of subjectivity 'alienates' itself into such material forms in order that, in recognizing and reincorporating itself within its own personality, it can define its uniqueness. In seeking commodities, that is to say, it seeks itself. It seeks itself with even greater urgency in the 'purest' of its alienated forms, as another person. To become a self the 'I' must engage itself in a relationship of mutual recognition.[77] It desires the desire of another, and becomes itself only by transcending all other, animal, needs.[78]

In spite of the groundless freedom of subjectivity, the personality 'emerges' by an immanent and absolutely necessary process of unfolding. Hegel's phenomenology is a philosophical description of this process. The self, in its completed, rational form (which in becoming philosophically possible becomes also, for Hegel, a practical reality) is after all a kind of duplicate of the 'world system'. It is, first of all, a system: in Hegel's terminology, pure mediation. The self is 'all of a piece', its various differentiations so many aspects of a totality recoverable from any starting point. It is therefore 'rational'. Its development through a series of dialectical steps or 'negations', its progressive 'otherness' is an inevitable, self-moving sequence of forms. It is a system understood in terms of a simple underlying principle, desire, rather than a fixed empirical arrangement of parts. Desire, we might say, is the gravity of the self; it is what lends weight to the personality and fills it with content.

This is not to suggest of course that there is no difference between a Hegelian and a utilitarian approach to the nature of individuated subjectivity. Such a contention would be absurd. Yet they are linked as two of the logically and existentially possible versions of bourgeois selfhood. Neither is comprehensible within the cosmology of fun or happiness. What the traditions of bourgeois psychology seek above all is confirmation of conscious selfhood.[79] And the route to conscious selfhood is through completion of the 'psychic' system. This system can be viewed in a number of different ways: as the complex interrelations of 'irritable' matter, the network of utilitarian calculations, the reflexive judgements of 'approbation', or the dialectics of 'otherness'. Each tends, in its own way, to progressively higher levels of internal organization that culminate in self-consciousness.

THE SYSTEM OF THE WORLD

The substantial unity of psyche and cosmos, whether as the undivided immediacy of fun or as the symbolic identity of happiness, was shattered by the new reality of capitalism. Two separate realms of being disentangling themselves from the confusions of the pre-modern world view confronted each other in suspicious silence. Nature at last exercised its claim to an independent existence, and in so doing exhausted the world of its material content. The human subject, no longer distracted by its implication in purely natural categories (its embodiment), counted this as a release rather than a loss.

Both cosmos *and* psyche, that is to say, were subject to an infinitizing movement. The boundedness and fixity of the symbolic universe, its minutely regulated hierarchy of parts, was overwhelmed by a new universality. The order of the world could no longer be conceived as a 'logic of place'. Rather, nature had to be seen as an inconceivably complicated mechanism whose coherence derived from its own indwelling lawfulness. Matter everywhere was the same. Nature anywhere, to exist as nature, had to conform to a few simple, formal rules. These rules could be expressed with perfect clarity in a mathematical form. The subject, similarly, to exist as subject, had everywhere to be the same. It was not just some vague 'left-over' from the realm of nature. Subjectivity, discovering its proper domain of 'inwardness', discovered its own infinity. And, in order to guarantee *its* coherence a similar lawfulness must regulate its interior interactions and provide it with its own, non-material mechanism.

It is misleading, therefore, to view the classical bourgeois picture of the world as split between, on the one hand, a 'rational', objective, and universal nature, and, on the other, an 'irrational', subjective, and particular psyche. Cosmos and psyche, it is true, become qualitatively distinct. All attempts 'scientifically' to reduce one to the other, or 'metaphysically' to force them to merge, failed. Yet this difference cannot be represented as the distinction between the rational and the non-rational world.

The 'mechanism' of nature could not be immediately apprehended but, in principle at least, could be easily understood. Matter, guided by ineluctable necessity, interacted in a complicated fashion to produce effects of which we became conscious as the phenomenal world. The psyche was not itself a division within this world of exterior nature. But its own inexhaustible inner variety was ordered by a 'mechanism' proper to itself. We experience this mechanism directly as the pursuit of pleasure. Pleasure is the spontaneous tendency and inherent organizing principle of the psyche. Its effects become visible, to ourselves and to others, as the individual and particular characteristics of personal identity. 'Reason' and 'passion' within the individual are opposed in just the same way as the complexities of real events may obscure or even 'contradict' the direct consequences of a simple 'law of nature'. The mechanism of pleasure, however, cannot, any more than the law of gravity, be revoked.

The pursuit of pleasure has nothing to do with sensuous gratification. It is a mechanism of individuation; the principle by which the self emerges from the flux of consciousness. And the laws of nature are not themselves 'material', but the principle in terms of which we 'make sense' of matter. Neither directly apprehends the pure 'substance' of either nature or consciousness. The organizing principles of mechanism and egoism are therefore 'rational' abstractions.

The 'theoretical' sciences of the bourgeois era, uncomfortably conscious of a gulf between object and subject, inadvisedly sought to span the abyss with a yet more abstract *system* of philosophical categories. But there was no need. The social relations of capitalism, the formation of the commodity world, had done the job for them. The commodity was itself an abstraction, an object drained of qualities. It was *its* universality and necessity that the scientist and the psychologist had rediscovered in the rational mechanisms of nature and subjectivity.

The commodity world furnished the sciences with a new social logic; one of individuated 'objects', differing only by quantitative criteria, and interacting upon a universal 'market' of pure dimensionality, according to strict rules of causality. Consequently, the philosopher who sought to reunite the disparate realms of object and subject succeeded only in hypostatizing the commodity. And since the commodity 'thinks for itself', there was a real sense in which philosophy had become redundant.

Yet, however much the bourgeois sciences made sense only within the context of a new universal social logic, the initial steps beyond its system of the world took the form of dismantling the commodity's metaphysical *alter ego*.

Part Four

EXCITEMENT

Actuality cannot be conceptualised.

Søren Kierkegaard

How should explanations be at all possible when we first turn
everything into an *image*, our image!

Friedrich Nietzsche

Chapter Ten

TOWARDS A CRITIQUE OF BOURGEOIS PSYCHOLOGY

The classical bourgeois world view can be understood as a process of individuation, as the pursuit of pleasure. The pursuit of pleasure is the pursuit of the self; and the self, like the cosmos, is a system of relations tending towards a unique equilibrium. This has long since ceased to be a plausible view of either psyche or cosmos. The pre-eminence of the classical bourgeois world view (mechanism and egoism) can claim an historical validity indeed for little more than 150 years.[1] The cultural history of capitalism since the mid-nineteenth century has been one of disintegration. It is a process which, still continuing, does not allow us to adopt a position 'beyond' its own apparently aimless course. There is no vantage point from which to observe its bewildering succession of styles, no way to avoid being caught up in (or caught out by) its next convulsion.

We are paradoxically therefore still firmly rooted in the very world view we have come to despise – or, which is the same thing, it is still firmly rooted in us. No longer convinced of the possibility of a complete and consistent 'system of the world', we none the less have to behave as if we were. All our criticisms turn into self-accusations and dissolve. The bourgeois world view is, as it were, carried forward into the heart of the modern world, the world which should have replaced it, and finding it heartless took root again. Strangely, the bourgeois world view has come to contain its own future. Perhaps we should not be surprised; it contains, after all, both its past (happiness) and its opposite (fun).[2]

More oddly still (modern criticism cannot have too much of this sort of thing), its future is contained in its own past! We find, in Montaigne or Cervantes, for example, at the dawn of the

199

bourgeois world, all the elements of corroded subjectivity we have come to love in its afterlife.[3]

The problem of identity, which is central to Montaigne is not, thus, a biographical struggle towards the 'realization' of the unique value of an individual. The 'self' emerges only to dissolve, existing separately and incoherently in the moments illuminated by an unreliable memory. Human personality has emerged as the privileged 'subject' of world history, but is still free of all historical determinants. The author exists then in a void; self-determining and self-deceiving.[4] There is no standard of validation against which the self can test itself. Any apparent norm is in the final analysis self-generating.

It was not in fact in this radically solipsistic form that the bourgeois ego emerged on to the world stage. It was rather as an uncomplicated union of reason and desire (as pleasure) that the ego made its appearance. It is almost as if Montaigne, anticipating 300 years of brutal self-abuse, had leapt to the farther side of the bourgeois world to describe the inner decomposition of what hardly yet existed. Of course it was just because the *order* of capitalism was not yet visible, while the collapse of feudalism was only too apparent, that Montaigne's psychology bears such a striking resemblance to the most advanced of modern writers.[5]

It is similarly tempting to read Giordano Bruno's *La cena de le ceneri* as an 'anticipation' of the theory of relativity;[6] as if, immediately prior to the long detour of classical science, a prematurely 'modern' cosmology had been stillborn. Indeed, from the perspective of the present, there is much that is familiar in the Renaissance – more, it often seems, than can be found in the exhausting 'progress' of reason during the eighteenth and much of the nineteenth centuries. But we should not be misled. Theirs is a contest between fun and happiness; a rediscovery of the inexhaustible transformative power of subjectivity released at last from the etiquette of a spiritual quest. The real struggle, however, was between happiness and pleasure, which could have but one result. 'Modernity' is neither of these; nor is it a competition between pleasure and something else. It does not challenge reality, it is not antagonistic. It is nothing but decayed pleasure and quite harmless. A certain aimlessness and love of paradox is hardly sufficient to join us once again to the pre-modern world. We notice these similarities only to convince ourselves of our own intellectual freedom.

We cannot therefore date the advent of modern cosmology from what appears, in retrospect, to be its initial spokesmen; any more than we might claim Democritus as the real founder of classical atomism. Nor should we begin with the spectacular changes in the physical sciences, as if they were propelled by some internal dynamo of their own. Both historically and conceptually, it is safer to look for the origins of modern culture in the 'human' sciences. Transformations of culture are always complex and many-sided. At this point we can simply continue with the story. In the history of human subjectivity, what has been the fate of the pursuit of pleasure?

DESPAIR

In the writings of Søren Kierkegaard can be found the first comprehensive and critical 'depth' psychology of the bourgeois era.[7] It is a critical psychology in a rather special sense. The general assumptions of his writings, as of his life, are just those institutionalized with the development of capitalism. An individualistic psychology is therefore as central to Kierkegaard as it was to Adam Smith. He takes quite seriously the obligation to become that 'single individual' which is his unique possibility.[8] So wholehearted is his commitment to the ideals of the age that his suffering for them exposes their cruelty. Of course he misjudges his age, importing into it the absolute demands of an older tradition.[9] For all his love of paradox, he fails to see one of the most fundamental of modern contradictions: the absolute demand to be hypocritical. Kierkegaard's 'error' was to take life seriously, to seek the realization of a value in himself. Modern values, however, are to be talked about rather than realized. For all that, indeed because of it, Kierkegaard is an unrivalled guide to the pitfalls of personal existence. Taking the urge to individuation to an extremity, he displays for us the formal impossibility of the 'self' and incidentally exposes the (less harrowing) difficulties faced by the mundane 'ego'.

What is at issue – for Kierkegaard, a life-and-death issue – is not whether from some other point of view the self-evident 'rationality' of bourgeois culture might appear grotesque or absurd, but rather, given this fundamental reality, what it means for an individual, personal existence.

Kierkegaard's astonishing insight into the incompleteness and contradictoriness of the bourgeois psyche has two obvious and equally important sources.[10] One is the philosophy of Hegel, particularly the *Phenomenology*, to which, in common with many other radical intellectuals, he responded during the 1840s.[11] The second is his own personal experience. The two are perfectly merged. His 'categories' cannot be understood as abstract philosophical terms, they unnervingly take on the qualities of personal existence. And equally his personal life existed as a philosophical gesture.[12] He tirelessly attacks both the 'abstractness' of the 'system' and the dialectical insufficiency of contemporary life.[13]

Hegel is, for Kierkegaard, the apotheosis of Enlightenment. He renders nature, history, and the human subject into aspects of an immanent, unfolding reason. The 'ego' is the personal inwardness of this process, continually expanding as the movement of human knowledge towards the absolute. While Marx countered Hegel's metaphysical comprehension with its 'reality' as the social world, Kierkegaard opposed both, as different versions of a totalizing rationality, with the unsystematic fragments of personal existence.[14] For him 'paradox' is not a sign of logical misbehaviour, nor of contextual confusion, it is the point at which life asserts itself against that part of itself (intellectual reason) which claims to speak for the entirety of human experience.

A convincing reply to the *Phenomenology* cannot therefore be conceived simply as an alternative 'system'. Hegel's system is (as Marx also argued) the definitive statement of bourgeois rationality. It can only take the form of a personal document, an autobiographical essay, in which metaphysical tendencies are rigorously suppressed in favour of the hard, irreconcilable differences of life itself. The reality that Hegel captures in his thought is the truth of reason. But that is only part of life; 'If Hegel had written his whole logic and had written in the preface that it was only a thought experiment', claims Kierkegaard, 'he undoubtedly would have been the greatest thinker who has ever lived.'[15] He is not incorrect, he is incomplete, which is worse: 'As it is he is comic.'[16]

Either/Or was conceived as a polemical autobiography against Hegelianism and the 'misunderstanding of the age'. This strange book, however, cannot be understood without reference to its purely personal content. It contains the 'story' of Kierkegaard's unhappy love for Regine Olsen. This 'private' matter provides perfectly sound subject matter for a philosophical work; it is indeed passion alone which can reveal the only interesting philosophical topic, the self.[17]

The prospect of marriage precipitated a crisis in Kierkegaard's life. Having hastily contracted an engagement he found himself in an impossible situation. The genuinely decisive nature of the situation aggravated his peculiarly melancholic anxiety: 'If you marry you will regret it; if you do not marry, you will also regret it.'[18] But more than this, it caught him in an irreconcilable contradiction between his intense and secretive life with his father, and his conception of marriage as an open relationship.[19] He could not marry, but to extricate himself from the situation, or rather to extricate Regine from the situation, justified a melodramatic subterfuge. That she might not harbour any lingering hopes in him he affected a dissolute public life. By appearing to be an unscrupulous libertine he also satisfied his own guilt and left Regine unattached and without blame, so that, in the small and intensely pious atmosphere of bourgeois Copenhagen society, her future prospects of marriage might not be damaged.[20] The situation became scandalous. It had been what Kierkegaard had intended, yet he could not help trying to 'explain' himself, to Regine and to the world. When *Either/Or* was published he had a copy sent to Regine. His use of a pseudonym fooled no one and the *cause célèbre* again became a topic of gossip as his book was hastily read for any new insights into the circumstances it might offer. As the first volume ended with a narrative entitled 'The Diary of a Seducer' and seemed to be a thinly veiled account of the entire 'affair', there was every hope that public curiosity would be satisfied.

His book, however, was far from straightforward. Whatever the significance of the 'Diary', what was to be made of its other extraordinarily diverse contents? It seemed to have no central focus; it even claimed to have been written by two different people. *Either* was apparently the work of a 'young man' of extraordinary literary and critical gifts, while *Or* was ostensibly the composition of an older friend of the author of *Either*. The *Or*, in the form of long letters addressed to the 'young man', thus forms a critical personal commentary upon the *Either*. This complex 'indirect communication' was not intended only as a 'confession' or as a covert 'explanation' for Regine's sake, but more generally as a description of the 'real' psychology of the bourgeois age.

Philosophy must grasp life directly, and the real content of modern life lay in just those disconcerting and 'contradictory' movements the metaphysician could not allow. Each individual is a chaotic history of internal states and dispositions that can be formed into a unity only through the deceptions of memory. In truth the individual – at least, the bourgeois individual – cannot become united with himself because he altogether lacks the guidance of a transcendental norm. He is left to define himself and is undermined by the endless self-doubt of existence.[21] The coherence of the individual is only an ideological assumption, and is destroyed in the moment we try to live 'seriously'. This is not an 'intellectual' problem. Reason as the necessary order of immanence, unfolding in our experience by a process of inner development, is powerless to help us. Kierkegaard can describe his predicament, but he cannot by some process of rational reflection 'decide' how to act. Human activity is not a question of 'rational' decision, but of 'choice', which is quite another matter. Kierkegaard realized therefore that his failure to marry could not be understood as a decision. The account of the circumstances which he offered to the world, and to himself, were at first self-flattering. The insurmountable contradiction of remaining 'faithful' to his father and to his own melancholy, while at the same time binding himself absolutely to Regine, was nothing more than a particularly testing example of the oppositions that went to make up everyday experience. He had lacked nothing in intellectual skill and discrimination. Like the 'young man' of the *Either*, Kierkegaard was well aware of his own talent for reflection; what he had lacked was the courage essential to 'choice'.[22]

Bereft of any 'natural' unity, the individual is not without a sense of direction, and Kierkegaard's psychology is in large measure an elaboration of the 'stages' through which the self *might* pass. It comprises an existential map of unprecedented (and almost unsurpassed) precision and detail. His aesthetic works in particular offer extraordinary insight into the nature of the modern experience of the self.[23] *Either/Or* is a description of the first two 'spheres' of existence, the aesthetic and the ethical.

The peculiar book is after all not chaotic; it turns out in fact to be a perfectly coherent, but deliberately unsystematic, account of these spheres of existence as possible modes of selfhood. From this perspective, we can see that the *Either* contains as a loosely connecting theme the nature of aesthetic enjoyment. This becomes more apparent in the light of the retrospective commentary written by Judge William who lives under the radically different determinants of ethical values.

The aesthetic sphere is 'immediate' existence. The 'young man' aspires to nothing beyond the enjoyment of the world presented to him. He seeks pleasure as the only meaning of his existence.[24] He is not without cultivation. He hopes to find pleasure in the finest artistic achievements. Immediacy does not mean sensuousness. But his superb education is to no avail. He finds only boredom. The more pleasure becomes the sole object of conscious endeavour, the more certainly it eludes the grasp of the longing subject. Part of the difficulty is the sheer number and variety of possible sources of pleasure. No sooner is one envisaged as the ultimate and final satisfaction than its attractions pale, to be superseded by some yet more beguiling objective.[25] Pleasure is a relation of desire, a movement towards the completion of the self. But the self is infinitely extendable, so the object world continually takes on new attractions. It is a movement which cannot be completed. We cannot find, in pleasure, the cessation of desire. But this is what the 'young man' desires above all. His fantastic projects and fits of enthusiasm always end in boredom and self-disgust because they fail to provide a real sensuous object in which he can 'take hold' of himself. The pursuit of pleasure is always the desire for the self. But the 'self' cannot become actual in 'immediacy', and remains accidentally dispersed throughout the world of experience.

Momentary pleasure is arbitrary and chaotic, it can provide neither the 'foundation' nor the 'image' of selfhood. To the degree to which the 'young man' identifies himself with these hectic adventures he also becomes 'volatilized' and chaotic. In a radical rejection of Hegelian philosophy and all other traditions from which bourgeois psychology had sprung, Kierkegaard begins his description of the individual as an *accidental* relation.[26] The characteristic inner feeling of existence under the sway of aesthetic 'categories' is thus neither pleasure nor frustration (which requires the clarity of a fixed objective), but boredom. The 'young man' consequently 'hovers above existence',[27] possessing only the melancholy which Kierkegaard himself had known so well since his childhood.[28]

The 'young man' is in 'despair'; that is to say, he is failing to be himself. Indeed, in a purely aesthetic existence, despair takes the extreme form of the failure to be *any* self, let alone the 'single individual' each person might become. But in suffering melancholia he begins to despair 'seriously', for melancholy is already a movement of the *self*.[29] It shades into ironic detachment from the world, and it is this deeper form of despair which brings him to the edge of aesthetic existence.[30] In it the intuition of another form of existence, entirely different from the arbitrariness of the immediate senses, comes to life. In melancholy and irony the 'young man' possesses despair, which is the starting point of 'real' philosophical existence, not intellectual doubt but a doubt of existence.[31]

There is no necessary, essential, or immanent movement which leads from boredom to melancholy to irony; far less any inherent force of self-development driving him to make the 'repetition' beyond all these 'qualifications' of immediacy into an 'ethical' life. The fundamental error of Hegelian psychology is revealed as the absurdly optimistic belief in the *continuous* development of reason. Real existence is made up not only of contradictions (the impossibility of grasping pleasure, for example), but also of brutal discontinuities and disjunctions. It is indeed only from the farther side of such a 'gap' in existence that its having a farther side can be known.

The ethical begins, then, not with a rational demonstration of its superiority, or by blind necessity, but by a self-motivated 'leap'. It is only in making a decisive 'choice' that the individual can exist in a different way. But once such a 'leap' has been effected, miraculously the individual finds that he has not left or destroyed the aesthetic, but has taken it with him – and now in a form in which pleasure finally does become a realizable goal.[32] The decisive choice 'crystallizes' the self into ethical categories. The real meaning of existence is revealed as the pursuit and expression of some 'value'. Ethical existence is therefore polarized between judgements of 'good' and 'evil'. However, just as the ultimate pleasure sought within the aesthetic turned out to be the individuality which it could never contain, so all the choices within the ethical, in being choice made by the self, are really choices *for* the self. The decisive choice is not in favour of one good rather than another, or one good in preference to evil, but between life conceived as falling within such choices and one (unaware of the 'higher' possibility) that abandons it to the chaos of immediacy.[33]

Judge William's example of the ethical life is his own individuality, and in particular the decisive choice of marriage. His rather touching and naïve account of bourgeois marriage as the paramount ethical life-choice has a particular relevance, of course, in the context of Kierkegaard's own failed engagement. It would be misleading, however, to regard the 'young man' as Kierkegaard or to 'explain' the latter's failure to marry by an inability to break free of irresponsible aestheticism. Kierkegaard's own difficulties lay deeper, and the choice of marriage is only one of a number of possible examples of the ethical.

The ethical organizes the inner life into a coherent and continuous unity. But within it divisions and contradictions reappear. Just as aesthetic existence fails to grasp the pleasure which is its only goal, so the ethical life cannot in fact realize the good as free expression of the self. Our limited knowledge, both of the world and the complexity of our own nature, the enormous scope for self-deception and misunderstanding, all implicate us in evil. As soon as we attach ourselves ethically to the world, we cannot evade responsibility for the consequences of our choice.[34] The typical inner feeling of the ethical existence is once again one of failure, this time of guilt. Guilt, however, is a 'higher' qualification of the individual than boredom; it has a precise quality of individuation and a longer time perspective. It maintains and builds a unique vision of the self.

To undergo a second 'repetition' into the higher existence of religion, stretched between the polarities of faith and sin, requires an extraordinary effort. Beyond the highest 'potentiation' of the ethical (the comic), lies a dreadful gulf that few can summon the spiritual resources to challenge. It was from the edge of this particular precipice[35] that Kierkegaard wrote his pseudonymous works. Gripped by a powerful intuition of the religious 'solution' to the problems of personal existence, he yet could not for a long time summon the energy to leap into faith.[36] There is in Kierkegaard's psychology here an extraordinary affinity with the great mystical writers of the twelfth and thirteenth centuries. There is just as obviously, however, a fundamental difference. The entire tendency of the Kierkegaardian 'steps' or 'stages' in life is towards the absolute inner certainty of faith as an experience of the *self*. The eventual attainment of religious categories grasps the profound paradox of personal existence,[37] and holds it in an absolutely inward and secretive relationship with God. It is the very opposite of the annihilation of all personally distinguishing features sought by St Bernard or William of St Thierry. It is the ultimate confirmation of the self as a 'single individual', rather than the ultimate release from the self as that stubborn pride which blocked genuine spiritual growth.

Even Kierkegaard's religious writings are therefore works of modern psychology. The religious terminus to the process of self-development is in fact deceptive. It is an unattainable goal. No one, other than Christ, could in fact become a Christian.[38] The reality of selfhood in the modern world, the duty and promise at the heart of bourgeois society, are in fact boredom, guilt, and spiritual terror. The 'person' never emerges from the trials of existence; he is incomplete, contradictory, and unknown to himself. In the face of this psychological reality, the Church has the bad taste to offer childish piety. It ministers to the superficial coherence of the bourgeois ego, which had ironically been the starting point also of Kierkegaard's subversive orthodoxy. To overcome man's struggle with himself, Kierkegaard had to resort to a solution as abstract as had Hegel. Christ, as the only living example of fully developed existential unity, is a 'theoretical' rather than an 'existential' possibility.

Modern life as the experience of individuality is an ideological impasse. And bourgeois psychology is exposed as a 'false-consciousness' of a peculiarly tenacious kind. It is the failure to achieve the beatification of the 'single individual' which turns Kierkegaard's life (as well as his writing) into a critical instrument, dissecting the hypocrisy of bourgeois 'idealism' and 'piety'. In this he is far more successful than Max Stirner, whose 'absolute ego' remains a grandiose extension rather than a critical negation of the bourgeois psyche.[39] Despair, which is the 'sickness unto death', is the incurable disease of modern subjectivity, the inescapable mark of its corrosive self-consciousness.

Though cast in traditional, even reactionary, terms as the search for rational individuality, Kierkegaard's psychology in fact dramatizes the decomposition of the modern self. In striving for unity it is driven deeper into despair, farther away from itself. The contradictions of modern life do not stop with the division between subject and object; subjectivity is itself fragmented, each 'sphere' threatening the self with its own impossible vision of the world. The synthesizing power of reason has somehow evaporated from everything human; poured out of the self and into the world of 'real', objective relations, into the commodity form, and into the scientific conception of nature. The human subject, thus 'liberated' from its constraint, finds itself in a world of unlimited inner freedom.

DREAD

It is from the point of view of the decomposed subject that we can grasp the significance of dread as a specifically modern phenomenon. Kierkegaard devoted his most obscure work to an essay on the subject.[40] It is a notable 'anticipation' of many of Freud's key ideas on anxiety, the term sometimes used indeed to translate Kierkegaard's title.[41] In Kierkegaard's text, however, the terror of an inordinate and unreasoning passion is more transparently present. Dread is to be simultaneously attracted and repelled by the world; immobilized by fright, confined by an inexplicable loss of freedom.[42] It is fear, but of a peculiarly objectless sort. It is fear of nothing, a fear of the next empty moment. Dread possesses the self and makes the simplest action an impossible risk.

It is in this very objectlessness of dread that its terror lies. Here an intense feeling is experienced, as it were, in a vacuum; in the absolute purity of emptied space. The pursuit of pleasure, or virtue, or faith, is conceivable as the search for selfhood. It is to discover ourselves in the object world that we act with an underlying consistency of motive. When this process of self-discovery becomes undermined by the intuition of the nothingness that lies at its end, then we become filled with dread. And where we might normally invest the world with the attractiveness of our own self-love, projecting our 'wants' into objects to render them delightful, the decomposed self – preoccupied with its own inner fragmentation – can project only unrecognizable images of itself. In dread, the 'man without qualities' recognizes himself, clothing the world of appearance in his authentic selfhood; that is, in nothing.

Kierkegaard's lingering ideological sentiment (he considers the rejection of selfhood as the core of life to be a pathology of the personality), does not prevent him proposing a genuinely critical psychology.[43] It is a psychology which is repeatedly rediscovered in the following hundred years. He does not so much inaugurate a tradition which, developing from his fundamental insights, elaborates and refines his 'categories', as propose a point of view which is subsequently rediscovered, independently, in a number of different contexts.[44] A fuller examination of dread, not only as objectless fear but as objectless passion in general, can be found, for example, in the major novels of Dostoevsky.

The extraordinary realism of Dostoevsky's characters can be readily grasped in Kierkegaardian terms. There is nothing artificial or ideological about their coherence. Whatever their author's original intentions, each of his major characters, as if strung along the *Stages in Life's Way*, might be viewed in the light of one of Kierkegaard's existential images. The aesthetic, the ethical, and the religious spheres of existence are all represented by absolutely convincing personalities.[45] Each one presses his or her own claims on reality with the shaping power of a real human presence. It is characteristic of Dostoevsky's cycle of mature novels that the contradictions among and paradoxes within these characters are not resolved by an authoritative, 'transcendental' act of synthesis on the part of the author. The complete independence of each character overcomes the residual ideological sentiment in Kierkegaard.[46] Each existential position is accorded an equivalent status, a possible mode of individuation. There is no inherent order linking such possibilities, no development, whether from inner necessity or by the irrationality of the leap, leading from one to another, and no privileged final state which realizes all the inner potentialities of the individual. Raskolnikov is converted at the close of *Crime and Punishment* only to reappear, more terrifying than ever, as Stavrogin and Ivan Karamazov.[47] The arguments

among the various protagonists are never-ending and inconclusive. Reason is powerless to reconcile their differences. Equally,
however, they cannot undergo 'repetitions' into progressively
'higher' categories that carry the hope of overcoming the
inconsistencies of all particularities.

It is only as readers that we are allowed the privilege of the leap
from one enclosed vison of reality to another. Where his
characters develop, it is towards a 'deeper' subjective realization of
their own reality. Any movement from one decisive possibility of
life to another is unsystematic, arbitrary, and frequently reversed.[48]

In the prelude to his major novel cycle *Notes from Underground,*
Dostoevsky introduces one of his central characters. He is 'a sick
man', suffering the sickness unto death.[49] He is in despair, the
despair of 'weakness'; he is incapable of acting and confines his
existence to abstraction, he is only an idea of himself.[50] He lives
aesthetically but, like the 'young man' of *Either,* is disgusted by his
own trivial and meaningless existence. He is incapacitated by a too
highly developed self-consciousness. His disease is an inability to
forget himself, even for a moment. This is a deliberate inversion by
Dostoevsky of the orthodoxy sanctified by the secular mythology of
the Enlightenment.[51] There the human is identified with self-
consciousness, which is the unique atttribute of man's species
being, and in the *Phenomenology* the human develops generically
only to the extent to which this power of reflexivity can take
precedence over man's instinctual life. The growth of inwardness,
which is an aspect of the dialectic of desire, is held to be the special
virtue of the human. But in the 'Underground' self-awareness
becomes an affliction. The cosmopolitan European has become
transparent to himself and cannot cease questioning his own
motives. He becomes frozen by self-conscious doubt over the
wisdom of any action. He is possessed of self-loathing – that is to
say, the loathing of not being himself. He envies those whom
intellectually and spiritually he despises but whose blissful lack of
refinement allows them the exhilaration of continuous activity.[52]

There is more here than the 'melancholy' temperament of the
artist and intellectual. 'Underground Man' is a general type of
modern humanity presented in its most 'intellectual' form. But
less sophisticated versions are continually met with in the
proliferation of neurotic symptoms, where dread, rather than
expressing itself in general terms, fixes itself in some objective

form. The intolerable objectlessness of the passions is temporarily denied by discovering some appropriate 'object' for the sensations of guilt, boredom, self-loathing, and so on.[53] Dostoevsky presents, however, the purest case – a case in which the highest development of European culture, a culture which began in a frenzy of spiritual liberation and confident self-assertion, is exposed as cancerously solipsistic.

Only the 'man of action', the thoughtless and insensitive fool, can tolerate the modern world as it is. Underground Man, in being stuck in immediacy, is in fact withdrawing from the world as it is, the world of ideological delusion and hypocritical values. A more dramatic attempt to avoid both the anodyne complacency of conventional culture and the corrosive self-consciousness of aestheticism is exemplified in Underground Man's immediate successor, Raskolnikov.

Crime and Punishment in fact marks a decisive movement beyond the conventions of bourgeois psychology.[54] In order to 'realize' his potential uniqueness, Raskolnikov essays a fearful leap. Now, however, given the inherent incompatibility but existential equivalence of any of life's spheres, the leap into faith can just as easily be the leap into self-justifying violence. By killing an old woman money-lender Raskolnikov aspires to become a 'single individual'. The novel exhaustively exposes and rejects all possible 'rational' motives for his crime. No utilitarian or moral calculus can comprehend the meaning of such an act.[55] It is not an action at all, but a form of inner reflection. Only accidentally related to the external world, his crime aims at an interior transformation. It is a technique of 'repetition' which will precipitate the personality in a more highly determined form.

Only a profoundly immoral act can satisfy Raskolnikov's thirst for authentication. He must define himself. No external norm can validate or limit an absolutely freely chosen interior selfhood. If actions flowing spontaneously from his personality were to conform to the conventional moral expectations of his family and friends, they would serve only to irritate his relentless self-doubt.[56] The despicable murder, a 'teleological suspension of the ethical' is a true test of uniqueness, a 'pure' act expressive of himself alone. Yet he chooses his victim with a certain amount of care and cannot wholly shake off a conventional responsibility for his motives.[57] He has caught himself in a dreadful contradiction. To remain inactive is to condemn himself to the lingering sickness of Underground Man. But to become the authentic Raskolnikov demands nothing less than an act of arbitrary criminality. Yet an arbitrary act, just because it is void of genuine expression, is powerless to bring his hidden self into existence. He cannot act in the world, nor can he withdraw from it. He has intensified rather than resolved Underground Man's subversive despair. And here there is no saving leap into a higher existence. Raskolnikov's 'conversion' is the afterthought of the novel and not its denouement.[58]

The superficiality of his conversion follows on the failure of his second 'technique' of self-realization, namely confession. By unburdening himself to Sonia, he hopes still to salvage something of his original intention. The guilt he feels is not remorse over his crime, but pure self-loathing, which communicated to Sonia is the tenuous form in which he can appear to himself.[59] If he can despise himself he must after all be a particular, despicable individual, someone worth despising. Sonia, a genuinely tragic heroine, shoulders the guilt he ought to feel. She has the feelings he lacks, and being unable to understand his brutal act, none the less forgives him. Worse, she loves him. Instead of hating him and thus confirming his perverted form of self-expression, she 'spoils' his confession, by loving him.[60] Nor can he confess to Porfiry, who knows he is guilty but treats him as the charming personality at the heart of a psychologically 'fascinating' case.[61]

Raskolnikov is trapped in his own subjectivity, in the paradox of aestheticism deepened to schizophrenic detachment. The theme of confession is used by Dostoevsky to show the reader – and more importantly to demonstrate to Raskolnikov – another version of himself. By receiving Svidrigaylov's confession Raskolnikov can view himself in another light. His fascinated horror of Svidrigaylov is an exact description of dread. He cannot shake off the obsessive curiosity, the awful attraction, of Svidrigaylov's amoral sensuousness. He recognizes in him his own double.[62] Dread is the fear of nothing, the nothing which is the core of selfhood; in Svidrigaylov it is the terror of pure self-determination. Raskolnikov's revulsion has nothing to do with conventional moralizing. Svidrigaylov as an 'objectification' of his own self-loathing shows him the ultimate tendencies of bourgeois inwardness. An active, sensuous immediacy, rather than a reflective, abstract aestheticism, is just as incapable of escaping the circle of self-doubt and dread. Sensuousness remains frozen in narcissism, or its close relative, incest; and intensifying in the only direction open to it, culminates in its one 'positive' act of self-expression; suicide.[63]

Dostoevsky is a negative Kierkegaard. His existential precision is not conditioned by an implicit or explicit order. His novels are loosely organized around conventional stories comprehensible in terms of everyday notions of morality. The intensification of normal existence which we find in his characters tends, however, to invert and pervert the original meaning of the narrative. The normal psychology of bourgeois egoism as the rational pursuit of pleasure (the self) is overthrown. Without a transcending norm each character is responsible for creating himself, a process which ends in self-destruction or madness as the ultimate forms of 'authentication'; the proof that he is, as he suspected, worthless. The 'logic' of personal autonomy is turned against the self. The realization of a unique individuality is in fact a self-defeating illusion.

Bakhtin has suggested that Dostoevsky's disregard for the formal niceties of the classical novel, his original forms and themes, his psychological depth, are all aspects of his revival of Menippean satire, the original literary form of carnival.[64] None the less, however indebted he may be to purely literary prototypes, his novels appear strikingly modern. He has broken through the veil of bourgeois conventions in another direction. They do not recall pre-capitalist ideals of order and disorder, hierarchy, and release. His characters are caught nicely in the act of withdrawing from the 'object relations' which define the 'normal' individual. This withdrawal, catching them out with its slippery logic, precipitates a bizarre and unpredictable world of experience.

This is not fun, but excitement. The transformations of mood and feeling form the foreground to the grey and dismal existence of boredom. A welter of events and unruly passions are concentrated into his books, yet they stand out, even there, as unusual and never claim the inclusiveness or 'naturalness' of fun. They exist primarily to be read, to be 'consumed' by a 'normal' reader delighting in the shock, surprise, and melodramatic teasing of his partially hidden wishes. A good deal of the time Dostoevsky's characters, like his readers, are hanging around waiting for something to happen.[65] The hectic narrative overlays a backdrop of a decadent ruling class consumed by its own inactivity. The provincial boredom depicted particularly by Chekhov is implicitly contrasted, not with classical bourgeois self-assertion, but with the further decomposition of such composure, with its breakthrough into irrational cravings for new, untested forms of experience.

DEFIANCE

Despair is to be other than the self. Dread is the fear of being nothing, of not being able to become a self. Dread is therefore a generalized form of despair, an inability to actualize any image of the self. It is the purest form of what Kierkegaard termed the 'despair of weakness', a somewhat misleading term in the light of Dostoevsky's unconventional characters. 'Weakness' embraces not only those who have thoughtlessly embraced bourgeois complacency, but also those engaged in the most intense inner spiritual struggle over themselves. Yet despair may be potentiated in another way, transforming comfortable hypocrisy into defiance. Within each sphere of existence there is the possibility of identification with the 'wrong' choice. There are a series of negative characters, deliberately in pursuit of unpleasure, or evil, or sin. There is a powerful demonic element in such personalities, a perverse willing not to be the self; a strange power of self-possession directed against itself.[66]

Such individuals – and Raskolnikov hovers on the verge of becoming such a one – are shut up within themselves.[67] Myshkin, for example, and in a more sinister form Stavrogin, remain unapproachable in the perfection of their inner individuality. But while Myshkin and Alyosha Karamazov approach the ideal 'positive' character, the Kierkegaardian religious type,[68] with whom Dostoevsky never felt totally at ease, Svidrigaylov and Stavrogin provide the more powerful image of uniqueness, as defiance. Their existence is an insult hurled against the western philosophical tradition and the values it claims to support. In the place of philosophical certainty conceived as knowledge, Dostoevsky claims the authentic singularity of personal truth.

Philosophy, however, was not slow to fight back. Nietzsche, towards the end of the nineteenth century, attacked the philosophical orthodoxy of the classical scientific world view head on. He proposes a fresh philosophical approach which does not, once it has laid claim to modern reality, lead away from it into 'abstraction'. Nor is it just another voice complaining of the obstacles that speculative metaphysics has placed in the path of real self-knowledge. His is not, as in Kierkegaard or Marx, a critique of philosophical language, but a philosophical critique of modern life. His writings are not conceived as exercises in a critical reason which places itself beyond existence to comment upon the shortcomings of the present.

Nietzsche generalizes Kierkegaard's *Attack upon Christendom* to modern bourgeois culture in general.[69] The core of this culture is not the scientific knowledge of the world to which it lays claims but the specific way of life espoused in its values. The search for 'objectivity' in philosophy is one such value and one which reveals the general inversion and turning away of bourgeois culture from a 'natural' disposition towards life. Nietzsche is primarily a moralist, not because the discourse over values can be more 'rational' than the discourse over epistemology, but in recognition of its more fundamental sense of reality.[70] Knowledge of the world, that is to say, is intimately linked to a particular way of life which it expresses. Modern science admits as much in its acceptance of the provisional nature of its results, and the conventionality of its theories.[71] It has abandoned the search for 'truth' in favour of the eradication of 'error'; an apparently slight transition which turns out to be momentous. Any view of the world is just that, a perspective whose claim to validity is supported ultimately by the 'will to power' which it expresses, and with which it is one.[72]

The discourse over values has itself become corrupted, and in just the same way as the discussion of knowledge. The search for the good has been replaced by the avoidance of evil. This is Nietzsche's general statement of the character of modern asceticism.[73] He traces the characteristic 'transvaluation of all values' to the formation of early Christian morality. The powerlessness of early Christian groups to realize their vision of life in the face of a powerful and oppressive foreign ruler led to an 'interiorization' of their values. They came to identify their own weakness with these values. Humility, passivity, forgiveness became the perverse aims of a profoundly self-denying form of existence.[74] The subjective world is organized by *ressentiment*, by a host of repressed feelings that express themselves as their opposites.

Modern bourgeois morality in Nietzsche's view is modelled on the Christian prototype of transvaluation.[75] The symptoms of *ressentiment,* anxiety, guilt, and self-loathing everywhere abound. Overtaken by these pathological forms of subjectivity, the modern individual seeks relief from them in the very processes that occasion their formation; by subjugation to an external authority, by confession, the neurotic substitution of real wants by acceptable but unsatisfying playthings.

The formal coherence of philosophy and Christianity make them both prime targets for Nietzsche. The distinction is not, as for Kierkegaard, between the cruel abstraction of the former and the existential vitality of the latter, but between the world of experience rendered as a systematic totality and the world as open possibility. The moralist cannot any longer espouse a positive value. The world is already rotten and corrupted, there is nothing in it but repetition, the 'eternal recurrence of the ever same'. Against the perpetual grimace of demonstrative reason, the inner world of the human subject opens to reveal an undiminished plenitude. The subjective becomes the exclusive realm of the aesthetically and morally valuable; not as a dignifying principle of individuality but as the home of freedom.[76]

Nietzsche's demonic passion is comprehensible in the context of Kierkegaard's psychology. He is the defiant character, turned inwards, relentlessly stripping existence of every conventional positive value. He takes the absolute freedom of the inner self to be its only principle. Indeed, the notion of the 'self' is revealed as illusory. Subjectivity cannot be formed and constrained by any such individuated 'structure'.[77] Personal identity cannot rise above the incomplete and contradictory fragments of immediate experience. It is the fundamental error of the western philosophical tradition to aspire to identity, completion, integration, wholeness, and systematic unity where none can exist. Personal integration, as much as a systematic metaphysics, is a denial of reality, a rejection of life.

The nobility of life, which is the heart of genuine philosophy, cannot limit itself to something (logic, value, pleasure) less than itself. Nietzsche's predilection for the cosmology of fun is evident. That it is life in its undiminished fullness need not be doubted. But here again we are faced with something new. Fun has already been suppressed and cannot be recovered.[78] It cannot be deliberately sought or made the object of rational action. At best we can hold ourselves in a state of readiness to catch its unpredictable vibrations. Excitement, an internally decomposed state within which the 'self' cannot locate itself, is the enjoyment of modernity and the modern form of enjoyment. It is inherently unpredictable and related to the 'object world' in an arbitrary fashion. It is the choice of defiance, a love of nothing rather than a fear of nothing; a demonic attachment to everything incomplete and unfinished. In an excited state, the psyche no longer 'knows itself' and manages to throw off every pretension to seriousness.

The world of pleasure, the bourgeois world, is composed of a system of relations (commodities) differentiating and linking the 'self' and the 'world'. The distance between the two is felt as desire. The psychology of excitement resolves these relationships into projective fantasies. Objects exist for it only as aspects of its abiding narcissism. Any external stimulus might trigger off the entirely free, internal, and irrational process which is the sanctuary of human freedom. Excitement is a kind of nostalgia over fun, as pleasure is ultimately the forgetting of happiness. The hope of excitement is a voracious consumer of novelties. As enjoyment is no longer predictably related to the nature of the 'object world' and is something entirely inward, experimentation becomes continuous. Something which triggers a keen enjoyment on one occasion may be impotent at the next.

The Newtonian mechanism finally disintegrated towards the close of the nineteenth century. The system of forces it defined and used to describe the 'system of the world' ceased to have an unambiguous and consistent meaning. This is not just a matter of a 'relativizing' tendency in the theoretical languages used to describe nature and society. These tendencies, admittedly, are important. But a more general issue emerges and becomes paramount, first within a loose 'tradition' of bourgeois thought which might be called 'philosophical psychology'. As against all 'totalizing' modes of thought Kierkegaard, Dostoevsky, and Nietzsche, most powerfully, but not in isolation, proclaim the priority of a paradoxical, contradictory, and fragmented reality.[79] The consciousness of the bourgeois world did not founder as a consequence of attacks mounted upon it from without so much as crumble from within in the response of its most devoted philosophers to the life it had promised them.

Chapter Eleven

NATURE'S STRANGENESS

The critique of bourgeois psychology is by no means an exercise in 'negative thinking'. It provides a description of the real world: the modern world. Kierkegaard, Dostoevsky, and Nietzsche lay bare the reality of decomposed subjectivity. They reconstruct the world from its point of view, or rather from its varied and incompatible points of view. The *experience* of modern life cannot be conveyed, they claim, by any system of rational concepts. Incompleteness and contradiction (rather than ignorance and error) are its fundamental conditions, however much it succeeds in rendering the 'objective' necessity of nature as a rational order. The standard of personal truth could never be subsumed within a hypothetico-deductive 'system of the world'.

The implication, which Nietzsche fully realized, was in fact to deny to nature as well as to personal 'identity' the rational form of a logical system. The consequent attack on bourgeois meta-physical pretensions must be taken up within science itself.[1] This line of thought is in one sense a curiously negative confirmation of the ontological argument which stands opposed to the metaphysical tradition, as it were from the other side. Both Anselm and Nietzsche deny the possibility of a reasoning faculty separate from the reality upon which it reflects. Indeed, our thoughts do not lead us hopelessly astray merely because they can never achieve the 'objectivity' supposedly inherent in a wholly 'detached' view. Our conspicuous failure to solve the puzzle of self-identity offers a genuine insight into the real nature of the self as paradox.[2] We must trust this intuition and resist all shallow forms of rational abstraction. But if we are not in fact constituted as isolated 'egos', how can we hope to understand 'nature' as if it were composed of nothing but the relations among isolated particles? Just as our own experience, if taken seriously, will undermine any rational psychological theory, so will it prevent the completion of the classical scientific world view. The frozen psychotic stare of rational mechanics gives way, then, and with dramatic suddenness, to a new appreciation of nature's strangeness.

Although central to our understanding of it, Nietzsche's philosophical insight can hardly be held responsible for the transformation in scientific thought. Scientists, in formulating their own problems, however, were subject to the same conditions, and had to use as their intellectual resources the same modes of experience as those he described. The relativism, discontinuity, and incompleteness which was held to characterize the experience of modernity also came to depict the basic properties of nature. We can thus find 'modernity' described, precisely but incommensurably, in the different languages of physics and philosophy, as well as in the more common idioms of painting, literature, and music.[3]

INVISIBLE MECHANISMS

The crisis of rational science lies just as deeply hidden in the nineteenth century as does critical psychology. 'Anticipations' of the new relational and morphological concepts of nature have in fact been detected as a continuously present, undeveloped alternative to Newtonian mechanics, as in the writings of Leibniz[4] or Boscovich[5] or Goethe.[6] In the present context, however, the first stirrings of the movement – which subsequently developed into a distinctively modern as opposed to classical scientific world view – might conveniently be dated from Sadi Carnot's *Reflections on the Motive Power of Fire*, which appeared in 1824, but was not well known until much later in the century through the work of Rudolf Clausius,[7] and from the invention during the 1830s of non-Euclidean geometry whose full significance was likewise not immediately apparent.[8] It is worth noting, none the less, that the *scientific* ideas which would come in time to render positivism absurd were contemporary with the publication of Comte's major works.

The aim of the classical scientific world view was to complete

Newton's *Principia*, subsuming all natural phenomenon and their regulatory forces within a system of rational mechanics. Two different views of the underlying mechanism of nature were developed as part of this endeavour. One focused upon the 'inherent' properties of matter and attempted to deduce from them the characteristics of the phenomenal world. The other, regarding matter as fundamentally 'passive', sought to isolate and describe the variety of 'forces' held to be responsible for the cohesion of and interaction among bodies.[9] Either approach, through progressive abstraction and mathematization, tends towards the statement of a law from which can be derived (given specific initial conditions) expressions descriptive of the 'real world'. Such laws are typically expressed in the form of equations. From such systems of equations might be calculated, for example, the position of Mars or some other planet in relation to the earth at some particular time. Such procedures, central to classical mechanics, are always *reversible*. This formal feature has important physical implications; it means that the fundamental theories of classical science are *isotropic* with respect to time.[10] The classical laws do not explain why the entire mechanism of the universe might not be put into reverse, and the sequence of planetary motions, eclipses, and conjunctions run backwards. Classical mechanics is in fact indifferent to time; it does not distinguish before and after, but establishes the ideal regularities of a purely conceptual world. Its fundamental laws are therefore conservation laws, and its cosmos is governed by principles of identity and exchange. Every physical process can be conceptualized as an exchange of equivalents, nothing is ever gained or lost. Mechanism can be viewed as an ideal market upon which pure, alienated objects perpetually circulate.

Classical science succeeded by *ignoring* rather than explaining nature. Meyerson expresses this idea forcefully: 'we only attain laws', he points out, 'by violating nature ... by isolating more or less artificially a phenomenon from the whole'.[11] Newton's extraordinary success in deriving laws of motion from a general principle and using them to account for the observed motions of the planets was fortuitous. Few aspects of nature, in reality, approached that degree of conformity to 'timeless' repetition. Carnot's memoir, one of the first directly inspired by the technology of the Industrial Revolution, introduced an entirely new element into scientific thinking. Rather than establish an identity, Carnot's principle expresses nature's inherent propensity to change.[12] In defining the idea of the efficiency of an engine, Carnot points out that, 'wherever there exists a difference of temperature, motive power can be produced',[13] but that, in the transfer of heat from one body to another which this involves, a certain amount of energy was necessarily dissipated through frictional and other forces. As all natural processes involve such energy transfers, nature as a whole must exhibit a similar 'cooling' effect. Meyerson, again, is to the point: 'In opposition to the illusion of identity to which mechanical theories, the principles of conservation, and even the form of laws in general give rise, Carnot's principle stipulates that the whole universe is modifying itself in a constant direction.'[14]

Made analytically precise and quantitatively measurable as *entropy* by Clausius, this new theoretical viewpoint played a central role in the development of physical ideas during the latter part of the nineteenth century. Where physics had previously been concerned with ridding its conception of nature of all 'subjective' and intuitive ideas, its most advanced branches now began to develop in response to an ineradicable aspect of our immediate experience of the world. The exact relationship between the second law of thermodynamics and our own inner certainty of duration has remained a matter of dispute.[15] The ambition to express physical laws in a time-sensitive form, however, has become commonplace. The idea of *entropy* seemed to provide a new principle of cosmic reasoning. If the entire universe were considered to be an isolated system, rather like a sealed container filled with gas, then all physical events detectable within it must tend towards a state of 'equilibrium' and end, ultimately, in a condition of complete disorder.[16] We can grasp this intuitively, since we know that machines wear out, their parts rust and become useless, that all living things perish and disintegrate, that the processes of nature are rarely in fact reversible.

Carnot, and more significantly his belated followers, were concerned primarily with the laws governing the diffusion of heat in fluids, both liquids and gases. In spite of the inherent novelty of the enterprise, this did not appear to pose any real threat to the integrity of the physical sciences or the validity of their foundation in rational mechanics. Such investigations indeed began as a confident extension of the classical laws, as nothing more than an application of basic principles long since established within the Newtonian tradition. Heat was a form of energy and must therefore be associated with the motion of particles. If liquid were heated, or a gas compressed, the activity of its molecules would be increased. If a hot liquid were poured into a container partially filled with cooler liquid, the more active particles would quickly lose heat through collisions with their colder, more sluggish companions, resulting in a uniform temperature. This kinetic theory, preserving an underlying 'mechanism' for heat transmission and diffusion, proved only partially successful. Quite apart from the difficulty of reducing a mass of experimental data to a satisfyingly simple law, the large-scale mechanics of elastic collisions had never been satisfactorily understood.[17] More important still was the realistic complexity with which any theory had to deal. There could be no question here of isolating a single particle and considering its properties. The level of description had to refer to the fluid as a whole and therefore to the behaviour of millions of interacting particles. New statistical methods were developed to deal with the general problem of description, but the presumed 'underlying' elementary processes remained obscure.

The kinetic theory of gases attracted some of the most able physicists of the second half of the nineteenth century, among them William Thompson (Lord Kelvin), James Clerk Maxwell, and Ludwig Boltzmann. The mathematical problem proved to be overwhelmingly complex unless the action of each individual particle was assumed to be *independent* of every other. If given a physical interpretation, this meant that a strictly mechanical view had to be abandoned.[18] The observation of the so-called 'brownian motion', the *random* movement (presumably as a consequence of elementary molecular agitation) of fine particles of pollen suspended in a fluid, even offered unlikely empirical support for such a drastic step.[19] But this made the Second Law all the more mysterious. If elementary particles were not governed by a strict mechanism, how could the properties of fluids, on a molar scale, be so clearly defined? Why should heat *always* be transferred from warmer to colder regions? Why were there no instances of fluids at a uniform temperature spontaneously dividing into regions of differing temperature? Such problems were to recur, in a more acute form, at the beginning of the twentieth century, but even without new 'discoveries', the development of classical mechanics had, by the 1890s, run its course.

Difficulties over the kinetic theory of gases were not an isolated problem. Increasingly, the 'loose ends' of Newtonian science became interconnected in their recalcitrance to classical solutions. From the mid-nineteenth century, they centre on the 'newer' sciences dealing with electrical and magnetic phenomena. Well-known electrostatic and magnetic effects had never sat easily within the Newtonian picture of the world. Repulsive as well as attractive forces were involved, and their operation was not easily reducible to a universal law. And later, with Faraday's brilliant experimental demonstrations of induction, a mechanical account seemed all the more distant.[20] Hence it was to the 'non-corpuscular' theory of light that physicists turned in search of instructive analogies. In spite of Newton's own professed preference for a uniformly 'corpuscular' view of nature, some of his own contemporaries, notably Huygens, and in the succeeding century, after the apparently decisive arguments of Young and Fresnel, almost everyone, conceived of light as some form of vibration within a luminiferous ether.[21]

The 'wave' theory of light was part of, rather than an anomaly within, the classical picture of mechanism. Light was held to be a physically real elastic deformation of the 'subtle fluid' which constituted the ether.[22] As the medium of propagation supporting natural forces, the ether was clearly not a 'material' substance in the usual (even usually scientific) sense of the term. Pervading space, it apparently offered no resistance to the passage of planetary bodies through it. But just as clearly, by virtue of propagating light, gravitational and possibly other forces, it was not simply another name for extension or continuity. It was possessed, that is to say, of 'mechanical' as well as purely geometrical properties.[23]

Faraday, however, in visualizing electrical and magnetic effects in terms of 'lines of force', broke with the implicit mechanism of the ether theory. There was no hint here of a 'physical' phenomenon underlying and uniting the heuristic images offered as interpretations of his results.[24] And if Faraday could for a time be ignored by mathematically sophisticated physicists, James Clerk Maxwell could not. Deeply impressed by Faraday's demonstrations, Maxwell set out to extend and complete his work by establishing a general and precise framework for the analysis of all electromagnetic phenomena. In establishing a mathematically faultless 'field theory', Maxwell quite self-consciously took the physical sciences beyond the confines of classical mechanics.[25] Maxwell's physical intuition, which Einstein so much admired, was entirely liberated from mechanical prejudices. Others often found his work difficult because of this and preferred to treat his physics as no more than a mathematical convenience. But this was never his intention. He always insisted upon the necessity of *analogy* in scientific thinking,[26] and treated mechanism as just one among several of its possible forms.[27] The aim was always to connect the unknown and unfamiliar with the known and familiar, to present the unity of nature in a series of connected images. Maxwell, indeed, gifted with extraordinary powers of synthesis, felt the more pressing methodological difficulty to be the establishment of difference rather than the forging of identities. 'If all that we know is *relation*,' he confesses, 'and if all the relations of one pair of things correspond to those of another pair, it will be difficult to distinguish the one pair from the other, although not presenting a single point of resemblance.'[28]

Electricity can be connected with magnetism, and both can be considered as analogous to light. They can be represented as varying states of fields of force whose local conditions remain qualitatively unlike the large-scale physical interactions known to classical mechanics.[29] Maxwell's electromagnetic theory, the foundation of the modern physical sciences, brought to an end the progressive mechanization of the world picture which had been at the heart of the bourgeois mode of thought.

ECCENTRIC OBSERVATIONS

Clerk Maxwell, expressing a view common to many physicists during the latter part of the nineteenth century, remarks that 'all our knowledge, both of time and space, is essentially relative'.[30] Ernst Mach and Henri Poincaré indeed went so far as to apply this principle of relativity to classical mechanics.[31] The former, in a brilliant analysis of Newton's original 'demonstration' of absolute space, argued that, far from establishing an invariant framework for physical description, such experiments showed rather the unrestricted interaction of all matter in the universe.[32] We could not imagine a physically real location from which we could observe the 'actual' motions of matter in the cosmos. Every observation had to be made from some particular place, and was subject to the 'distorting' effects of its own motion. It was an illusion to suppose that there were any physical experiments we might perform which would free us from the effects of our own particular perspective. The interpretation of classical mechanics as the true 'system of the world' was therefore an unwarranted generalization. Classical mechanics, rather like Euclidean geometry, exercised an overwhelming attraction on our common sense but, in the final analysis, could be regarded as nothing more than a convention that might be expressible equally well in quite a different form.[33]

A 'relational' and 'conventional' view of scientific knowledge stimulated a wide-ranging methodological debate that most scientists were happy to ignore.[34] Such reflections seemed to offer no novel *physical* ideas. Even in the context of ingenious new experimental investigations, it prompted at best an *ad hoc* criticism of classical ideas.[35]

Scientific interest centred on the development of Maxwell's theory and in particular on the search for a definitive demonstration of the reality of an ether as the quasi-physical seat of fields of force. The most celebrated and refined of such experiments was conducted by Michelson and Morley in 1887. They developed an ingenious apparatus to compare the velocity of light along two orthogonal paths.[36] In theory, its speed measured along the direction of the earth's motion should be greater than that found for the transverse path. No difference was in fact observed, and all efforts to detect the 'drift' of the earth through the ether failed.

These developments form the immediate context of Einstein's Special Theory of Relativity, which appeared in *Annalen der Physik* in 1905. They do not furnish us, however, with Einstein's real point of departure, which was – to him – the evident contradiction between Newtonian science and Maxwell's electrodynamics.[37] He begins his historic paper thus: 'It is known that Maxwell's electrodynamics – as usually understood at the present time – when applied to moving bodies, leads to asymmetries which do not appear to be inherent in the phenomena,'[38] and only later mentions the failure to detect the movement of the earth through a 'luminiferous ether'. The Principle of Relativity which he introduces to resolve these difficulties is, unlike those of Poincaré or Mach, a physical postulate. The abandonment of absolute space, and a privileged point of observation within it from which to construct a 'system of the world', necessitated a much more thorough revision in our basic physical concepts than had been realized.

Ultimately, it required a scientific theory of much greater complexity than any envisaged within the Newtonian tradition. Generality rather than simplicity must become the standard of formal elegance. Einstein makes this clear in his 1916 paper on the General Theory of Relativity: 'The general laws of nature are to be expressed', he insists, 'by equations which hold good for all systems of co-ordinates, that is, are co-variant with respect to any substitution whatever.'[39] The realization, that is to say, of the necessarily 'subjective' and 'relational' viewpoint from which we observe the world does not mean that science is restricted to empty formalism and the artificiality of a purely conventional viewpoint. Properly understood, it provides us rather with the starting point for a new and more secure scientific realism. Its method inverts the, largely implicit, assumption of classical science that 'reality' is more complex than the theory through which we understand it. Now 'reality' is absolutely simple but remains unknowable in its simplicity.

Einstein's science begins, then, with a postulate of constancy rather than a relativistic hypothesis. Physical observations might be made from any point within an indefinitely large range of possible systems of co-ordinates. There is no absolute space, and there are no stationary points, and any observational platform must be considered in motion relative to any arbitrarily chosen point in the universe. Even so, assuming *uniform* relative motion of two such 'inertial frames', we can say that the laws of mechanics (however expressed) must hold good for that observer and, what is really implied in such a statement, the velocity of light 'is always propagated in empty space with a definite velocity c which is independent of the state of motion of the emitting body'.[40] The uniformity of nature, as we have seen, is the fundamental postulate of the scientific revolution, and Einstein's Special Theory can be understood as a thorough analysis of the meaning of this conviction. Just as Kierkegaard or Dostoevsky exposed the depths of bourgeois self-deception by embracing rather than rejecting the fundamental reality of the ego, Einstein does not base his science on a radically new principle but on a careful consideration of all the implications of the existing principle to which he remains fully committed.[41]

Given these two fundamental postulates, the phenomenal world must be 'deformed' in relation to them. If we know that the velocity of light must remain constant for any observer, then observers moving relative to one another must disagree about such apparently 'universal' things as the measurement of length or time. And as any measurement must depend upon the sending and receiving of light signals, this becomes a general theoretical, rather than a technical, problem. Suppose an observer A, situated at the centre of an unrealistically rapid train (travelling at an appreciable fraction of the speed of light), switches on a lamp just as he passes a companion B on a platform stationary with respect to the moving train. A observes (that is, his instruments record for him) that light impulses from the lamp reach either end of the carriage simultaneously. B, however, records that light impulses reach the rear of the carriage before reaching the front. This is a simple consequence of their relative motion. In the brief moment it takes for the light to reach the rear of the carriage, the carriage itself has moved forward, so that, from B's point of view, the light still has some distance to cover to reach the new position of the front of the carriage.[42] Simultaneity, that is, does not reside in nature, and in comparing events in different reference frames, is a meaningless concept. There is no absolute time any more than there is an absolute space.[43]

The constancy of the velocity of light forces even more bizarre conclusions. Imagine that A and B conduct a more refined experiment and each measures the velocity of the light impulses emitted from A's lamp. Before the experiment takes place, A and B have satisfied themselves as to the dimensions of the carriage while at rest relative to each other and have synchronized their clocks. Apparatus is set up on the platform to replicate exactly the position of mirrors and clocks within the stationary carriage. Again, as the train speeds by B, A switches on his lamp. For A there is no difficulty, it does not matter in which direction he takes measurements as he is part of the inertial system containing all the relevant apparatus. For B it appears, at first sight, that if he measures the velocity in the direction in which the train is moving, he must arrive at a higher value than if he were to measure it in the direction against the motion of the train. But in fact he cannot 'compound velocities' in this way. The velocity of light is not augmented or diminished by any relative motion in its source. Intuitively we can understand that this must be the case. If we could, by moving very rapidly, 'overtake' light signals as we can sound waves or a ball thrown ahead of us, we could see events which had taken place *before* we began to move. We could move observationally into the past. The incoherence of such a possibility is far more disturbing to our sense of reason than are the non-intuitive consequences of the constancy of the velocity of light.

Suppose B measures on *his* watch the time it takes for the light to reach the forward end of the carriage. Because of the forward motion of the carriage, the light travels farther, relative to his 'stationary' position, than it does for A. However, since velocity is given by distance traversed divided by time elapsed and the velocity *must* be the same for both observers, A's clock must run 'slow' in relation to B's. Similarly, if A and B agree on the elapsed time, then they must disagree about the length of the carriage. It is important to note that these relativities are not 'subjective' phenomena but physical consequences of relative motion. Clocks remain synchronized and measuring rods remain of equal length only so long as they are confined to the same inertial frame of reference.

Measurements of space and time could no longer be considered as independent and absolute magnitudes. Minkowski, introducing his own formalization and extension of Einstein's theory in 1908, remarks that 'Henceforth space by itself, and time by itself, are doomed to fade away into mere shadows, and only a kind of union of the two will preserve an independent reality'.[44] This marks a decisive shift in the Galilean viewpoint. The physical universe cannot be understood from the point of view of an isolated, observing ego. All physical observations become comparisons among separate observers, and physical laws become the rules by which their differing 'results' can be rationalized. In Einstein's physics we are not asked to 'imagine ourselves' in a particular location or situation; we are asked, rather, to 'imagine someone else' in a particular set of circumstances. It is the other's observations, when compared to our own, or to a third party's, that provide the theorist with his basic material.[45] His is the viewpoint of the decomposed subject. The observing ego must be distributed into a variety of locations and times if it is to overcome its tendency to project into nature the illusion of its own rational order. And reason is powerless to recombine these disparate images into a unified and coherent picture of the world. The generality Einstein sought in his theoretical work is an expression of powerful physical intuitions which, in satisfying formal criteria of consistency, proved difficult to visualize. In extraordinarily lucid, non-technical presentations, however, the same difficulties arise. If by 'rational' we mean the completion of a single Newtonian deductive 'model' of the cosmos, then the 'absolute simplicity' of nature cannot be rendered into the form of a rational intellect.[46] Nature *must* appear strange.

These difficulties become much more acute when the inter-relation between space-time as an undifferentiated continuum, and matter, are considered. Minkowski, realizing that the classical language of point-mass, velocity, force, and so on had finally been overthrown, suggested a new geometric approach. Any 'substantial point' in the universe can be assigned a space-time co-ordinate and its 'everlasting career' represented by a *world-line*.[47] Then, 'physical laws might find their most perfect expression as reciprocal relations between these world lines'.[48] It was only at this point that the enormous *physical* implications of developments in non-Euclidean geometry became clear. The new geometric description of space-time would be an unimaginable four-dimensional and non-Euclidean *manifold*.[49] Only then would the 'simple' physical phenomena of accelerated motion, gravitational 'force', and field effects be fully comprehensible.

As early as 1870, William Clifford had suggested that many of the phenomena we associate with 'ponderable' matter might be better understood as properties of space. Might not 'physical variations', in fact, be 'effects which are really due to changes in the curvature of our space'?[50] It was not until 1916, however, with the publication of Einstein's General Theory, that such spec-ulations were made rigorous. The extension of the Special Theory led not only to a deeper understanding of motion, but also to a new conception of matter. The technical problem was deceptively simple. The Special Theory had found ways to express the invariance of the laws of nature across reference frames in uniform motion relative to one another. But what about reference frames *accelerating* with respect to one another?

In a non-technical presentation, Einstein introduces the problem by asking us to imagine someone inside a closed lift from which he can make no observations of outside space. The lift, transported to a distant region of space, is in 'free fall'. For the observer enclosed within this 'inertial system', the laws of mechanics hold good and are directly observable in their classic form. Everything, including himself, is 'weightless', and objects remain where they are in relation to the boundaries of the lift, unless the observer interferes with them in some way, for example by pushing them, in which case they will persist in their motion until colliding with another object or the walls of the container. The 'mass' of any object can be determined by its resistance to motion. A 'large' object requires more of a push to achieve a specific velocity than a smaller object.[51]

Now suppose the lift is acted upon by an external 'invisible' force. A rope is attached to its 'top' and a distant energy source used to accelerate it through space. For the interior observer things are now quite different. He feels a downward pressure forcing all 'heavy' objects to the base of the lift. He becomes conscious of his own 'weight'. And now he can determine the relative masses of the objects around them by a weighing device, such as a spring balance, which measures the 'force' acting upon it. The larger the object, the greater the force. He might well explain such fundamental physical features of his world, Einstein suggests, by supposing the base of his lift to be endowed with some quality such as gravity which, 'acting at a distance', accelerated objects towards it.

The internal observer assumes that the lift is at 'rest' in relation to some absolute frame of reference. An external observer, however, does not make this mistake and attributes the 'gravitational' effects within the lift to its acceleration relative to the space surrounding it. Again, this is not a purely formal relativism. It persists only because of the apparently fortuitous equivalence of inertial and gravitational mass.[52] This equivalence, however, allows Einstein to treat 'gravitational' forces as accelerating reference-frames, and suggests, furthermore, that light and other forms of electromagnetic radiation must be 'bent' by gravitational fields. From the perspective of an accelerating reference-frame this is easy to grasp. Suppose a small hole is bored in the lift and light from an 'outside' source introduced into the lift. The illuminated point on the wall opposite the hole will not be directly in line with the outside source but displaced due to the movement of the lift during the interval of its travel across the interior. For the interior observer, this effect will also be attributed to the 'gravity' he has discovered governs all other displacements.[53]

Since there is no kinematic difference between a gravitational field and an accelerating reference-frame, Einstein, in generalizing the classical laws to expressions invariant under *any* transformation (including acceleration), arrives at a new conception of gravitational 'force'. Instead of invoking a 'force', either acting at a point or distributed throughout a field, to account for acceleration, Einstein suggests that change in velocity can be fully characterized by the geometrical properties of space. Accelerated motion does not betray, any more than does uniform motion, the action of an external 'force'. In Einstein's physics, all motion is 'effortless', all objects drift through space, irresistibly 'falling' along their world-lines. Acceleration is simply a bending of space, a contortion in its 'normally' flat metric. All motion has finally become 'natural'.[54]

The naturalness of motion, however, is no longer predicated upon the properties of matter. Now, space and matter are geometrically linked. The character of space is, first of all, 'deformed' by the matter it contains. 'Our world is not Euclidean', Einstein warns. 'The geometrical nature of our world is shaped by masses and their velocities.'[55] Large concentrations of matter curve space, and world-lines running through such regions converge, thereby accelerating objects. Equally, however, such concentrations of matter can be viewed simply as 'dense' regions of space: 'Matter is where the concentration of energy is great, field where the concentration of energy is small.'[56] The categorical distinction between matter and space is dissolved. 'We could regard matter', claims Einstein directly, 'as the region in space where the field is extremely strong'[57] – a statement which might be taken as the logical implication of generalizing Maxwell's field theory. The cosmos, it appears, is composed of nothing but space, or rather space-time, variously condensed and warped: 'A thrown stone is, from this point of view, a changing field.'[58] The relativizing of absolute space, that is to say, implied the overthrow also of classical conceptions of matter as substance separate from the 'empty' extension, or ethereal medium, containing it.

HIDDEN PERVERSITIES

Relativity Theory interpreted the sensible world as a specific instance of more general laws, the variations in which were observable, for us, only in exceptional circumstances. The immediate experience of the world exposed us directly to only a small segment of its relations. Inconceivably high relative velocities estranged us from this commonsense reality. Reason led us, when these 'extreme' situations were considered, to bizarre results. A rational image of nature no longer seemed congruent with the 'system of the world' as a mechanism; the relational system of universal physical causality which, in the bourgeois world, had become 'intuitive' broke down under the stress of internal contradictions.[59] These inconsistencies arose not only in Relativity Theory, but appeared, with even greater force, when mechanical ideas were applied to insensibly minute atomic interactions.

In 1900 Max Planck succeeded in showing that the electromagnetic energy distribution within a closed system not only tended towards an equilibrium condition in accordance with Boltzmann's formulation of the gas laws, but did so in a series of discrete 'jumps'.[60] Energy was absorbed and emitted by elementary particles of matter in a discontinuous fashion; in *quanta* whose value was a universal constant, independent of any variation in the empirical circumstances of the system as a whole.[61] Boltzmann had resorted to a statistical approach to thermodynamics as a mathematical convenience, and supposed that, in reality, an 'underlying' mechanism provided a continuous network of physical causes wholly determining the motion and energy level of each particle. Planck argued, however, that Boltzmann's theoretical results, which were extendable to all forms of radiation, could be supported only by inferring a real physical discontinuity in the process of absorption and emission.

Einstein similarly argued that light must be considered a discontinuous form of radiation. In a paper published within a few weeks of his 'Electrodynamics of Moving Bodies', he proposed a 'quantum' interpretation of the photo-electric effect.[62] When a piece of metal is illuminated, it periodically expels electrons in rapid motion. The velocity of the emitted electrons depends on the wavelength and not the intensity of the incident rays; it is independent, that is to say, of the distance of the light source. If light is considered as the wave-like deformation of an elastic ether, then its energy level, its capacity to impart motion, would progressively decline with distance. If, however, light is considered as a stream of 'photons' or light quanta, the constancy of the photo-electric effect becomes comprehensible.[63] Einstein was also forced to the conclusion that the metal absorbed and emitted radiation in discrete amounts or 'packets', He was able to show that experimental results had been reached which were consistent with such a view and quite inconsistent with the classical wave theory.

In spite of the serious logical difficulties into which it seemed to lead, the quantum theory made rapid strides. Discontinuity and duality appeared everywhere and refused to be unified or integrated at some 'higher' theoretical level. The more successful it became in predicting experimental effects, the more mysterious did the theory appear.[64] Systematic problems were pushed into the background for some time with the success, in 1913, of Bohr's quantum view of atomic structure.[65] A growing body of experimental work, such as that on the photo-electric effect and on radiation particularly connected with Rutherford, necessitated a revision of the classical conception of an atom as an indivisible, impenetrable, and simple body. The atom was itself a structure of simpler particles, but if this were the case, what accounted for the stability of this structure? Bohr developed the planetary view of the atom with a positively charged nucleus and a number of negatively charged electron satellites. He supposed that atoms of a particular element possess a finite number of discrete energy states, 'stationary states', in which neither emission nor absorption of energy can occur. Atoms may 'jump' from one stationary state to another with accompanying emission or absorption. He felt, at this point, that classical mechanics were adequate to describe the 'stationary states' but broke down when applied to the 'transitions', which remained fundamentally incomprehensible.[66]

Bohr's theory had considerable success in accounting for the spectral patterns associated with different elements, and even more in providing a rational foundation for Mendelev's periodic table and, consequently, for chemical theory.[67] The discontinuous nature of energy states, the acceptance of which Bohr's theory finally made respectable, posed acute and fundamental problems that could not be ignored for long. Imagine a piece of radioactive material. It is unstable, its large complex atoms tend to break down into simpler and more stable structures, in the process emitting a variety of energized particles. It is a strange instability. One atom, mechanically identical to every other in the sample, suddenly transforms itself. It has existed in its original state for perhaps several thousand years. Another atom may persist in its original condition for several thousand more years before it too follows its companion. In terms of classical science, there seemed no possible explanation of either the 'how' or the 'when' of this process. Yet, as a process within the sensible world, it followed its own laws. The precise number of such 'jumps' in a defined period of time, the nature and intensity of the resultant radiation could all be predicted with admirable accuracy.

The same difficulty had arisen in the statistical approach to thermodynamics. There the mathematical formalism had been justified by an appeal to an as yet unknown 'underlying' mechanism. Even so, it had been embarrassing. Lorentz admitted, in an opening remark at the first Solvay Congress at Brussels in 1911, that 'We do not understand why a lump of iron does not glow at room temperature'.[68] Just when physicists had reached down, as it were, to a deeper level of nature, nature itself seemed to dissolve into incomprehensible processes. It was not simply a matter of a more ingenious application of existing ideas. Using conventional concepts Bohr claimed that, 'a description of atomic processes in terms of space and time cannot be carried through in a manner free from contradiction'.[69] From the point of view of the 'ordinary description of physical phenomena', indeed, 'the quantum theory represents an essentially irrational element'.[70]

The generalization of quantum theory during the 1920s by de Broglie, Heisenberg, Born, and Schroedinger intensified rather than resolved this 'irrationality'. The wave-particle duality was taken by de Broglie to be characteristic of any physical process whatever. Everything we called a 'particle' had associated with it wave-like characteristics, and every wave phenomenon was associated with particular particles.[71] If the General Theory of Relativity had united matter and space in a new intimacy, quantum theory bound them together in the same paradox. The fundamental characteristics of matter and energy, taken separately or in interaction, were inexpressible through a single, logically uniform, and physically meaningful set of concepts. Progressive formalization, particularly in Schroedinger's elegant 'wave mechanics', did nothing to resolve the difficulty.[72]

Heisenberg, in particular, offers a ruthlessly uncompromising formulation of the conceptual problem. 'Quantum theory', he says bluntly, 'does not allow a completely objective description of nature.'[73] Classical physics had been founded on the possibility of 'detached' observation of nature. The fundamental 'variables' of any mechanical system could be accurately and *independently* measured. The limitations on our knowledge of physical events were purely practical. In the subatomic world of quantum physics this was no longer the case. Our knowledge of such systems is, in principle, limited. Heisenberg introduced his principle of indeterminacy as a problem of measurement. Any experiment we might perform to determine the position of, for example, an electron would, by the Compton effect, alter its momentum in an unpredictable way. And any procedure which might accurately measure its energy would be temporally imprecise.[74] We could not know the future state of a system because we could not completely know its present state. In Heisenberg's view this was something more than a limitation on our knowledge of the world. The Indeterminacy Relations, as he preferred to describe these limitations, were all that we *could* know, and therefore to ask if an electron could none the less be conceived as a definite but unknowable magnitude associated with a specific location made no sense.[75] Even if the classical laws held, we could never at the subatomic level determine physical variables with sufficient precision to apply them.

Bohr's view of Heisenberg's derivation of Indeterminacy Relations underwent considerable changes over a period of years.[76] From an early view of quantum physics as a 'limiting condition' of classical mechanics, he developed a methodological postulate of Correspondence and finally of Complementarity. By the late 1920s, the unreasonableness that had been originally confined to highly specialized studies of radiation – namely, that 'there were two incompatible models . . . each well confirmed; yet evidence for the one constituted counter-evidence for the other' – was widely felt in all fields of physical enquiry.[77] Bohr, influenced perhaps by Höffding to a Kierkegaardian admiration of paradox, argued that the fundamental dualisms of quantum physics did not stem simply from different conventions of description.[78] They did not arise as the arbitrary but incompatible points of view from which nature as a whole might be analysed, nor did they originate in differences in the type of phenomena selected by each as central to physical structure. Each view, one associated with measurement of momentum and energy (particles), the other with determination of frequency and wavelength (waves), was essential to the description of *all* physical phenomena. The logical inconsistencies in physics were no more than a realization of the fundamental discontinuities of nature. Any physically realistic theory must contain logical incompatibilities. Rather than aim at the construction of a uniform and complete 'system of the world', or be satisfied with either a vague or a purely formal 'algorithmic' approach, the physicist should aspire to the construction of internally consistent, precise, but contradictory *images* of nature.[79]

The 'Copenhagen' interpretation of quantum physics was generally accepted within the community of physics, but in a 'harmless' form. The mathematical formalism, itself quite free of vagueness or contradiction, was held to be perfectly adequate, and the 'philosophical' problems raised by Bohr were regarded as inessential complications, arising from an unfortunate insistence on attempting to picture the processes described by the theory. Bohr's views, however, were coherent and systematic.[80] And the issues were central to the *scientific* understanding of the natural world. The indeterminacy relations, the deep unpredictability of matter, the failure of strict causality, the irreducible 'weirdness' of the quantum world could only be grasped if we supposed that, prior to its termination in an observable 'event', the fundamental constituents of nature existed in their own peculiar play world; both 'here' and 'there', both 'now' and 'then', only relinquishing the freedom of inherent possibilities when forced to do so by the experimenter's conscious probing. Even the more positivistically inspired Heisenberg expresses himself in Kierkegaardian language on this fundamental issue. The mathematical formalism of probability theory introduces 'something standing in the middle between the idea of an event and the actual event'. It is a 'strange kind of physical reality just in the middle between possibility and reality'.[81]

Chapter Twelve

THE CHARM OF DECADENCE

Within the classical bourgeois world, cosmos and psyche tend to draw apart. The success of the physical sciences justifies attempts to explain the psyche in a mechanistic fashion. But the failure of such attempts encourages a superficial distinction between an 'irrational' domain of subjectivity on the one hand, and a 'rational', objective order of nature on the other. In time, however, the social logic of universality, individuation, and exchange relations which is implicit in both becomes expressed, in apparently incompatible forms, in both domains.

With a conspicuous display of equifinality, both travel by different routes through disillusionment, to paradox, to solipsism, and acquire the exotic trappings of 'modernité' at about the same time.[1] Neither realm can develop to that point of systematic self-closure which would 'complete' them. Mechanical accounts of action decline and finally give way to irrational 'leaps', to outbursts of refractory and disenchanted energy. And each domain becomes more fully 'decomposed' when the elementary processes, from which in some sense they must be constructed, are found to be outrageously free of constraint. The 'sciences' of both subject and object become infected with playfulness; an ineradicable arbitrariness is rediscovered in their foundations.

The intellectual revolution of modernism does not imply the demise of the classical bourgeois world view, far less the commodity form in which the latter is most completely embodied. Modernity is a cultural spasm, a radical but ineffectual transformation of classical rationality. The completeness and profundity of its break with the science of pleasure has to be balanced against its exoticism and lack of seriousness.[2] In recapturing a sense of the primordial fun of experience, it isolates it in fact at an extremity we can visit only in our imagination. Einstein and Bohr, like Kierkegaard or Nietzsche, are describing reality, but in revealing its truth they place it safely beyond our grasp. We must live in a mundane and pleasant world; non-relativized, non-quantized, free of paradox. Just as we exclude from this 'system of the world' any direct apprehension of our own transformative powers (fun), or any symbolic expression of estrangement (happiness), we erect, ahead of it so to speak, an unattainable world of another kind, one filled with fascinating arbitrariness (excitement).

It is easy to believe, then, that quantum physics, and all the other decorative absurdity of modern culture has little to do with our immediate experience of the world. Truth has eluded us and leads its own incomprehensible life beyond our grasp.[3] What interests us, however, is more immediately comprehensible and still, thankfully, obeys the laws of causality and non-contradiction.

Yet if we have found a clue to the separate but linked cosmo-logical schemes of fun, happiness, and pleasure in the related but different social relations of the bourgeois world (freedom from labour/dependence, mutual obligation/hierarchy, alienated labour/individuation),[4] can we not discover another social form from which *excitement* is given reckless birth, a social relation peculiar to its captivating perversity?[5]

In earlier chapters the notion of 'reason' was distinguished through a series of putative oppositions: the reasonable, civilized, and self-controlled adult over against the unreasonable, uncivil-ized, and emotional child or savage; the rational western mind as distinct from the clouded and irrational superstition of the primitive. These distinctions were associated with the development of capitalism, first of all with an ability and willingness to work, and second with a facility for exchange; with the general social distinction between production and circulation.[6] In the modern period, however, such distinctions lose their exclusive and oppositional character. Reason and unreason come together in an intolerable *purification* of experience.[7] This process is associated particularly with the sphere of consumption. It is consumption which creates the realm of the modern irrational, and therefore the realm also of the modern rational.

CONSUMPTION

In order to elucidate the connection between modern forms of the rational and irrational and between both and the process of consumption, it is helpful to look again at Freud's psychology, but this time in the context of twentieth-century science rather than nineteenth-century biology or anthropology.

Freud's analytic viewpoint is admirably relativistic. The framework of absolute internal space and time (the utilitarian matrix of pleasure and identity) is completely dissolved. The individual no longer exists as a point of *integration* of his own experience, or *telos* of his own uninterrupted recollection. The normal individual is bewildered by himself, as much as he is by the world. He is no more a privileged observer of himself than he is of the cosmos. The only constant of the psychic world is an arbitrarily chosen value, a self-identity, a groundless 'I', which cannot any longer impose itself upon the flux of experience with the vigour of the bourgeois ego. Yet even in the modern world the constancy of the 'I' 'deforms' the space and time of classical subjectivity. Both perceptually and in memory, a fitful selfhood 'distorts' the previously certain framework of utilitarian expectations.[8] This, as in Einstein's physical theory, is not observable in 'normal' life, but can be detected only in 'absent' moments when the 'I' is absorbed into the ever-changing metric of immediate experience. The parapraxes and jokes, the dreams and neurotic symptoms that we still tend to think of as the uncontrolled distortions due to an inattentive ego, are better viewed as the psychic flux within which the 'I' is a temporary 'deformation'.

Freud's 'relativism', in other words, is nothing to do with moral indecisiveness. It is more akin, as is Einstein's theory, to a Nietzschean 'transvaluation'. What had previously been conceptualized as a 'framework' or as the dimensions *within which* objects could be defined and events observed, ceased to have any absolute meaning. The psychic 'material' of the self could no more stand apart from its own experience and memory than could a physical object exist as anything other than a particular wrinkle in space-time. The *separateness* and substantiality of the bourgeois ego, its differentiation from an infinitized and homogeneous inner space of 'pure' subjectivity, can no longer be maintained. The persistent illusion of 'foreground' and 'background', of *things* standing out from non-things, is exposed.

There is more here than a vague analogy. There is a formal, though not a rigorous, relationship between the 'objective science' of nature and the 'subjective interpretation' of the psyche. Freudian psychology and Einsteinian physics are alike in both rejecting the classical market model of reality. The 'frame of the world' no longer appears as the self-evident, unchanging, and homogeneous *dimensionality* of classical mechanism,[9] but is made an aspect of its substance. The inner world, its primordial immediacy sunk beneath the threshold of experience is, like the cosmos, fragmented and relativized. The 'self' no longer constitutes a 'logical' totality; it appears, rather, in a bewildering variety of incompatible and contradictory experiences.

There are, furthermore, conspicuous quantum elements in Freud's analytic writings.[10] The psyche is conceptualized as an 'energized system' capable of assuming a fixed number of positions. His early 'Project for a Scientific Psychology', in spite of its evident indebtedness to the mechanistic traditions, bears a resemblance to Planck's struggle with the gas laws. Something approaching thermodynamic equilibrium is assumed as the norm for the 'psychic apparatus'. A certain level of external 'stimulation' (energy) is 'absorbed' as perception. Beyond a particular level, however, any additional stimulation puts the apparatus into an 'excited' state, in which perceptual processes are 'converted' (emitted) as motor activities, allowing the system to return to a condition of equilibrium.[11] Freud, more readily than Planck, gave up the assumptions of mechanism. As an interpretive guide, however, rather than a causal theory, the notions of 'excitation' and 'conversion' continued to play an important part in Freud's work.[12]

Even more significantly, Freud was well aware of the impossibility of the 'objective' observation of psychic life. Here again, his ideas changed very considerably over a number of years. In his early analytic work, he held firmly to a faith in Enlightenment. To elucidate the inner meaning of neurotic symptoms *was* to cure them.[13] Later, however, he came to view the 'transference relation' as the key to the analytic situation.[14] The analyst could not remain apart from the world he was trying to investigate. Nor could his 'interference' be allowed for by some mechanical calculation. The significance of the observer was not that he 'disturbed' the system he wished to investigate, but, more importantly, the interrogation itself brought to light the reality he half-suspected lay hidden in the patient's psyche. The analyst collaborated with the patient to realize fully those elemental impulses which, until they were provoked into a visible form, existed in an indeterminate but still potent world of possibility. Freud, one feels, would have been quite at home with the 'Copenhagen interpretation'. Observation cannot be 'objective', but is not for this reason rendered meaningless. There is, in other words, no fundamental difference between an act of observation and any other act; just as in physics there is no fundamental difference between an observation event and any other event.[15] The protean world of possibility terminates, in an unpredictable fashion, in 'events', and the scientist has no option but to participate in such events.

In attempting to describe the psyche, Freud is driven to the same extremities as the quantum physicist. Quite apart from the apparently disconnected (or connected but incompatible) elements within the psyche, each well-formed part on closer examination was found prolific of contradictory images. The 'simplest' dream proved to be an inexhaustible source of competing and paradoxical meanings. No neurosis would relent in the face of an 'explanation' of its symptoms. A casual joke both conceals and reveals violent and ambivalent feelings. Psychic images cannot be given a simple 'rational' translation because reason is itself just one of these images; a differentiated segment of the primordial world to which it remains fatally attracted.

Freud proposes a relativized 'microphysics' of the psyche, and, correspondingly, Bohr adopts a psychoanalytic approach to nature.[16] Both, that is to say, having rejected the determinateness and endless causality of mechanism, are prepared to accept incomplete, arbitrary, and contradictory *images* of reality. Both find in reason itself, or rather in reason 'diffracted' in another, tiny clues to a vast and perplexing world that lies hidden within the conventions of our normal experience. The unconscious is the quantum world of the psyche. It 'knows no contradiction'.[17] Each of its elementary images can refer simultaneously to different and incommensurable systems of meaning.[18] Many of its elements are interchangeable.[19] It exists as a continuous flux from which, inexplicably, the settled world of everyday perceptions and experience is distinguished – yet never *quite* distinguished.

In the bourgeois world, the psychic 'constant of action' had been as effectively masked as had been its natural counterpart. But once admitted as a possibility, the irrationality of the 'self', its fragmentation, and incompleteness became perfectly obvious. The philosophical problem of 'quantum psychology' is thus rather different from that raised by physical questions. The discontinuities within the psyche are directly experienced and the everyday world cannot escape them. Freud's psychology may be logically unsatisfying but it certainly 'makes sense'. We are willing to admit irrationality of ourselves – indeed, only the mad refuse to do so. But we still cling to a rational image of nature as the inherent truth of things, and consequently find nature's elemental processes profoundly strange.[20]

Rational individuation is therefore only one 'model' of human activity. The obligation to impose upon primordial experience an ever more clearly defined self-identity is subverted by a tendency to give way to its elemental incongruity. In Freud's psychology, the pursuit of pleasure oscillates with the hope of *excitement*. The 'ego' can no longer be understood as the necessary outcome of a continuous 'development'. Kierkegaard had already exposed the emptiness of such a view, but Freud went much further in his analysis of 'illusory' selfhood. The ego, like matter, is differentiated from the 'primary process' by the observation of a *rule* rather than the application of a mechanical 'force'.

What gives matter its impenetrability, its hardness, when its elementary constituents are tiny in relation to the 'space' which they appear to occupy? The answer offered by Pauli takes the form of a grammatical rule rather than a physical 'cause'. The Exclusion Principle prohibits any two electrons sharing the same 'quantum number'. This results in a 'quantized' space around each nucleus, providing for the more complex elements a characteristic set of physical and chemical 'properties'.[21] The 'solidity' of the ego is similarly constructed. A series of prohibitions organizes the continuous flux of the primary process into an interior 'space' upon which can be 'mapped' the features of the 'outside world', and an objective 'self' which can move and act within it. An Exclusion Principle defines for each 'self' a set of unique features which preserves the personal identity of each inner world. The psyche's exclusion principle, however, is a mere convention. It is far less difficult to envisage, and in doing so actually to invoke, the primordial inner freedom of unlimited psychic interchange, than to imagine the undifferentiated particle 'soup' that was nature before time and space had any meaning.[22]

The fact that cosmologists now regularly describe the 'polymorph perversity' of the very early universe, and seek to understand the present structure of the cosmos in relation to this singular origin, should not be understood simply as the result of 'objective' scientific progress.[23] The process of 'self-understanding' is structurally similar, and conforms to a common 'grammar'. This is not a matter, as it is for the cosmology of happiness, of a direct comparison of macrocosm and microcosm. The rejection of mechanism and the formation of new images of space and time derive in both cases from changes in the social relations of capitalism – changes, that is to say, in the direct experience of the world.

The classical bourgeois world was experienced through a network of alienated relations. It was just this which lent it its logical coherence and overwhelming objectivity. The psyche's relation to an 'outside world' was first of all of need and then of desire. In expressing itself as desire, it differentiated from itself 'objects' the appropriation of which would 'develop' its potentiality for absolute individuation and completeness. The world of objects was primarily a world of commodities. Objects bound to one another through the (logical and practical) process of exchange. Their circulation was governed by a classical rule of mechanism, a principle of conservation which insisted that in every exchange no value was lost or gained. In such a world, production was concealed as a 'natural' process and consumption 'subjectivized' and personalized as desire. Circulating commodities ideally formed a distinct realm within which the mechanistic illusion was born. Every exchange was reversible, and linked with every other by an uninterrupted chain of causes. The 'laws' of exchange were binding and irreproachable. Consumption was the process through which, in satisfying his various needs, the individual could 'attach' himself to this ideal system. Consumption, beyond the level of mere survival, was the process in which inner uniqueness realized itself in the world. Desire was the determination of the self, and endowed particular commodities with the special attractiveness of anticipatory satisfaction. To consume was to possess *particular* commodities, those which had received and reflected the inner movement of the self as desire.[24] Consumption was therefore a process of individuation; it was above all others the sphere of self-realization.

The peculiarly distinctive character of the *modern* sciences, in addition to their special subject matter, describes a new relationship of self and world, and a new form of consumption. The *distance* between subject and object[25] is suddenly overcome. The self no longer exists as the desire stretched between itself and the distant image of itself which it had attached to an object. The self could exist only as a continuous flux.[26] The object world, or rather the commodity world, draws close. The ego ceases to be defined by a *force* (desire), but falls effortlessly along its path of least resistance, the geodesic of continuous consumption. To consume does not distinguish and individuate, but joins and merges.[27] We no longer consume 'rationally' in the pursuit of pleasure (the form of self-realization), but 'irrationally' in the hope of excitement.

Excitement, unlike pleasure, expresses itself openly and outwardly. The hidden and inward 'maieutic art' opens into a direct appeal to others.[28] Excitement, which might occur anywhere, is never contained. It communicates itself directly and does not require pre-existing social relationships to 'channel' its meaning. In an excited state strangers become friends through mutual consumption. The anonymity of the modern metropolis is therefore, as Freud as well as Simmel noticed, the privileged setting for modern, exciting experience.[29] The metropolis is exciting just because it is anonymous and unsettled; capable therefore of receiving ever new impressions and reflecting them. The metropolis is the ideal social 'black-body', in perfect equilibrium, absorbing and radiating each new wave of fashion, each unpredictable surge of opinion.

'Object-cathexes' are no longer sought as 'tests' of the ego against the world. There is no strenuous effort here to bridge the gap created by desire. In principle *any* commodity *might* prove exciting, and in an excited state any commodity might be consumed. There is no guarantee that something found satisfying once will prove satisfying a second or third time. The distribution of excitement in relation to the commodity world is *randomized*, so that the attainment of subjective satisfactions can never become the object of precise calculation. The ego has a new *accidental* relation to the object world.

The ideal commodity therefore ceases to have any personally distinguishing features. Properly speaking, commodities should become identical one with another, or reduced at least to a few fundamental 'types'. As the possibility of excitement is no more than an arbitrary relationship, we cannot choose what to consume on the basis of the commodity's own distinguishing features. If they are all the same, the chance of success is manifestly equalized. No promise is made. The commodity will not necessarily bring pleasure, but in consuming it there is a thrilling possibility.[30] Consumption in being freed from the cycle of need, want, and satisfaction, is made permanent. Paradoxically, it is as its satisfactions have become episodic that consuming has become an ideally continuous process. [31]

The only really important distinguishing characteristic of the commodity, for the consumer, is its *novelty*. It is newness that is craved as the most probable stimulant of excitement.[32] This fundamental condition allows the homogeneous and isotropic commodity world to be inherently expansive. As distinct from all mechanisms, an 'arrow of time' marks its most superficial appearance.

Freud's patients suffered from the diseases of consumption.[33] The neurotic is, literally, excessively excitable. He does not consume because every potential 'object-choice' has aged before it can be possessed. The 'cathexis' has become so superbly mobile that it keeps too far ahead of the ego and is distributed too 'thinly' over the object world, which consequently takes on a uniformly drab and uninspiring appearance. The psychotic, conversely, is not excitable enough. The 'cathexis' never leaves the 'muffled interior', and he remains indifferent to any possibility. The psychotic consumes himself; the neurotic lacks the self with which he might consume the world. As types, they serve to define a model of regulated insatiability; the ideal modern consumer or, better, the ideal consumer of modernity. In their open acceptance of the ephemeral and insubstantial, they celebrate the 'arbitrary, fleeting and transitory' as the accidental relationship of selfhood. The *Man Without Qualities*, as Musil does not tire of repeating, is the same as 'qualities without the man'.

The notion of *excitement*, then, like the other 'subjective' terms, *fun, happiness,* and *pleasure,* describes both a form of immediacy and a picture of the cosmos. These terms stand, so to speak, on the boundary between nature and the psyche, and from them the appropriate descriptive languages for each realm have been developed. This is possible only because these terms initially describe fundamental social relations. Excitement is not confined, therefore, to the process of consumption but can be rediscovered in any social sphere.

CIRCULATION

Since the eighteenth century, a close association between the social process of circulation and the attributes of reason has been maintained. It is the sphere of circulation which appears most readily to conform on the one hand to the metaphysical requirements of pure thought,[34] and on the other to the principle of mechanism. This is particularly the case when the general process of circulation is represented by *money* rather than by any particular class of commodity. It can then be described both as a conceptual and as a natural system. From the first perspective its 'timeless' qualities become its most significant characteristics; its universality, its absolute mobility and fluidity, its power to represent *any* value, and to bring together all commodities, no matter how distant in time and space.[35] In the second perspective the 'material' character of money becomes more evident. Its interaction with all other commodities, which defines the relationship among commodities themselves, follows a principle of *conservation.* In any exchange in which money acts as a universal equivalent, value is conserved. Money seems to purify the process of exchange of all accidental and extraneous elements. It was just in this 'condensation' of conceptual and physical principles that the classical notion of reason existed.

It is all the more significant then that Simmel is able to find in the *modern* money form all the relativities, discontinuities, and contradictions celebrated in the new scientific imagination. Money embodies a new social logic. It is a logic discovered, however, in a more consistent and a more general application of already existing social assumptions. Just as Einstein brings to light the non-intuitive implications of an orthodox belief in the *universality* of nature, and Freud uncovers the radical consequences of viewing the *self* as a rational pursuit of pleasure, Simmel describes the new social world that grows, uncontrollably, from the apparently unprovocative idea that money constitutes a general *equivalent* of all values.[36]

The general process of exchange, Simmel points out, tends to 'objectify' value. It 'converts' purely subjective judgements into 'objective, supra-personal relationships between objects'.[37] And it is only in exchange that 'objects' can take shape from 'the chaotic material of our images of the world and the continuous flux of impressions'.[38] Money as the 'autonomous manifestation of exchange relations'[39] is therefore the general medium through which our 'images of the world' become ordered and rationalized. It is its very lack of particular qualities[40] which allows money to become a general 'signifier' of exchangeability, and therefore of all that is most fundamental to rational social life. 'The philosophical significance of money', he argues, 'is that it *represents* within the practical world the most certain *image* . . . of all being.'[41] And as it 'represents pure interaction in its purest form; it makes comprehensible the most abstract concept.'[42]

In addition to being the general representation of all relations, and therefore the immediately available form for the conceptualization of reality, the development of money as a general social mechanism enormously extends the range and power of 'rationality'. Money is a pure *means* in social life, and its general development lengthens the chain of purposive activities that we can project into the future, and provides a common language into which the plurality of qualitatively different values can be translated and compared.[43] It is, as Max Weber also argued in a somewhat different context, the foundation of calculability in social relations.[44]

The indifference of money to any particular value, its characterless 'dimensionality', and its pure instrumentality also make it, however, susceptible to new forms of irrationality. The ultimate goal of any chain of purposive action tends to become detached from the sequence of means.[45] Money, in this respect, is only the most extreme example of a general process inherent in the growth of social complexity. But just as money most completely embodies the central rationalizing tendencies of capitalism, so it most openly expresses the subversive tendencies of modernism. Simmel is acutely conscious of the inherently contradictory notion of rationality as 'pure instrumentality'. All particular values, in finding a rational form of representation, are in fact made less realizable. Life's 'ultimate purpose is floating above the teleological sequence'.[46]

Money's prodigality of interrelations leads the individual into a network of connections within which he becomes lost. An earlier sense of worth and dignity which came from the power to use money is overtaken by a kind of 'dispersal' of the ego over a seemingly infinite variety of possible 'projects'.[47] It seems preferable to possess money rather than the specific values for which it might be exchanged. In money *possibility* is stored, and in possibility, apparently, lies freedom. We have returned to the familiar psychology of the *Either/Or*. The loss of reality Kierkegaard so brilliantly described through the literary experiments of the 'young man' finds its social meaning as a money fetish. We have returned also to the still unfamiliar world of a later 'Copenhagen interpretation'. Money is a 'quantum phenomenon'. Its 'reality' is somehow hypothetical and provisional. Initially nothing in the social world seems more precise than money. Yet its effect is to render reality strangely vague and dreamlike.[48] The preponderance of intellect over sentiment in modern times, which is itself 'the consequence of money's character as a means',[49] should not be mistaken for a general 'rationalizing' process. In fact, as the human 'content' of social life exists primarily in 'sentiments', the formalism of money has the effect of distancing the psyche from the direct apprehension of reality.

The perversions of money relations are, therefore, typical rather than otherwise of the modern situation. Simmel writes strikingly of the stimulation of *wishfulness* in the money relation.[50] In its extreme forms of greed and avarice, we see a 'pure' relation to money. An overwhelming wish for money, for an object *without qualities*, is only the realization of money's real nature. And the possession of money is an adult and legitimate fantasy.

Simmel hints in these passages at a new psychology of excitement. If satisfactions are only accidentally related to the possession and consumption of commodities, then money itself becomes, for the first time, a genuine commodity. The acquisition of a characterless 'equivalent' can bring no more than an anticipatory pleasure, but to possess modern money might be truly exciting. Money has just the right 'texture' for the ideal commodity, it is a divisible substance, homogeneous and isotropic; it does not divert its possessor with superficial particularities. In greed and avarice Simmel identified a new distrust of commodities, or what, from a different viewpoint, might be termed 'secular despair'. The commodity is described by a 'probability function': the chance that it might precipitate a state of groundless excitement. But frequently it will fail to do so. To attach to money itself the hope of excitement, to make it *glamorous*, avoids the recurrent disappointments of 'real' consumption. Each 'quanta' of money is identical to the last, but above a magic and wholly indeterminant threshold, its possessor will glow with excitement. In the 'saving' in anxiety a wish for money, rather than any specific value, displays a certain rational force.

More directly, Simmel illustrates the new psychology by the inverse phenomenon. *Extravagance* is not genuine purchase or consumption so much as a form of *radiation*. It is no more than excitement in spending.[51] Everything hard and 'impenetrable' in the self has been dissolved. The ego, identifying itself 'rationally' as means, is wholly 'volatilized' into the sphere of circulation. This sphere, however, in perfecting itself as pure instrumentality has lost the synthetic unity which was the foundation of the Enlightenment's claim to reason. Its inner world consists, then, in a multiplicity of inherently meaningless 'projects'.[52] The 'ego' is fragmented. Its most strenuous efforts to 'realize' itself in the world issues in a series of disconnected 'adventures', each once again expressive of nothing more than an ephemeral excitement.

Simmel argues that a less florid but more widespread cynicism, and a characteristically metropolitan blasé attitude, are indicative of the same dissolution. There is a typical 'reduction of the concrete values of life to the mediating value of money'.[53] Money is fundamentally antagonistic to all distinguishing characteristics. Its general indifference imposes a new kind of anonymity and erodes the social form of the personality.[54]

The continuity and fluidity, the unbroken mediation of the circulatory process on closer inspection is seen to be sustained by irrational and incommensurable acts of abandonment. The contrast between classical and modern science is replicated – or, rather, that antithesis is itself a rediscovery of the fragmentation of the sphere of circulation. And as with the breakdown of a single, coherent 'system of the world', the contradiction upon which it founders proves to be intractable to any 'theoretical' solution. It is not simply a contrast between, on the one hand, a 'statistical' description of an 'individuated' micro-world, and on the other, a causally complete and deterministic explanation of the gross qualities of a macro-world. Rather, the elementary processes – themselves neither micro nor macro – that make up the sphere of circulation prove recalcitrant to rational analysis.

The 'irrationality' of circulation has two distinct aspects. First, as with any formal logical system, it can never be complete.[55] The rationality of any system of social interaction depends upon its content. This content, however, cannot be expressed within the system itself, which is composed only of interconnected means.[56] By investing money, which is pure instrumentality, with the gravity of an 'ultimate end', the rationality of means becomes transformed into the irrationality of disconnected psychic adventures. And second, the primordial reality in which circulation takes root, when it does expand into a general form, loses the sharpness required of rational distinctions. Its fundamental relations, flaunting their arbitrariness, become decadent with excitement.

PRODUCTION

The universe appears to be homogeneous and isotropic.[57] The 'self' seems to be composed of its own images. It is somewhat odd, therefore, that we should be so convinced of duration, that an 'arrow of time' imparts a particular direction to the order of nature, and shrouds naked selfhood in a biographical narrative. This is not the case in the cosmology of fun or of happiness, and only with qualification can it be said of the cosmology of pleasure.[58] How can relations among identical elements take on the character of an immanent 'development'?

If cosmos and psyche are viewed as images, of each other and of particular types of social relation, then we can understand this curious 'asymmetry' as a representation of the historic role of production in the creation of modern society. The world of commodities, like the 'developing self' and the 'expanding universe', if it is to exist at all must continually extend its boundaries. From the ultimately irrational 'singularity' of north-western Europe around 1600 (the date of Bruno's execution), the capitalist mode of production irresistibly expands. The unity of this mode can be grasped only as a picture of continuous and uninterrupted development.[59]

This was first of all a generalization and extension of production *within* a larger 'space' of the 'undeveloped' world. Both in principle and in fact this expansion was limited by simple geographical exhaustion. It none the less provided for almost 300 years the 'cosmographical' framework of capitalist production. During this period it was always possible to visualize the *market* expanding independently of production itself. Capitalism sent ahead of itself (so to speak) an ideological halo. The benefits of a 'rational' civilization were to be bestowed upon the world at large. And as a consequence of this act of farsighted charity, newly-made 'rational' workers and consumers were drawn into a system of commodity exchange. New processes of production 'naturally' followed.[60]

The inherent tendency to equilibrium within such a cosmographical scheme was clearly expressed in Newtonian science.[61] When capitalism expanded to fill the availab.e space of the world it was to 'settle down' to the stability of the cosmological model, and look forward to the slow decay of inevitably increasing entropy. This of course did not occur, and it is the endlessly inventive dynamism of the commodity which invests the revolution in modern cosmology with its social significance. Capital accumulation, without regard to the 'classical' limitations of the market, found new expandable forms. The 'absolute' dimensionality of the market and its independence from the commodities exchanged upon it turned out to be illusory. The market is itself an aspect of production, locally 'deformed' by the mass of commodities in it and directly expandable, perhaps infinitely so, through the creation of new exchange-values. Production is best understood relativistically. The 'flat' Euclidean and independent space of the market gives way, that is to say, to the exotic complexities of 'imperfect competition'.

It is not simply a matter, then, of large corporations monopolizing production and, by ideological means, manipulating the 'need' for a particular commodity.[62] Modern commodities are no longer subservient to the market; they are not tied 'rationally' to needs or desires. The wishfulness they might satisfy is intrinsically renewable and unlimited. The 'space' for their exchange is created in the same moment that the commodity is produced.

Continuous production, however, does not make the world dense with commodities because any commodity, to be exciting, must be *new*.[63] Novelty is the only distinguishable feature of the ideal commodity, and perpetual novelty is the commodity world's own 'arrow of time'. Excitement breaks the symmetry of all classical equilibria and precipitates the dizzying advance of modernism. Exciting commodities can be endlessly produced and the sphere of production, in principle still constrained by the fiction of a 'market economy', is best described in a formula common to Nicholas of Cusa and the modern cosmologist as 'finite but unbounded'.[64] There is always room for something new because novelty is a 'time-like' rather than a 'space-like' transition in the commodity's world-line.[65] The overriding tendency within classical thought towards the 'spatialization' of all categories brought expansion to a halt. But modern thought, in 'temporalizing' space, confirms the possibility of endless development.[66]

Production is primarily, then, the creation of new values. This of course places an insignificant constraint on the homogeneity of commodities. Newness must become recognizable, it has to be represented by some, preferably minimal, stylistic feature.[67] Once classical stability had been swept aside,[68] the appealing idea of production as perpetual creation was given a direct cosmological interpretation.[69] The 'steady-state' model, as its name implies, preserved a fundamentally spatial notion of stability (the universe had always been much as it appears to us), and accounted for the Hubble expansion on the hypothesis of the continuous creation of matter.[70]

A deeper cosmographical picture of modernity emerged in the so-called 'Standard Model'.[71] The large-scale spatial structure of the universe appears to be very nearly isotropic. This was given impressive confirmation with the discovery of a universal background radiation of about 3 degrees Kelvin measured in any direction.[72] The most economical account of this distribution of matter and of its presently observed expansion is to suppose a time-like variation. By reversing the process we are led back to a singular origin, some fifteen to twenty billion years ago, from which sprang the entire detectable universe in its present form.[73] Modern cosmology seeks to account for the observable universe by a physics of this origin. The present 'laws of nature' cannot be simply generalized to the very early universe. In the first moments of the 'big bang' the few types of fundamental particles interacted freely with each other, and the 'forces' of nature were coalesced into a single interactive rule.[74] As the universe cooled and expanded, a series of differentiations occurred, prohibiting certain interactions and establishing certain basic distinctions.[75]

The correspondence between modern cosmology and modern psychology is fairly evident. Both place primary significance on *origins*. More than this, they *interpret* the present in terms of a primordial, undifferentiated reality. They define the structure of cosmos and psyche as the spontaneous breaking of symmetries inherent in such a primordial reality. This correspondence should not be viewed as an extension or extrapolation of a 'theory' of the commodity. It is not ideological in form; it is, rather, an interpretation of the commodity itself.

The commodity world indeed is heavy with its own history. Each act of production is a small theatrical recreation of the origins of capitalism. Through it a primordial equivalence of human labour to all other commodities is established and then concealed. In alienating itself, human labour takes on the general characteristic of all commodities, and becomes indistinguishable from them. Labour simply exists, alongside other commodities, and enters into an indefinite sequence of exchanges with them. But as commodities are 'known' primarily through consumption, the entire process of production becomes, as a result, 'invisible'. New commodities continually appear, effortlessly extending the social and psychic space of modern life. They appear as so many stimulants to excitement. It is only by an act of critical reflection that we can discover the human content 'frozen' within them, and discover from this the peculiar circumstances of their birth.

This is a peculiar 'history'. Its narrative content does not unfold as a necessary sequence of cause and effect. Rather, for any arbitrarily chosen point, a common origin is postulated and 'proved' as its essential 'precondition'. 'Events' are linked together through their embodying in different ways these preconditions. The 'origin' of the commodity is covertly carried forward into modern society in much the same way, therefore, and our early experience is disguised and protected by the adult psyche, or the initial conditions of the hot 'big bang' are embedded in the present distribution of matter and energy.[76]

It is so deeply embedded that production no longer requires

justification; its existence is not in doubt. Indeed, so secure has it become that, no longer fearing criticism, it can allow subjectivity the freedom to seek excitement. This, in its turn, encourages the 'experimental' consumption essential to the continuous accumulation of capital. The political genius of capitalism has always lain in its principled defence of 'freedom': initially from the constraint of 'tradition' in the extension of market calculability; then to allow the 'rational individual' the unhampered pursuit of his own interests; and finally to release the psychic energy of wishfulness. Each step is founded upon a more highly 'energized' state of capitalist production. Only in the initial stage is an 'ideological' control over production essential[77] and a close relation between production and consumption established. Thereafter the bifurcation of 'object' and 'subject' becomes possible as 'wants' are expressed and acted upon in conformity to a causally effective 'rationality' common to both. In modern society even the fiction of reason, placing a theoretical limit upon the self-expansion of production, has been dispensed with. The accidental relation to the commodity, which at an earlier 'stage of development' would have been hopelessly disruptive, proves the best guarantee of 'growth'.

Classical science, as the intellectual and practical mastery of the world of immediate experience, is also secure. It is still the 'real' science of the engineer.[78] Beyond the realm of the technologist, however, we are free to speculate. The theoretical sciences, infected with this new freedom, are able to indulge, albeit within an enlarged framework of reason, the most exotic images of reality. The marriage of particle physics and cosmology seeks in the origin and foundation of things a bizarre other world upon which the coherence of our own more prosaic world depends: a world filled with 'virtual' events – with particles born in the first moment of time only to spend the rest of eternity, as the cosmos cooled beneath the temperature at which they could interact with any other particles, to no 'purpose'; with quarks, both charmed and strange that can never now be liberated from their tiny hadronic prisons; with multi-dimensional superstrings, wrapped in themselves, to make the elementary quanta of space as baffling as its large-scale structure.[79]

The 'givenness' of the commodity world encourages the decadence of theoretical imagery and allows reason to escape the constraint of production. Practical mastery of nature is assiduously renewed, but becomes tangential to the life of reason. Indeed, the rational and the irrational become in some circumstances indistinguishable.[80] Reason becomes 'infected' with a beguiling strangeness.

THE SYMMETRY OF CREATION

The decomposed subject experiences quite a different world from that of the integrated and rational self. Nature no longer confronts him as a 'self-enclosed world of bodies'.[1] Cosmos and psyche, during the classical period, having been separately and independently constructed from the laws of nature, were thrown together again, giving rise to a series of fragmented and 'mixed' images.

These images take on a meaning only in relation to the classical 'system of the world'. They offer their separate, impractical, and effective opposition to the completion of scientific rationalism as the exclusive carrier of bourgeois consciousness. *Excitement* is a partial cosmology, then, and relishes the fact. It exists as a defiant gesture against the philosophical abstractedness, facile individualism, and totalizing mechanism of the classical sciences. It is a perversion of, rather than an alternative to, the cosmology of pleasure. And as modernity exists only as a perversion of capitalism this could hardly be otherwise.

Not that excitement lacks anything in profundity or originality. It is nothing if not radically perverse. Indeed, it is only through its subversive thoroughness that we have come to appreciate the true character of the classical world view.[2] The 'cosmology' of excitement is best viewed, in fact, as the consequence of preserving the most fundamental principles of the classical picture. Nature must everywhere be the same; and in order that this should be so, the more superficial uniformities of Newtonian science must be 'relativized'. Once this is accepted, the Laplacean ambition to render the cosmos into a single equation is thwarted. The reversible relations of the classical system are then found to be an incomplete and inadequate representation of a moment torn from the real flux of nature.

This transition was 'simultaneously' effected in the decomposition of the self. The rationally integrated person fragmented into a collection of discrete experiences, each containing a separate and incompatible 'intuition' of the self,[3] and any one of which was likely to undermine the synthetic unity of consciousness.

The systematic totality of both cosmos and psyche dissolved into partial and paradoxical images. The causal mechanism which had been central to both had, at the limits of observation, broken down. The fundamental simplicity and symmetry of the rational structure which had been their common property was therefore pushed back into the *origin* of the world. Reason was itself transformed in this process. The rational interconnectedness of differentiated elements having been conceptualized as a causal mechanism, this mechanism could be expressed as quantitative conservation laws; now in their turn the conservation laws could be viewed as but imperfect replicas of the symmetries constitutive of primordial nature. The reality of cosmos and psyche thus became comprehensible as the spontaneous symmetry breaking of a primordial undifferentiated flux. In this primary process, the origin and transcendence of all difference, reason, and unreason merge.[4]

The parallel decomposition of cosmos and psyche, their mutual rejection of a totalizing mechanism, is not to be understood as the generalization of a single 'idea' originating in one particular discipline. Nor should it be interpreted as part of a general cultural decay.[5] Just as the 'mechanization of the world picture' can be seen as the mode of thought most appropriate to the development of the commodity world, so the perverseness of post-classical modernism should be viewed, wherever it arises, as expressive of a new relationship to this commodity world: to the appearance of a new commodity relation. The language of relativities and quanta, as of dream interpretation or symptom formation, invokes a new excitement; the excitement inherent in the arbitrariness of the modern commodity form. And in common with cosmos and psyche, the commodity is understood 'historically'. A singular process of symmetry breaking gives rise to the differentiations of the social world, establishing the spheres of production, circulation, and consumption upon which both reason and its perversities are founded.

Excitement, by its very nature, cannot be a comprehensive, exclusive, or exhaustive cosmology. An accidental relation to the commodity is possible only provided that, at another level, production is organized still on the basis of classically conceived rational necessity. In liberating wishes into the sphere of consumption it is assumed that 'authentic' needs continue to be met within the framework of everyday calculability.

CONCLUSION:
INTIMATE COSMOLOGY

The bourgeois world believed itself, for a time, to be in possession of a uniquely 'rationalizing' power. The scientific revolution provided it with a means to eradicate the ignorance and to subdue the errors corrupting less fortunate views of the world. Today, we are only too ready to congratulate ourselves upon outgrowing such rash optimism. Yet, for the most part, we uncritically accept the distinctions bequeathed to us by just such optimism. Science and reason can still appear as different expressions on the world's benign face. Object and subject go their separate, unrelated ways. We unhesitatingly 'explain' isolated events within the physical or the psychological world with an appeal to universal principles and causal laws.

It has been the purpose of this work to show that such optimism has *never* been justified. The project of bourgeois thought was not, and could never have been, carried through to its putative conclusion.

It is rather too simple to say that this is a *consequence* of a 'contradiction' inherent in capitalism; that bourgeois culture must betray in the incompleteness of its systematic sciences the incommensurability between use-value and exchange-value. The language of 'consequences' covertly accepts the exclusive validity of the very view of the world we wish to challenge. This is why questions of 'causality' and in particular of 'historical explanation' have been avoided. These problems are *part* of the classical world view, and remain the concern of what is here called the 'science of pleasure'. A critical reconstruction of the bourgeois world view, however, must now be made from a point of view which, so to speak, has been only half released from the totalizing ambition of all classical theory. The result is bound to appear, by outmoded but still authoritative conventions, unsystematic or even incoherent. As the alternative is to recycle a form of theory founded upon an illusion of rigour, this may be a risk worth taking. It might even stimulate a certain amount of intellectual excitement.

The central task therefore is not to explain but to discriminate. Here the 'official' distinction between object and subject takes on an air of ideological simplicity. The formal 'antinomies' of the classical world force the varieties of human experience, including the varieties of its modes of thought, into a single relation. Confronted with its stark choices, the 'correct' answer is always obvious; only the demonic clings to the negative. The superiority of the objective, universal, necessary, and rational is clear to all.

From the perspective of the present, however, it is clear that the bourgeois theorist was never so confident as he would have us believe. The rational cosmos and its equally rational psyche had to establish their supremacy, not so much over the recalcitrant material of their own imperfectly developed theories as against the resistance of relations which could not find a place within any of their imperative oppositions. These relations did not belong to the modern world and ought simply to be forgotten. But they were not forgotten, and now that the demands of 'rationality' have been relaxed somewhat their continuing presence can be more easily detected. Old distinctions re-emerge and contest the exclusive ground of reason. They do not make their challenge on the basis of any formal claim; to do so would be to admit defeat at once because, in those terms, science *is* more rational than magic or religion. They propose, rather, a new division of experience within which the undoubted superiority of scientific rationality can be *contained.*

The classical world view, therefore, is only part of bourgeois cosmology; that part which knows only the science of pleasure. In contrasting it in different ways with the cosmologies of fun, happiness, and excitement 'ideas' are not being pitted one against another. The key classificatory terms used here – 'fun', 'happiness', 'pleasure', and 'excitement' – are deliberately chosen as words rather than concepts. They are rich in associations; associations which in becoming more self-conscious separate themselves from one another to describe fundamentally different relationships.

The cosmology of fun is not a naïve or impoverished account of the same world as that described by the classical scientist or psychologist. It captures and conveys a world of its own, a world which the scientist can only ignore. A description of this world of metamorphic freedom leads directly to its intellectual elaboration as an order of signs. Similarly, a hierarchy of symbols emerges from an attempt to convey the inner meaning of the cosmology of happiness which lies equally outside the scope of classical reason. The 'system of the world', as a language of causality, is exclusive, in other words, to the classical picture of the world. 'Beyond' causality we cannot yet (should we try?) lay claim to a total vision of reality. But in the cosmology of excitement as the *enjoyment* of its incomplete and contradictory images we have decisively rejected the classical world's search for systematic order. This is as true of the 'modern' scientist as it is of the modern novelist or painter. The 'ideal' (rather than typical) scientist is now closer to Calvino's bewildered Mr Palomar than to Thomas Mann's self-confident Professor Kuckuck.[1]

Fun, happiness, pleasure, and *excitement* – that is to say, constitutive inner worlds of meaning the elements of which cannot be 'translated' from one to another: each is associated with a different apparatus of thought (order of signs, hierarchy of symbols, system of causes, and network of images), and different intellectual preoccupations (description, exposition, explanation, interpretation); each is 'expressive' of a different social relation (absolute dependence, personal subordination, market freedom, leisure relations); and each constructs a different 'object world' (toys, goods, commodities, images). The list might be arbitrarily extended, arranged into groups, cross-tabulated and diagrammatically consecrated. The temptation to formalism remains powerful, but can be resisted. Each distinguishing term, the centre of its own world and not just a means to trigger a series of cognate terms (terms that are divisions *within* the science of pleasure as much as they are distinctions *among* the different cosmologies), is better left unmolested. These distinctions, if they are real, invite their own forms of exploration; and if they are not, cannot by any amount of 'rational' analysis be made so.

NOTES

INTRODUCTION

1 See, for example, Leach (1982: 212-20).
2 On the distinction between 'everyday' reality and theoretical constructs, see Schutz (1976), and Schutz and Luckmann (1974: 3-18).
3 Or the discussion becomes almost exclusively methodological. For a recent example see Komesaroff (1986). There are of course some notable exceptions (see below nn. 1 and 2 to Chapter 8).

CHAPTER ONE

1 This includes many primitive societies. See Durkheim and Mauss (1963), Durkheim (1964), Lévi-Strauss (1968: 132-67), Needham (1973).
2 As diversely as, for example, in Swift's *Tale of a Tub* and Dostoevsky's *Notes from Underground.*
3 Gay (1966: 120-5); a theme developed generally in Turgot (1973), and Condorcet (1955), a good deal less complacently than in their nineteenth-century followers. See also Simon (1963).
4 Cassirer (1966: 5) remarks that '"Reason" becomes the unifying and central point of this century', and makes it plain that even as *intellect* reason had a much wider meaning that is common for us. 'Thought consists not only in analysing and dissecting but in actually bringing about that order of things which it conceives as necessary, so that by this act of fulfillment it may demonstrate its own reality and truth', ibid., viii. See also Hazard 1965: 37-55).
5 Notable particularly in Hutcheson's *An Enquiry into the Original of our Ideas of Beauty and Virtue*, a vigorous argument in defence of the autonomy of aesthetic sensibility. This line of thought influenced Hume; see Norton (1982: 92); Vereker (1967: 56).
6 Goldmann (1973), critically following Kant (1963), takes a somewhat narrow view in castigating the Enlightenment ambition to liberate 'thought' from 'reality'. See above, n. 4. Not all, of course, were as

subtle as Hume, Holbach, for example, declares bluntly: 'theology is only the ignorance of natural causes reduced to a system', quoted in Manuel (ed.) (1965: 58). While Reid, from a completely different standpoint argues none the less similarly that 'philosophers, pitying the credulity of the vulgar, resolve to have no faith but what is founded upon reason' (Robinson 1961: 139-40).

7 The implication of such a view is that *The World as Will and Representation* is no more than Locke's *Essay Concerning Human Understanding* a genuine 'precursor' of *The Interpretation of Dreams*.

8 The differentiation of social life into production, circulation, and consumption is more often than not simply taken for granted. The most impressive recent attempt to utilize these distinctions is to be found in the work of Braudel (1979, 1983, and 1984); see particularly Braudel (1983: 20-3).

9 This argument will be pursued below in chs. 9 and 12. The basic insight is contained in Sohn-Rethel (1978), Thompson (1961), and before them Lukács (1968) and Simmel (1978).

10 For detailed persuasion see Febvre and Martin (1976: 143-66), and more generally Braudel (1983).

11 Hume wrote more on history than on philosophy, Rousseau wrote on music and chemistry as well as everything else. And it was for his versatility that, as Cassirer notes, Diderot was idolized. See also Gay (1966, ch. 1).

12 A system as potentially complex as that sketched for the rational cosmos by Foucault (1970), who somewhat surprisingly oversimplifies the internal relations of the unreasonable.

13 Rousseau (1911: 1).

14 Ariès (1973: 60-2, 98-111). But note critical remarks by Pollock (1983: 33-67) and Hunt (1970: 32-51). See also Porter (1982: 284-6), Stone (1979: 254-303), Wrightson (1982: 108-18), Davidoff and Hall (1987: 335-56).

15 Rousseau (1911: 50). Children, he says, experience 'desires which are not true needs, desires which can only be satisfied with the help of others'. Childhood, as a consequence, 'has its own ways of seeing, thinking, and feeling', ibid., p. 54.

16 Ferguson (1983: 146-51).

17 Locke (1968: 143). Though Locke as representative of a declining aristocratic tradition in this respect was less wholehearted than were his followers (Bantock 1984, vol. 1: 277).

18 Rousseau (1911: 95).

19 Locke (1968: 232) remarks the 'Sauntring Humour' which he describes as a 'listless condition', that 'I look on as one of the worst qualities that can appear in a child'. See also Ferguson (1983: 139-40).

20 Rousseau (1911: 29).

21 Ibid., p. 32, typically inverting Freud's later formulation in which all satisfactions are felt as a single pleasure.

22 Ibid., p. 53.

23 Ibid., p. 72.

24 Ibid., Hunt (1970: 133-9).

25 Ariès (1973) especially stresses the 'roughness' of schoolchildren before the era of bourgeois discipline. The masters of the sixteenth century had to 'cope with armed revolts', ibid., pp. 302-15.

26 Locke (1968: 145) recommends that authority be established 'as *soon* as he is capable of submission'. And as 'Children are not to be taught by rules', which will only be thwarted by 'the perverseness of their Will', discipline must precede instruction. Even Rousseau recognizes the difficulty: 'No doubt [Emile] ought to do what he wants, but he ought to want to do nothing but what you want him to do,' quoted in Bantock (1984, vol. 2: 11). See also Ferguson (1983: 123-4).

27 See particularly La Salle's *Conduite des Écoles Chrétiennes* (Avignon, 1720), reprinted in *Cahiers lasalliens* (n.d.), which is the most precise of the teacher's manuals of the period and makes clear the intimate relationship between reason and *conduct*; see also Foucault (1977: 184-94). Such disciplinary systems were, of course, ideal; the reality in many instances was pedagogically less fussy; see, for example, Cole (1950: 444-8).

28 Elias (1978, vol. 1) remains the fundamental work.

29 Rousseau (1911: 118), believing the consumption of meat predisposing to cruelty, recommends a vegetarian diet for children.

30 Donzelot (1980: 48-95).

31 Ariès (1973: 48-59). Bakhtin (1968) has completely transformed our view of the carnival tradition.

32 France (1988: 108-12), Pascal (1960).

33 Gay (1969: 89), notes a general unease evident in a number of French writers in this regard: 'The relationship between reason and the will, and between the will and the passions, that Condillac developed is symptomatic of the paradox of the Age of Reason ... in which reason unexpectedly turned upon itself and, by its own rigorous application, struck at the very foundation on which it rested – an intellectual development of baffling circularity and continuing relevance.' See also Cassirer (1955: 93-134).

34 Braudel (1984: 24-70, 115-30, 207-34) insists that the European economy was always a world economy. See also Scammel (1981). Ashtor (1983: 367-82) points out that maritime insurance was developed as early as the mid-fifteenth century.

35 Hodgen (1964: 117), Lovejoy (1960: 144-83).

36 Hodgen (1964: 213), Montaigne (1958: 105-19), was exceptional.

37 Lord Monboddo (James Burnet) (1773: 204), holding that there was 'no natural propensity to enter society', proposes a 'natural history' of human society. Many non-human species, he suggests, display characteristics wrongly considered to be exclusively human and social – for example, 'herding' together – and inversely many human societies lack 'all social arts'. In particular, language makes a relatively late appearance in his developmental scheme; Schneider (1967: 271-87). Burnet's approach owed more to Buffon than to Rousseau, who had begun his influential *Essais sur l'origine des langues* with the

categorical statement that 'La parole distingue l'humane entre les animaux' (Rousseau 1969: 27). As 'la prémière institution sociale', language must be derived from a non-rational source, from sounds which are at first mere 'vives et figures', ibid., p. 41; see also Cassirer 1968, vol. 1: 139-55).

38 The approach was effectively criticized by Henry Home (Lord Kames), and by Adam Ferguson. Kames's *Sketches of the History of Man*, continually revised and enlarged, was, however, part of the same popular tradition: 'a veritable speculum mirabilis, reflecting even the popular interests and tastes of the time – the current interest in "natural history", in the manners of remote peoples, and in at least a semi-scientific effort at social and historic interpretation' (Lehmann 1930: 213).

39 Kames, in Schneider (1967: 257).

40 James Beattie (1974), for example, following Rousseau, traces the origin of language to animal cries as the 'audible expressions of the passions', ibid., p. 236, and explicitly links intellectual 'development' to 'progress' in language.

41 Diderot, in Gay (ed.) (1973: 394).

42 Which is not to say that the institution of slavery can be explained wholly in terms of such beliefs. See, from different viewpoints, Davis (1970: 187-48) and Miles (1987: 73-93).

43 'During the Dutch East India Company's two centuries of existence, the directors sent out to the East and maintained at their own expense a total of nearly 1,000 Calvinist *predikants*, and several thousand lay readers and schoolmasters' Boxer (1965: 133); they presumably felt it a worthwhile investment.

44 *Gulliver's Travels*, exceptional in its satirical intent, provides only a partial model. Diderot, also critical, was more openly exotic and popular in form. See particularly Chinard's 'Introduction' to Diderot (1935).

45 Turgot (1973: 41-59). Condorcet distinguishes ten such epochs. Neither assumes, however, that the process terminates in the present: 'Nature has set no term to the perfection of human faculties; the perfectability of man is truly indefinite' (Condorcet 1955: 4).

46 Rousseau (1984: 78).

47 Ibid., p. 70.

48 Ibid.

49 Ibid., p. 89. A point which quite escaped Burnet or Beattie, but not Hume, whose distinction at the outset of the *Inquiry*, into the *active* and *reasonable* aspects of human being anticipated it. Hume indeed recognizes that the two are never fully united. In the most 'civilized', 'The feelings of our heart, the agitation of our passions, the vehemence of our affections dissipate all conclusions, and reduce the profound philosopher to a mere plebian' (Hume 1854, vol. 4: 3).

50 Rousseau (1984: 90).

51 Ibid.

52 Ibid.
53 Ferguson (1966: 6).
54 Ibid., p. 82.
55 Ibid., p. 87.
56 Ibid., p. 90.
57 Ibid., p. 89.
58 Ibid., p. 93.
59 Cassirer (1966: 41) argues that, as nature is 'now an original formative principle which moves from within [it] signifies the integration of all parts into an all-inclusive activity and life'.
60 Foucault (1965), particularly pp. 65-85. Note also Doerner (1981), Rosen (1968), and for some useful critical comments Sedgwick (1982: 125-49).
61 Foucault (1967: 58).
62 Foucault is by no means unique in this. Very generally the social history of madness has been coloured by later conceptions of the relation between lunacy and reason. Porter, for example, in an interesting essay on Samuel Johnson, ascribes to the period a view of madness as defective *reason*, rather than disturbed *appetite* (Bynum, Porter, and Shepherd 1985, vol. 1: 76-7).
63 As expressed particularly by Adam Ferguson (1973, vol. 1, ch. 1, and vol. 2, ch. 1). Ferguson, rejecting the synoptic 'histories' of Condorcet and others, defines man's 'progressive' nature in a complex fashion.
64 Diderot (1966). He exemplifies the ideal *philosophe* as 'a cultivated man, a reputable scholar and a scientific amateur', (Gay 1966: 14).
65 Diderot (1966: 34).
66 Ibid., p. 37.
67 Ibid.
68 Wain's comment on Savage, in Johnson (1976).
69 Johnson (1963: 377).
70 Ibid., p. 378.
71 Ibid., p. 379.
72 Quoted in Skultans (1979: 58).
73 Pinel, in Zilboorg (1941: 330). Sade is the most complete modern madman; see Barthes (1976).
74 Klibansky, Panofsky, and Saxl (1964); also Yates (1979: 49-59), Babb (1951, 1959), Orrù (1987: 64-93).
75 Foucault (1970: 117-35); Hunter and Macalpine (1963), extracts by Willis, Whyte, Battie, Cullen, and White.
76 'We are at length persuaded . . . that the distemper named from the womb, is chiefly and primarily convulsive, and chiefly depends on the brain and nervous stock being affected', Hunter and Macalpine (1963: 190).
77 Doerner (1981: 106). Tuke 1882: 109), reports John Wesley's treatment of both melancholia and mania; for the former 'juice of ground ivy with sweet oil and white wine as an ointment' is recommended.
78 Whytt, in Hunter and Macalpine (1963: 407). Wesley recommends for

'raving madmen', 'binding' then 'set the patient with his head under a great waterfall, as long as his strength will bear' (Tuke 1882: 109).

79 The Scottish intelligentsia were united as much by manners and 'style of life' as by ideas, though many subscribed to a form of Stoicism; see Kettler 1965: 33-9.

80 And not with women, who begin to go insane in an interesting way only when they are recognized as rational consumers in their own right (Skultans 1979: 77-97). Similarly, in more recent times, the interest of children to psychiatrists, rather than developmental psychologists, marks their integration into bourgeois society as consumers.

81 Porter, in Bynum, Porter, and Shepherd (1985: 73).

82 Mead, quoting Hale, in Hunter and Macalpine (1963).

83 Johnson, *Rasselas*, p. 190.

84 Ibid., p. 191. Rousseau (1979: 91) agrees, less timidly: 'and during these wanderings, my soul rambles and glides through the universe on the wings of imagination, in ecstasies which surpass every other enjoyment'.

85 Swift (1909) is both an exposition and an example; a genuinely manic work full of disjointed vigour and restless energy, it has all the appurtenances of modernism; see particularly pp. 93-7, 'A Digression in Praise of Digressions'.

86 Scull (1982: 76-125); Bynum, in Scull (1981: 35-58); Bynum, Porter, and Shepherd (1985); Parry-Jones (1972: 168-281); Jones (1955: 49-66); Tuke (1882).

87 An insight subsequently overlooked, or deliberately ignored, in the professionalization of psychiatry.

88 See, for example, the striking analysis of Fitzhugh's writings by Genovese (1971: 118-245).

89 Davis (1970: 221-48).

90 Even if the figure of 96,000 visitors a year to Bedlam is a gross overestimate, there is no doubt the figure was large and continued at a high level over a lengthy period; Alderidge, in Bynum, Porter, and Shepherd (1985: 17-34).

91 Foucault (1970: 322-8).

CHAPTER TWO

1 France (1987: 5-8).

2 They remain also, of course, among its chief exponents.

3 Weber (1978, vol. 1: 24-6) uses the term *Zweckrational* to refer to a particular type of *social action*. Rationality in the present context is restricted to a particular mode of thought.

4 The philosophers were only expressing abstractly what was already known to practical men. Andrew Ure's *Philosophy of Manufacture* (1825) appeared before Comte's *Cours*.

5 *The History of the Evolution of the Organism* consists of two kindred and closely connected parts: 'Ontogeny', which is the history of the evolution of individual organisms, and 'Phylogeny', which is the history of evolution of organic tribes. 'Ontogeny' is a brief and rapid recapitulation of 'Phylogeny' (Haeckel 1874: 1-2). See Gould (1977) for a comprehensive and critical study.

6 Manuel (1962), Charlton (1959: 11-12), and generally Kolakowski (1972: 20-59).

7 Comte (1883, vol. 2: 149).

8 Ibid. (vol. 1: 3).

9 All history is '*development*', which 'brings after it *improvement*', assumptions which he declares, 'we may consider to be admitted as facts' (ibid., vol. 2: 89).

10 Ibid. (vol. 2: 159).

11 The term 'fetishism' had been coined by Charles de Brosses in 1760. See Evans-Pritchard (1965: 20), van Gennep (1914, vol. 5: 161-78), Jevons (1896: 166), Schmidt (1931: 56-7).

12 Comte (1883, vol.2: 186). Later this 'subjectivity' was taken as evidence of the disordered state of the primitive mind. Thus Jevons (1896: 16): 'one primitive man's experience must have consisted of a stream of events as disjointed and disconnected as the successive incidents in a dream.'

13 A point of view which proved remarkably persistent. Kelsen (1946: 1) claims that the primitive mind is dominated by 'the emotional component', which leads to 'ideas which neither describe the world nor satisfy our curiosity and desire for knowledge'.

14 Comte (1883, vol. 2: 151) (emphasis added).

15 Ibid., p. 152.

16 See particularly Turner (1974 8-37); also, Burrow (1966: 65-100), Peel (1971: 131-65), Moore (1979: 153-73), Simon (1963: 19-72), Charlton (1959: 86-157).

17 A species of intellectual pornography that was more refined but hardly less hypocritical than the moralizing that was so much a part of the period. See, for example, Marcus (1974), Turner (1974: 50-9).

18 Turner (1974: 17-24).

19 'The savage state in some measure represents an earlier condition of mankind, out of which the higher culture has gradually been developed or evolved' (Tylor 1913, vol. 1: 32).

20 Ibid., vol. 1, p.74.

21 Ibid., vol.1, p. 237; Tylor, as Comte, borrowing here from de Brosses.

22 'Just as no one is likely to enter into the real nature of mythology who has not the keenest appreciation of nursery tales, so the spirit in which we guess riddles and play at children's games is needed to appreciate the lower phases of language' (Tylor 1913, vol. 1: 237).

23 Ibid., vol.1, p. 284.

24 Ibid., vol. 1, p. 285. Compare Jevons (1896: 11): 'The savage imagines that even lifeless things are animated by a will, a personality, a spirit, like his own.' Either might have been quoting directly from Comte.

25 Tylor (1913, vol.1: 288).
26 Ibid., vol. 1, pp. 297-300. It was just such a 'deep consciousness of nature', of course, that the philological school attempted to bring to light. See, for example, Müller (1887: 77-86) and 1898, vol. 9: 58-62, 102). The specific mechanism of myth creation, the 'confusion of language', is not far removed from the uncritical 'association of ideas' that Tylor held to be of central importance.
27 Tylor (1913, vol. 1: 299).
28 Ibid., vol. 1: p. 303.
29 'In psychology the savage is he who (extending unconsciously to the universe his own implicit consciousness) regards all natural objects as animated and intelligent beings' (Lang 1887, vol. 1: 31).
30 Tylor (1913, vol. 1: 116).
31 Ibid., vol. 1, p. 306.
32 See particularly Jevons (1896), and for a comprehensive review, Schmidt (1931), Evans-Pritchard (1981: 91-5, and 1965: 20-32). The important exception is the work of John H. King (1892), in which the most 'primitive' categories are 'pre-animistic', impersonal forces of 'luck' and 'ill-luck'. His work was ignored, but Schmidt (1931: 151) goes so far as to claim that 'King's account of the real origins of the idea of magic is still the best'. Tylor's influence survived the development of ethnographic methods. It is still evident in the work of Lowie (1924), Murphy (1927), Allier (1929), and many others including, in spite of a somewhat crude 'materialistic' approach, Radin (1937). But compare Rivers (1924) for a more specific ethnography of the medical context of magical practices.
33 The arguments had been rooted more in academic politics than anything else. 'Theoretically' the savage and the ancient were equivalent, but anthropologists in establishing their university credentials needed to monopolize a specific field of study. Frazer's universalism was born of greater professional confidence.
34 Frazer (1900, vol. 3: 290). The combination of potent narrative and forbidden modes of thought proved highly popular and exercised a lasting influence over the development of anthropology; see Stocking (1983).
35 Frazer (1900, vol. 1: 403).
36 Ibid., vol. 3, p. 290.
37 Ibid., vol. 1, p. 61.
38 Ibid., vol. 1, p. 62.
39 Comte (1883, vol. 2: 63).
40 Frazer (1900, vol. 1: 75-6).
41 This is contradictory on Frazer's part. Having argued that magical thought is a result of the savage's failure 'to recognise those limitations to his power over nature which seem so obvious to us' (ibid., vol. 1, p. 128), he should not then claim that the origin of religion can be traced to intellectual disillusionment in such a belief. Marett (1936: 78) repeats the argument, attributing it to Tylor: 'Loose and illogical as man's early reasonings may be, and slow as he may be

to improve them under the check of experience, it is a law of human progress that thought tends to work itself clear.'

42 Frazer (1900, vol. 1: 71).

43 Ibid., vol. 1, p. 77.

44 Frazer (1922, vol. 3: 134).

45 Ibid., vol. 3, p. 43.

46 A viewpoint which owed almost everything to Tylor: see Burrow (1966: 228-59).

47 Radin (1957: 52), quoting Rasmussen's report of a dialogue with an Eskimo respondent: 'What do you believe? We do not believe, we only fear.' Or Murphy (1927: 178): 'What primitive man does not understand, while yet it forces itself upon his notice, he fears.' And Allier (1929: 92) claims that for the savage, 'uncertainty is the dominant feeling' and 'uncertainty inevitably begets fear'.

48 Feuchtersleben (1847: 83).

49 Doerner (1981: 64-97).

50 Ellenberger (1970: 57-85).

51 Feuchtersleben (1847: 110). In such states, he admits, 'the mind is a conjuror'; however, 'it is so only at the moment, and involuntarily, which entirely deprives it of all merit'.

52 Ibid., p. 111.

53 Scull, in Bynum, Porter, and Shepherd (1985, vol. 1: 103-50).

54 Feuchtersleben (1847: 246). It is 'the erroneous combination of manifold ideas often united with the patient's own inclinations without his being aware of the error, or being able to overcome it' (ibid., p. 211).

55 Ibid., p. 270.

56 A characteristic elaborated in another classic of psychopathology, Dostoevsky's *House of the Dead*, and under the term 'degeneration', was made popular by Morel. See Ellenberger (1970: 281).

57 Robertson (1910: 22), for example, declares that 'American Mythology and Religion are alike but aspects of the general *primitive psychosis*', (emphasis added).

58 The basic nosology had been established by Kraepelin; see Ellenberger (1970: 284-5). It is Bleuler's work, however, that 'remains until today the most complete description of the symptoms of the schizophrenic from a phenomenological point of view' (Arieti 1974: 11).

59 Bleuler (1950: 27).

60 Bleuler (1916: 81). Or again, Bleuler (1950: 16): 'Only the goal-directed concept can weld the links of the associative chain into logical thought.' He quotes as an example of undirected, pathological associations a short essay written by one of his patients, 'The Golden Age of Horticulture', of which the following is an extract:

> At the time of the new moon, Venus stands in Egypt's Autumn-sky and illuminates with her rays the commercial ports of Suez, Cairo, and Alexandria. In this historically famous city of the Califs, there is a museum of Assyrian monuments from

Macedonia. There flourish plantain trees, bananas, corn-cobs, clover, and barley, also figs, lemons, oranges and olives. Olive-oil is an Arabian liquor-sauce which the Afghans, Moors and Moslems use in ostrich-farming. The Indian plantain tree is the whisky of the Parsees and Arabs. The Parsee or Caucasian possesses as much influence over his elephant as does the Moor over his dromedary.

(Ibid., p. 15)

61 Darwin's 'A Biographical Sketch of an Infant' was first published in 1877 about thirty-seven years after its initial composition (reprinted in Gruber 1974). Froebel (1897) formulated his pedagogics within the tradition of Hegelian dialectics. Particularly indebted to the 'biogenetic law' were the works of Baldwin (1895) and Hall (1904).

62 Harris in 'Preface' to Preyer (1896, vol. 1. vii).

63 Most notably by Groos (1898 and 1901).

64 'Preface' to Preyer (1896, vol. 1: vi). For a fuller statement, see Harris (1898: 23-31) and Ladd (1897: 193-227).

65 Sully (1896: 35, 37).

66 Ibid., p. 70. A view shared by Rasmussen (1920, vol. 3: 105) and Stern (1924: 222-3). Compayré (1893: 209, 270-8), stresses the importance of positive education.

67 Sully (1896: 91).

68 Ibid., p. 70.

69 Problems anticipated to some extent by Compayré (1893: xxiii): 'quand il s'agit de l'historique de l'enfant, c'est-à-dire d'un être en formation, dont les facultés sont en train de s'organiser, chez qui les phénomènes divers ... constituent comme des couches successives, comme une série de palimpsestes superposés'.

70 Stern (1924: 148) treats speech development as non-associative: 'Early speech is innocent alike of grammar and logic.' See also Bühler (1930).

71 Piaget (1956: 27): 'The child does not in the first instance communicate with his fellow-beings in order to share thoughts and reflexions; he does so in order to play.'

72 Ibid., p. 43.

73 Ibid., p. 46.

74 Ibid., p. 79. 'The same phenomenon', he points out, 'is undoubtedly to be found in Dementia Praecox and other pathological cases' (ibid., p. 101).

75 Ibid., p. 135. Rather than comparing child thought to the active associations of reasoned memory, it is more revealingly likened to the *dissociation* of spontaneous recollection. Piaget refers here to Bergson; he might equally have mentioned Proust: 'the tendency of every memory to gather others around it can be explained by a natural return of the mind to the undivided unity of perception' (ibid., p. 132).

76 Ibid., pp. 140, 154, 159; and Piaget (1973: 49-109).

77 Piaget (1956: 159).

78 Bleuler (1916: 90). Reil had already talked of the 'ideal assimilation of the universe' as typical of pathological thought. See Feuchtersleben (1847: 85).

79 Bleuler (1916: 90).

80 Lévy-Bruhl (1928: 17).

81 Ibid., p. 19. It is characterized by 'lack of concepts rather than the contradiction among concepts'. Lévy-Bruhl (1975: 10). And with specific reference to notions of individuality he says, 'The participation of the individual in the social body is an immediate datum contained in the feeling which the individual has of his own existence' (ibid., p. 83).

82 Piaget (1973: 49). It is therefore thought which has 'not become conscious of itself' (ibid., p. 46).

83 Ibid., pp. 236-86.

84 Lévy-Bruhl (1928: 20). He thus 'sees no difficulty in metamorphosis'. In this he went far beyond the original account of *mana*. See Codrington (1891: 117-27).

85 Piaget (1973: 176).

86 Ibid., p. 156. Thus, 'the shadow, the likeness, the reflection etc. are confused with the man whose form and features they represent'.

87 Ibid., p. 80.

88 Ibid., pp. 350-2.

89 Lévy-Bruhl (1928: 21).

90 Schreber (1955).

91 Bühler (1930: 66), quotes Helmholtz recollecting an incident from the age of three or four: 'I can remember as a child passing by the spire of the garrison church in Potsdam and seeing people on the balustrade, whom I took to be dolls. I asked my mother to take them down for me, believing she could do this by merely reaching up with her hands.'

92 Lévy-Bruhl (1975: 3).

93 Arieti (1974: 231).

94 Ibid., p. 230.

95 Bleuler (1950), quoted in Arieti (1974: 231).

96 Werner (1957).

97 See particularly Kohler (1927), Buytendijk (1935), and most significantly Uexküll (1926).

98 Werner still regards the intellect as a mechanism of adaptation, but the notion of 'development' as a continuous function has all but vanished.

99 Kelsen (1946: 3). Constrained to think entirely in terms of personal motives they have no means of 'discovering nature' (ibid., p. 43).

100 Note also Piaget (1929: 294) on children's tendency to attribute to things 'such powers as they would need either to obey us or resist us'; and Lévy-Bruhl (1928: 19). For an eccentric exception, see Marett (1932).

101 For example, Lévy-Bruhl (1928: 70). The human individual 'exists by virtue of his participation in the essential principle of his group'.

102 Ferguson (1983: 139-46). Schreber (1955), for many examples. And Bleuler (1950: 118): 'They are dissected, beaten, electrocuted; their brain is sawn to pieces, their muscles are stiffened. A constantly operating machine has been installed in their head. Someone has injected something into their tear-ducts. Their eyes have been exchanged for those of an old woman.' It is the lack of affect in recounting such phenomenal nightmares that justifies the term 'schizophrenia'.

CHAPTER THREE

1 Marx, it should be remembered, presents his analysis of production as a *critique* of political economy. Piercing the mystery obscuring the source of profit is here something rather more than the provision of an adequate causal account of its origin. See Marx (1976, vol. 1: 163-77). Value 'transforms every product of labour into a social hieroglyphic' (ibid., p. 167). In this respect Marx might be considered a modernist. See Berman (1982: 90-115), Frisby (1985: 20-7).
 Freud and Proust, it is worth noting, developed a similar 'critique of bourgeois psychology', by uncovering the source of pleasure accumulating in the normal process of psychic exchange.

2 'Interesting', that is, in Kierkegaard's precise use of the term. See *The Concept of Dread* (1957: 16). The interesting always retains at its core something of personal significance; in being 'objectified' social relations therefore become uninteresting.

3 And, paradoxically, becomes *dematerialized*. Cf. Baudrillard (1975). Again, following Kierkegaard (1967), the 'systematic' is taken up into purely reflective categories and becomes metaphysical. Genuine materialism must, in a rather particular sense examined below (Chapter 11), remain unsystematic.

4 Thus Engels' well-known letter to Joseph Bloch: 'According to the materialist conception of history the *ultimately* determining element in history is the production and reproduction of real life' (Feuer, ed., 1959: 397-8).

5 Weber (1976: 181) notes that 'The Puritan wanted to work in a calling; we are forced to do so'.

6 See particularly Kierkegaard (1959: 169-224), Löwith (1964: 235-51).

7 Marx (1975: 243-58, 323-34). And for useful comments, Ollman (1971: 131-225), Mészáros (1970: 93-122). Schacht (1971: 65-112).

8 As in Lukács (1968) and Sohn-Rethel (1978).

9 See Abercrombie, Hill, and Turner (1980) for a vigorous argument. This is not to be confused of course with the once popular notion of the 'end of ideology', traceable to Saint-Simon, but in its contemporary revival owing most to Bell (1960).

10 Once again Kierkegaard provides the first and most compelling examples – particularly, in this instance, in the *Either* of *Either/Or*.

11 See below, Chapter 11.

12 For example, Lévi-Strauss (1973), Szasz (1961), Holt (1974).

13 A tradition analysed by Rorty (1980). Freud (S.E., vol. 14: 122), it should be remembered, defined 'instinct' (*Trieb*) as 'the psychical *representative* of the stimuli originating from within the organism and reaching the mind'.

14 Saussure (1966: 66), expresses it thus: 'The linguistic sign unites, not a thing and a name, but a concept and a sound-image. The latter is not the material sound, a purely physical thing, but the psychological imprint of the sound, the impression that it makes in our senses.' And Delacroix (1934: 131): 'Le signe linguistique, dans son essence, est incorporel; ce qui le constitue, ce n'est pas sa substance matérielle, ce sont les différence qui séparent son image acoustique de toute les autres'.

15 Saussure (1966: 120): 'in language there are only differences'. See also Barthes (1967: 38-9) and Martinet (1969: 20-2).

16 Cassirer's *Philosophy of Symbolic Forms*, vol. 1, p. 69, as a '"morphology" of the human spirit', which aims at 'the specification of pure subjectivity', is one of the most impressive efforts along these lines. It is a similarly 'morphological' perspective which informs Uexküll's romantic biology. For example:

> No attempt to discover the reality behind the world of appearances, i.e. by neglecting the subject, has ever come to anything, because the subject plays the decisive role in constructing the world of appearances, and on the other side of that world is no world at all.
>
> (Uexküll 1926: xv)

And again:

> In the world of the physicist there are only objects which react on one another through the medium of space; in the world of the biologist there are only appearances which react on one another through the medium of the subject.
>
> (Ibid., p. 31)

17 A limiting possibility approached, ironically, in the contemporary relations of production. The boredom of work has been commented upon for well over a century.

18 A distinction implicit in both the title and the bold opening statement to *The Interpretation of Dreams*: 'every dream reveals itself as a psychical structure which has a meaning'. It is equally evident in many of Simmel's finest essays – for example, Simmel (1971: 121, 179-86, 294-323).

19 Particularly striking in Jaspers' monumental *General Psychopathology*, a work which more than any other succeeds in convincing us of the profound arbitrariness in the systematic structure of what we take for everyday reality.

20 Lévi-Strauss (1968: 55-100). And making use of an older anthropological tradition, Onians (1951) and Benveniste (1973).

21 Koerner (1973: 166-74).

22 Bergson (1911: 24).

23 Ibid., p. 195. Thus: 'Perception is never a mere contact of the mind
with the object present' (ibid., p. 170). 'With the immediate and
present data of our senses we mingle a thousand details out of our
personal past experience .... In most cases these remains supplant
our actual perceptions, of which we then retain only a few hints, using
them merely as "signs" that recall to us former images' (ibid., p. 24).

24 Ferguson (1983: 51-7).

25 Freud (S.E., vol. 7: 174-5), asks, 'Why should our memory lag so far
behind the other activities of our minds?' – an anomaly that appears
even more striking when we realize that there is 'good reason to
believe that there is no period at which the capacity for receiving and
reproducing impressions is greater than precisely during the years of
childhood'.

26 The 'golden age' of childhood stretches approximately from the
publication of *Alice in Wonderland* (1857) to *The Wind in the Willows*
(1908).

27 The process described by Jaspers (1963: 348) as a 'release from the
obscure bondage of the undifferentiated'.

28 Compare an incisive passage from Nietzsche (1969: 57-8):

> Forgetting is no mere *vis inertiae* as the superficial imagine; it is
> rather an active and in the strictest sense a positive faculty of
> repression ... to make room for new things, above all for the
> nobler functions and functionaries, for regulation, for foresight,
> premeditation (for our organism is an oligarchy) – that is the
> purpose of active forgetfulness, which is like a doorkeeper, a
> preserver of psychic order, repose, and etiquette: so that it will be
> immediately obvious how there can be no happiness, no
> cheerfulness, no hope, no pride, no *present*, without forgetfulness.

29 Proust (1966, vol. 1: 58-62) is as devoted as Freud to these sensory
remains. Like Bergson, he sees the perceptual world as suffused with
memory, condensing into fleeting sensations 'the vast structure of
recollection' – recollections that become part of our 'body-image';
'Our legs and our arms are full of torpid memories', ibid., vol. 12; p.
2. See also Poulet (1977: 3-4), and Deleuze (1973: 39-50), who argue
that such recollections are, for Proust, the accidental intermediaries
through which the narrator gains access to a world of 'pure' signs.

30 Freud provides many examples. For a general statement of this
'fundamental rule', see Freud (S.E., vol. 22: 10-14), and for an early
example (ibid., vol. 2: 56). By following the paths of 'free association'
rather than 'rationally' reconstructing the patient's recollections,
Freud was also able to display his literary talent to great effect. We
need not accept at face value his claim to find it 'strange' that his case

histories 'should read like short stories' (ibid., vol. 2, p. 160).

31 Proust (1966, vol. 1: 61), also clearly distinguishes between these two types of memory: between a predominantly visual 'intellectual memory' which 'preserves nothing of the past itself', and involuntary recollection carried in chance encounters with the world of contemporary sensations, especially of taste and smell,

> but when from a long-distant past nothing subsists, after the people are dead, after the things are broken and scattered, still, alone, more fragile, but with more vitality, more unsubstantial, more persistent, more faithful, the smell and taste of things remain poised a long time, like souls, ready to remind us, waiting and hoping for their moment, amid the ruins of all the rest.

See also Sperber (1975: 115-19) and Deleuze (1972: 51-64).

32 Bergson (1911: 102).

33 Freud (S.E., vol. 2: 7).

34 Ibid., pp. 3-17.

35 Freud's major case studies provide the fundamental and unsurpassed demonstration of this contention.

36 Anzieu (1986: 234).

37 The argument was fully developed in the context of the 'wolf man' analysis (Freud, S.E., vol. 17: 29-47).

38 There seems little doubt that Freud himself exaggerated the opposition to psychoanalysis; see Ellenberger (1970: 783-4).

39 Analysis may therefore become 'interminable' (Freud, S.E., vol. 23: 216-54).

40 Ibid., vol. 12: 97-108, 157-74.

41 See note 27 above. Sulloway (1980), it seems to me, misrepresents Freud's work by viewing it exclusively within the context of nineteenth-century biology.

42 Proust traces it differently, to an intuition of the essence of things, to art (Deleuze 1972: 39-50).

43 Freud (S.E., vol. 14: 146-58). 'We recall the fact that the motive and purpose of repression was nothing else than the avoidance of unpleasure' (ibid., p. 153).

44 This is the major theme of Freud's *Three Essays on the Theory of Sexuality*. See particularly ibid., pp. 232-3.

45 'The symptoms *constitute* the sexual activity of the patient' (ibid., vol. 7: 163) (emphasis added).

46 Analysed first in relation to dreams (ibid., vol. 4: 279-304, 305-9). But quickly extended to jokes (vol. 8: 19-33), slips of the tongue and pen (vol. 6: 58-9, 61-2), and symptom formation (vol. 5: 671).

47 Saussure (1966: 123) defines syntagmatic and associative relations as follows:

> In the syntagm a term acquires its value only because it stands in opposition to everything that precedes or follows it, or to both ....

Outside discourse, on the other hand, words acquire relations of a different kind. Those that have something in common are associated in the memory, resulting in groups marked by diverse relations.

48 Rojek (1985: 18-33).
49 Anticipated, as with almost everything modern, by Nietzsche, himself an ideal madman. See Deleuze and Guattari (1977: 20-2).
50 At the turn of the century neurasthenia was more 'popular' than schizophrenia. Ellenberger (1970: 242-3), and more generally Sontag (1983).
51 Jaspers (1963: 309), in particular stresses the affinity between modern creativity and certain pathological states.

> Patients see into depths which do not so much belong to their illness as to themselves as individuals with their own historical truth . . . in psychotic reality we find an abundance of content representing fundamental problems of philosophy; nothingness, total destruction, formlessness, death. Here the extremest of human possibilities actually breaks through the ordinary boundaries of our sheltered, calm, ordered and smooth existence.

52 The point at which Marx and Kierkegaard, in different ways, parted company from Hegel; Löwith (1964: 137-61).
53 Jaspers (1963: 127): 'Why do you ask me – you know it already.'
54 Thus, Deleuze and Guattari (1977: 35): 'The schizophrenic deliberately seeks out the very limit of capitalism: he is its inherent tendency brought to fulfilment.'
55 A patient of Minkowski's believed, each night, that he would be killed and was never consoled by the discovery that he was still alive. For him quite literally, 'each day life began anew', (May, Angel, and Ellenberger 1958). The 'selfless', therefore, live spatially.
56 This is quite apart, of course, from their 'meaning' as 'signatures' (Freud, S.E., 16: 269), 'Neurotic symptoms have a sense, like parapraxes and dreams.'
57 A reflex which reason, as self-consciousness, tends to undermine. See below, Chapter 11.
58 Kierkegaard (1971, vol. 2: 161-75). The 'aesthetic' 'loses itself in the multifarious'.
59 Many of Freud's patients were markedly 'intellectual' in temperament (Sulloway 1980: 57). Any simple dualism here must be avoided, reason and sensuousness do not form a 'zero-sum' relation.
60 The abstraction and necessity of reason tends 'to bring everything to a standstill' (Kierkegaard 1971, vol. 2: 179). 'I can either do this or do that, but whichever of the two I do is equally mad, *ergo* I do nothing at all' (ibid., p. 174).
61 This is not to deny the frequent 'secondary gain from illness' (Freud S.E., vol. 18: 158-60).

62 Most notably by Freud (S.E., vol. 12), Canetti (1962: 434-62), Schatzman (1973).

63 'The logic of pure identity' (Gabel 1975: 155), following Minkowski and Bergson, regards as an unbalanced extremity of the normal tendency to view duration spatially. In psychosis there are no counteracting tendencies, so that thought becomes 'geometrism and morbid rationalism' (ibid., p. 79), quoting Minkowski. Jaspers (1963: 63) puts it succinctly: 'They have to live forever because time no longer exists.' It's hardly surprising that one patient should report: 'I felt lost, abandoned to the infinities of space, which in spite of my insignificance somehow threatened me' (ibid., p. 81).

64 Schreber (1955: 74).

65 As Canetti (1962: 440-1) points out Schreber's mastery of space makes him not simply the last human survivor, but the centre of the universe. This megalomania is, at the same time, the experience of utter helplessness.

66 And failing, retreat into 'the soft muffled gloom of the interior' (Kretschmer 1936: 161).

67 'I had the feeling that I was being pulled up to heaven and saw the whole earth under me, a picture of incomparable splendour and beauty stretched out under the blue dome' (Jaspers 1963: 295). And again: 'my body bears fruit . . . it is a world-body' (ibid., p. 296).

68 Minkowski, in May, Angel, and Ellenberger (1958: 132-3).

69 The neurotic and psychotic are equally 'abstracted', logical transformations of each other rather than distinct 'conditions'.

70 Gabel (1975). The neurotic holds to the truth of his own humanness, the psychotic to the truth of an objective social order.

71 See below, Chapter 11.

72 Durkheim and Mauss (1963: 5) assume an originally undifferentiated condition: 'It would be impossible to exaggerate, in fact, the state of indistinction from which the human mind developed.' Thus (ibid.): 'Animals, people, and inanimate objects were originally conceived as standing in relations of the most perfect identity to each other.'

73 Religion is *defined* through the sacred/profane distinction. And 'If religion has given birth to all that is essential in society, it is because the idea of society is the soul of religion' (Durkheim 1964: 419).

74 Durkheim and Mauss (1963: 11) explicitly reject the notion that the relation between the two can be reduced to a simple causal mechanism; to do so would be to reduce the act of classification to some ethereal form of pure 'thought': 'Society was not simply a model which classificatory thought followed.'

75 Bergson (1935: 18) remarks that in society 'only one thing is natural, the necessity of a rule'.

76 Durkheim (1964: 225) 'Religion ceases to be an inexplicable hallucination and takes a foothold in reality . . . this power exists, it is society.'

77 Durkheim (1964: 38). Robertson Smith (1907: 140), in his Burnett lectures of 1888, anticipated Durkheim to some extent. 'The

distinction between what is *holy* and what is *common* is one of the most important things in ancient religion.' He did not attempt, however, to formulate a general theory of religion based upon such a conventional distinction.

78 Durkheim (1964: 229).

79 Durkheim's 'functionalism' tends to obscure this point. Failing to observe his own dictum that society must be granted a reality *sui generis*, he persists in attempting to 'deduce' its cause from itself. For example, ibid., pp. 370-88.

80 For example, in Hertz (1960) and Needham (1973).

81 The primitive is not characterized by lack of distinction but by lack of differentiation among domains of distinction. Every 'aspect' of social life is 'embedded' in every other.

82 Mauss (1970), Hubert and Mauss (1964). Exchange forms a complex system. They exchange 'courtesies, entertainments, rituals, military assistance, women, children, dances and feasts . . . a system of *total prestation*' (Mauss, 1970: 3).

83 Ibid., p. 12: 'In perpetual interchange of what we may call spiritual matter, comprising men and things, these elements pass and repass between class and individual, ranks, sexes and generations.'

84 Mauss (1972: 76-9). 'Things affect each other only because they belong to the same class or are opposed in the same genus' (ibid., p. 78). And 'Magic becomes possible only because we are dealing with classified species' (ibid., pp. 78-9).

85 Its analogous form in bourgeois society is *money*, rather than science or religion. It constitutes, in other words, the *medium* of exchange and a mechanism for establishing equivalence among things exchanged.

86 Augé (1982: 91): 'one cannot find social activities that are strictly speaking outside the sacred'.

87 Though Durkheimian functionalism pervades his earlier work on kinship (Lévi-Strauss 1969: 478-81).

88 Brilliantly, for example, in 'The Story of Asdiwal' (Lévi-Strauss 1977: 146–97).

89 Lévi-Strauss (1970), (1966: 35–74). In spite of Lévi-Strauss's elegant approach, 'symbolic' interpretations of myth and ritual, such as Bachofen's (1967) and Jung's have remained popular. See, for example, Firth (1973). From either point of view it should also be borne in mind that the 'primitive' also reflects self-consciously upon his existence; see particularly Griaule (1965).

90 Lévi-Strauss (1966: 75–108), (1968: 67–80).

91 Particularly in relation to Propp (1968), who is 'the victim of a subjective illusion' (Lévi-Strauss 1973: 131).

92 This is particularly the case with myth, where Lévi-Strauss offers methodological advice similar to Freud's in relation to dream interpretation (Lévi-Strauss 1968: 210–15). See Wilden (1972: 31–62). And, like dream interpretation, 'There is no real end to mythological analysis, no hidden unity to be grasped once the breaking-down process has been completed . . . . Myths, like rites, are "in-terminable"'

281

(Lévi-Strauss 1970: 5–6); cf. Freud (S.E., vol. 23: 216–54).
93 Lévi-Strauss (1966: 1–34).
94 '... but with models which are built up after it' (Lévi-Strauss 1968: 279).
95 Ibid., p. 229.
96 Lévi-Strauss (1969: 12): 'I therefore claim to show, not how men think in myths, but how myths operate in men's minds without their being aware of the fact.' Lévi-Strauss's debt to Rousseau is even greater than his debt to Durkheim, it is through mythology that man's universal nature finally makes itself felt, though not in a 'rational' or 'self-conscious' form. In another context see Duerr (1985), particularly pp. 64-6.
97 Lévi-Strauss (1966: 13).
98 Lévi-Strauss (1981: 639). The analysis of mythology makes it possible 'to discover certain operational modes of the human mind'.

THE COSMIC BODY

1 Huizinga (1971: 25).
2 See particularly in this context Turner (1969: 80–118), Douglas (1970: 41–53).
3 Onians (1951: 13–17), and for a psychoanalytic viewpoint Melanie Klein (1932: 35–57), and Winnicott (1971: 1–30).
4 Bergson (1911: 171), writes: 'my present exists in the consciousness that I have of my body'. Schilder (1935), following Bergson, argues that our conception of space is intimately related to our body *image*, to the 'immediate experience of the human body' (ibid., p. 11) – an experience which is 'more than a perception' (ibid., p. 11) and has to be seen even in the simplest terms as a 'postural model ... in perpetual inner self-creation and self-destruction' (ibid., p. 16). More generally we are 'enshrouded by various body images' (ibid., p. 237). The body is a privileged object in being directly felt; it does not, however, constitute the foundation of a system of 'natural' signs. See also Douglas (1973: 93–112), Needham (1972: 138–51), and more generally Turner (1984).
5 For example, Crocker (1985: 21–30), Reichel-Dolmatoff (1971), Christine Hugh-Jones (1979: 235–74), Stephen Hugh-Jones (1979: 151–7).
6 In this context see particularly Schilder's neglected work. Schilder (1935).
7 Huizinga (1971: 25)
8 Examples drawn, of course, from Proust.
9 Most notably in *The Interpretation of Dreams*.

CHAPTER FOUR

1 As, for example, in Lea (1888, vol. 1: 5–6, 212, 218, 401); and, generally, Thorndike (1923).
2 Lea (1888, vol. 1).
3 Bloch (1965, vol. 1: xvii); Brunner, in Cheyette (1968: 32–61).
4 Ganshof (1952: xv).
5 Hintze, in Cheyette (1968: 24).
6 Mitteis (1975: 11), Poggi (1978: 17–21), Ganshof (1952: 5–6). In early German kingship, 'the king belongs to the people without the people "belonging" to the king' (Mitteis 1975: 31).
7 Ganshof (1952: 16), Myers (1982: 150).
8 Ganshof (1952: 96) defines the fief as 'a tenement granted freely by a lord to his vassal in order to procure the latter the maintenance which was his due and to provide him with the means of furnishing his land with the services required by his contract of vassalage'. The fief referred originally to movable property, particularly cattle (Bloch 1965, vol. 1: 165). Vassals could also be maintained by making them members of the household, a process which became progressively less important. Frequently vassalage would be instigated 'from below', by a free gift of land to the lord who subsequently granted it back to his vassal as a benefice. This was the method through which the Church brought itself under the protection of powerful patrons (ibid., pp. 166–75).
9 Ibid., p. 228.
10 Ibid., p. 232. From another point of view, of course, he was 'absorbed' into the personality of his superior.
11 Quoted in Bloch (1965, vol. 1: 233).
12 Ibid., p. 233.
13 Fourquin (1976).
14 Myers (1982: 156).
15 Bloch (1965, vol. 1: 148). This was of course an ideal. Painter (1961: 7), argues that 'as a political system pure feudalism was little removed from anarchy', its ruling warrior class was 'turbulent, savage, and utterly undisciplined'.
16 Bloch (1965: vol. 1: 147).
17 Fourquin (1976: 35–8). For examples, see Coulton (1927, vol. 2: 35–6), Knowles and Obolensky (1972: 32–56). The priest, indeed, 'could be regarded as a feudal dependent, and his office as a gift or a reward (*beneficium*) ibid., p. 53.
18 Bloch (1965, vol. 1: 251).
19 Ibid., p. 272.
20 Weber (1978, vol. 1: 257) remarks that 'the ideal extent of feudal authority has never been effectively carried out in practice or remained effective on a permanent basis'. Even Bloch (1965, vol. 1: 173) admits that it 'was never a perfect system'.
21 Mitteis (1975: 11–13). Myers (1982: 160):

In practice, a king in the feudal period either had to lead his armies himself, or personally see to it that they were well commanded. It is no accident that many of the most successful feudal monarchs, regardless of the century or country, were experts in contemporary military tactics and strategy.

22 Petit-Dutaillis (1983: 17) Myers (1982: 156):

The royal demesne was the aggregate of territories in which the king exercised on his own behalf the privilege of baron, or independent lord, and of justice; this last was the most important for it offered him an opportunity for constant intervention, and endowed him with a reality of power.

23 Weber (1978, vol. 1: 246–9; vol. 2: 1121–3, 1141–2) Ullmann (1961: 117–27), Mitteis (1975: 71):

It was a medieval commonplace that each must recognise obligation to some higher authority; every form of human society was viewed as overshadowed by transcendental mysteries.
Therefore the function of the state was also conceived in religious terms; its ruler was anointed by the Church.

24 Kantorowicz (1957: 44): 'Not only the bishop, but also the king appeared as a *persona mixta*.'
25 Ibid., p. 44.
26 Ibid., p. 48, quoting an anonymous Norman tract. See also Ullmann (1961: 121): 'Power came "from above" and was transmitted downwards through an act of concession'.
27 Ullmann (1961) argues that the original theocratic element was gradually overtaken by legalism.
28 Knowles and Obolensky (1972), Southern (1970: 38, 102–27). The Gregorian reforms, Southern claims, created a 'machinery of government as effective as any government could be before the late nineteenth century' (ibid., p. 121).
29 Adalberon (d. 1030), *Poème au roi Robert*, quoted in Fourquin (1976: 77).
30 Duby (1977: 90–1). The major distinctions were: (a) Monks, clergy, laity; three steps towards perfection on the basis of sexual purity – *virgines, continenti, conjugati*; (b) free and unfree; (c) *oratores* – aid to sovereign through prayer; *bellatores* – aid through arms; (d) the 'powerful' and the 'poor'. Duby (1980) makes clear the intimate connection between the revival of the 'three orders' view of society and the development of a new ideology of knighthood.
31 These relations were always personal. Southern (1970: 110) quotes the Archbishop of Vienne in 1119, 'Wherever I went there was almost no one of note who was not my nephew, or cousin, or neighbour, or vassal.'
32 Coulton (1923: 67–77), Boase (1972: 19–30). Emmerson (1981)

explores the eschatological imagery. But compare Ariès (1983: 97–9), who suggests the image of the Second Coming was as significant as that of the Last Judgement.

33 Time spent in prayer and penance was time lost to military training and service. The Church sometimes imposed 'unrealistic' penances (such as the forty days for each person killed in the battle of Hastings), to encourage the knight to discharge them through grants of land and other endowments.

34 Coulton (1923, vol. 1: 93), Ariès (1983: 183).

35 Southern (1970: 29–31), Brooke and Brooke (1984: 15–37).

36 Le Goff (1980), Landes (1983: 60–6).

37 Hilton (1975: 47), Murray (1978: 50–60).

38 For an authoritative summary see Wallace's 'Appendix' to Aquinas (1964–81, vol. 10). This was a hierarchy of virtue as well as position. Even as unconventional a thinker as Roger Bacon did not hesitate to argue from one to the other: 'But it is not natural that a nobler body is contained by a less noble body, and the orb of Mars is less noble than the Sun's orb; it cannot therefore contain the Sun's orb' (quoted in Duhem 1985: 147).

39 Bede's spheres were earth, air, ether, Olympian fire, the stellar sphere, heaven of the angels, heaven of the Trinity (Gurevich 1985: 71).

40 Complications arose particularly in relation to the movement of the planets. Their perfection demanded that they circle the cosmic centre in a regular manner, yet, from the centre their orbits appeared eccentric. Borrowing heavily from Arabic sources, the 'feudalization' of the cosmos consisted in introducing additional intermediary 'spheres' appropriately 'geared' to achieve the required result (Taton 1963), Gurevich (1985: 71–2), Nebelsick (1985: 121–6). The 'standard' later medieval text was Sacrobosco (see Thorndike 1949), Lovejoy (1960: 61–4), Haren (1985: 72–81), Klibansky (1982), Marenbon (1983: 43–80).

41 Le Goff (1964: 405). Nature was a 'great reservoir of symbols'. Not only was the material world saturated with symbols, it was also materially 'dense'. God's beneficence being unlimited, all that could exist did exist (Lovejoy 1960: 52, 68–77). There were no 'gaps' in creation; 'nature does not make [animal] kinds separate without making something intermediate between them' (Albertus Magnus, quoted in ibid., p. 79).

42 Duhem (1985: 139–268), Koyré, (1957: 1–38).

43 Gilson (1936: 129).

44 Lovejoy (1960: 68–77). 'There were no empty places, no gaps; nothing was superfluous or unnecessary' (quoted in Gurevich 1985: 69).

45 Higher beings, by implication, could be trusted always to choose 'correctly'. On the possibility of man being thus created, see Mackie (1982: 162–6).

46 Nebelsick (1985: 92).

47 Gurevich (1985: 73).

48 Particularly evident in the knightly quest. And further justifying the seclusion of the monastery as a *place* set apart.

49 Mâle (1958: 390).

50 Gurevich (1985: 56).

51 Quoted in ibid., p. 61.

52 Ibid., p. 59. Gurevich, however, fails to distinguish between identification and analogy: 'In the universe man saw the same forces at work as he was aware of in himself. No clear boundaries separated man from the world' (ibid., p. 56). It could be countered, as in the case of Lévy-Bruhl, that it is only because a clear distinction between the two has been drawn that a metaphorical relation can be established between them.

53 Mâle (1958: 133): 'God who sees all things under the aspect of eternity willed that the Old and New Testaments should form a complete and harmonious whole; the Old is but the adumbration of the New.'

54 For a detailed examination of a *Hexameral* text see Steneck (1976) and Stock (1972).

55 Le Goff (1980: 23–35). Special difficulties over usury which could be viewed as 'selling' time which, ultimately, belonged to God (Gurevich 1985: 141).

56 Ibid, p. 125.

57 Le Goff (1988: 332): 'Lapidaires, lists of flowers, and bestiaries, in which these symbols were catalogued and explained, occupied an important place in the ideal library of the middle ages.' Though Mâle (1958: 48–50) criticizes the mania for symbolic interpretation and prefers to see in the overabundance of decorative detail the freedom of the sculptor, 'looking at the world with the wondering eye of the child' (ibid, p. 51). Yet even Villard de Honnecourt's sketches, widely admired today for their realistic detail, are conceived as part of a purely symbolic order (see Lassus 1968, particularly pp. 63-4, Bowie 1959).

58 Knowles and Obolensky (1972: 91): 'the hierarchical arrangement of these decorations combined to produce a symbolic image of the cosmos'.

59 Mâle (1958: 139).

60 Ibid., p. 29.

61 Le Goff (1980: 238).

62 Distinctions more commonly expressed in the typology outlined above.

63 Southern (1970: 92): 'Documents were therefore drawn up in which the theories of the present were represented as the facts of the past.' Also Gurevich (1985: 177–8).

64 Ibid., p. 175.

65 Le Goff (1980: 230–68). See also Duby (1977), (1980).

66 Le Goff (1980: 241).

67 Ibid., pp. 245–6).

68 Ibid., pp. 282–6).

69 Ibid., pp. 242–3. Perella (1969: 12–13, 15–20).

70 Duby (1977: 175–80).

71 This view was expressed with particular force by St Anselm, but was common to the later Scholastic tradition. See particularly Evans (1978: 13–38), Gilson (1950: 229–47).

72 Anselm (1974, vol. 1: 5). The later orthodoxy of Aquinas did not subordinate human reason quite so rigorously to an act of faith. See Gilson (1924: 17–33).

73 Though, as Mâle points out, 'all medieval art was didactic', and the distinction between intellect and sentiment was far from clear.

74 Anselm (1974, vol. 1: 5).

75 Ibid., p. 6.

76 Ibid.

77 Gilson (1936: 55): 'He is perfect because He is' was the foundation of Christian philosophy, as opposed to the Platonic formula, 'He is because He is perfect'. The perfection of the Christian God 'is that perfection which is proper to being as being'.

78 Anselm (1974, vol. 1: 7).

79 Ibid., p. 9.

80 Ibid., p. 17.

81 The ontological argument has, after a long period of neglect, again found favour with some philosophers. See Hartshorne (1962), Plantinga (1974), Mackie (1982). Anselm's contemporaries, it seems, were more impressed by his systematic derivation of God's attributes than by the ontological 'proof' itself. See Eadmer (1962: 29).

82 Anselm (1974, vol. 1: 93).

83 Gilson (1955: 68): 'all save God might possibly not exist', and this 'radical contingency stamps the world with a character of metaphysical novelty of immense significance'. For the challenge of Judaism, see Evans (1980: 34–41).

CHAPTER FIVE

1 Lovejoy (1960: 69).

2 Ibid., p. 83: 'The one God was the goal of the "way up" of that ascending process.' See also Leclerq (1979: 51), Gilson (1924: 272).

3 Aquinas (1967, vol. 10: 119).

4 Ibid., p. 105.

5 Ibid., p. 109. Note the implied criticism here of over zealous asceticism, a critical issue throughout the thirteenth century (Lambert 1961).

6 Gilson (1924: 46–75) argues that the five 'ways to God' are really five versions of a single fundamental argument, and notes of this proof from the First Mover that it 'displays its full significance only on the assumption of an hierarchically ordered universe' (ibid., p. 58).

7 Aquinas (1911, vol. 1: 24-5), (1955, vol. 1: 86–96).

8 Gilson (1924: 272). Gilson again brings out the feudal imagery central to Aquinas's argument:

Precisely because every operation is the realisation of an essence and because every essence is a certain quantity of being and perfection, the universe presents itself to us as a society of superior and inferior beings in which the very definition of each essence places each directly into its proper rank within the degrees of this hierarchy.

9 On Llull see particularly *The Book of the Gentile*, in Llull (1985, vol. 1), Johnston (1987: 158–9), and Yates (1982: 9–77). On Silvestris, see Dronke (1978), Stock (1972), Curtius (1953: 110–11).

10 Auerbach (1961: 20). He continues: 'It incorporated the wild, crude power of the heroic legends, turned the feudal system into a symbolic hierarchy, and transformed God into the supreme feudal overlord.' And as Curtius (1953: 52), points out, 'The Middle Ages subjected profane authors to allegorical interpretation exactly as it did the Bible.'

11 The *trivium* (grammar, rhetoric, dialectic) was preparatory to the *quadrivium* (arithmetic, geometry, music, astronomy), together forming the seven liberal arts (Steenberghen 1966: 59).

12 Steenberghen (1966), Gilson (1955: 277–94).

13 Knowles (1949), Lawrence (1984), Leclerq (1978: 65–86), Cowdrey (1970), Knowles and Obolensky (1972).

14 Benedict, ed. by McCann (1960).

15 Hunt (1967: 30).

16 Benedict (1960: 163). Though well regulated, the spirit of Benedictine monasticism was far from severe: 'we hope to ordain nothing that is harsh or burdensome' (ibid., p. 13). See particularly Butler (1919: 35–45).

17 Hunt (1967: 31).

18 Its only serious rival was St Denis (see Klibansky 1979). Its Abbot, Suger, like the Cluniacs, conceived decorative magnificence to be spiritually uplifting.

Thus, when – out of my delight in the beauty of the house of God – the loveliness of the many-coloured gems has called me away from external cares, and worthy meditation has induced me to reflect . . . then it seems to me I see myself dwelling, as it were, in some strange region of the universe which neither exists entirely in the slime of the earth nor entirely in the purity of Heaven; and that, by the grace of God, I can be transported from this inferior to that higher world in an anagogical manner (ibid., pp. 63–5). It was a standard defence of opulence; see also Idung of Prüfening (1977: 42).

19 Hunt (1967: 111) remarks that 'such building was, in a sense, a concrete expression of a certain theology'.

20 Southern (1966) notes additionally the importance of Lanfranc in making Bec the more attractive to Anselm.

21 Though manual labour had already been confined, by tradition, to domestic tasks (Knowles 1949: 466).

22 Southern (1970: 220–70). Knowles (1949: 191–226). Halinger, in Hunt (1971: 45), in fact detects at Cluny 'the first traces of individualistic affective devotion', a tendency that became increasingly evident in all the reform movements. Thus Knowles and Obolensky (1972: 194): 'The monk, from being primarily an intercessor for society and a discharger of liturgical services, had become an individual seeking the fullness of evangelical perfection'.

23 Cluny pioneered adult monastic recruitment, see Leclerq (1979: 9–10), Cowdrey (1970: 67–75), Hunt (1967: 68–88). Peter the Venerable complained of indiscriminate recruitment: 'Yokels, children, old men and idiots have been taken in such numbers that they are now near to forming a majority' (quoted in Hunt 1967: 87).

24 Butler (1919: 195). Hunt (1967: 43) notes that 'the gradual accumulation of lands drew Cluny into a network of economic relationships which modified that overall independence of the monastery in secular matters which the founders had hoped to secure'.

25 Knowles (1949: 216). The fact that we now find their abandoned English sites 'picturesque' is a measure of changing aesthetic attitudes towards 'nature'.

26 Cowdrey (1970).

27 Knowles (1969: 93).

28 Donkin (1978: 21–9), Knowles (1969: 24), Duby (1977: 4).

29 William of St Thierry (1971: 107-8): 'For although reason sends us to you, O God, it cannot of itself attain to you.' See generally, Vandenbroucke (1972: 25–50), Leclerq (1978: 65–86).

30 William of St Thierry (1971: 106). 'For the soul's sense is love', and when love 'reaches out' it is 'transmuted into the object loved'. It does not become the 'same nature' as the object, 'but by its "affections" it is conformed to what it loves'. Cf. Déchanet (1972: 52).

31 William of St Thierry (1971: 37). And as Bernard of Clairvaux (1940: 71) remarks: 'Just as faith leads to full knowledge, so desire leads to perfect love.'

32 William of St Thierry (1971: 37). Ascent requires progress in the 'animal', the 'rational', and the 'spiritual' aspects of being. Perfection of the first is reached 'when the habitual exercise of virtue becomes a pleasure' (William of St Thierry 1971a: 27).

33 It draws on the traditions of both *eros* (seeking perfection), and *agape* (humility in the face of God's transcending love) (Nygren 1953: 609–58).

34 William of St Thierry (1971: 172-3): 'The steps that I set up for myself are three; first a great will is needed, then an enlightened will, and thirdly a will upon which love has laid its hand.'

35 Ibid., p. 56. 'My ardent longing wearies and grows shattered in the quest and from your heights I fall back to my depths' (ibid., p. 145).

36 Ibid., p. 56.

37 Ibid., p. 104.

38 Modelled on St Benedict's *Rule*. Bernard (1940) distinguishes twelve steps in humility, each defined by its opposition to a particular temptation: 'Curiosity ("the beginning of all sin"), Frivolity, Foolish Mirth, Boastfulness, Singularity ("zealous for himself, indifferent to the community"), Conceit, Audacity, Excusing Sin, Hypocritical Confession ("they condone their guilt by confessing it, they conceal it by revealing it"), Defiance, Freedom to Sin, Habitual Sinning'.

39 Ibid., p. 149. William's approach is more direct. There are two basic temptations, two 'weaknesses congealed in nature . . . the yearning of the Flesh and blasphemy of the Spirit' (William of St Thierry 1979: 36).

40 See, for example, Gilbert of Hoyland (1978, vol. 1: 27, 135), Aelred of Rievaulx (1981: 82–3).

41 See generally Leclerq (1978), Hallier (1969), Bell (1984).

42 In this sense it was also a form of serfdom. 'It is for others to serve God, it is for you to cling to him' (William of St Thierry 1971a: 14).

43 Obedience is man's principal labour (Bernard of Clairvaux 1971, vol. 2: xiv). Those seeking God 'must turn their eyes to the earth rather than up to heaven' (ibid., p. 17). Bernard is again invoking Benedict's *Rule* (1960: 7): 'That by the labour of obedience thou mayest return to him from whom thou hast strayed by disobedience'.

44 In this connection it is worth noting the significance of the kiss in Cistercian spirituality. Their favourite text was *The Song of Songs*, which they expounded as an ontology of love in the sense of *agape*. See, for example, Gilbert of Hoyland (1978), (1979), Bernard of Clairvaux (1971), (1976), (1979), (1980), William of St Thierry (1970). Dronke (1984: 209–34) notes the secular significance of the text.

45 The friars should rejoice 'when they find themselves among mean and dispossessed persons, amongst the poor and weak and infirm and the lepers and those that beg in the street' (quoted by Lambert 1961: 39). It is not so much poverty as worthlessness which is embraced, hence the Franciscan 'love of nature' is an identification with a degraded rather than an innocent category. See Little (1978).

46 Duby (1980: 296–306), (1977: 175–80).

47 Chrétien de Troyes (1982: 5).

48 Chrétien de Troyes (1966: 123).

49 Ibid., p. 75.

50 Ibid.

51 William is to be taken literally: 'The more perfect charity becomes in us, the more necessary it becomes that the flesh be lessened until it (charity) be perfected there, where there is no yearning for anything but God' (William of St Thierry 1979: 11).

52 Chrétien de Troyes (1966: 123).

53 Llull, in Herlihy (1970: 310).

54 Ibid.

55 Ibid., p. 312.

56 Ibid.

57 Ibid.
58 Ibid.
59 Ibid., p. 313; Duby (1980: 299).
60 Duby (1977: 80, 181).
61 Bernard of Clairvaux, in Herlihy (1970: 290).
62 Duby (1980: 306): 'The forest . . . was the dominion of the wild, the untamed, the weird dangers that had to be faced alone'.
63 Llull, in Herlihy (1970: 329–30).
64 Auerbach (1968: 133), Artin (1974).
65 Auerbach (1968: 134–5).
66 The distinction is drawn from Weber (1964: 325–6), (1965: 166–83).
67 Auerbach (1968: 135), Artin (1974: 28).
68 Auerbach (1968: 140): 'Only two themes are considered worthy of a knight: feats of arms, and love.'
69 Andreas Capellanus (1959), Herlihy (1970: 44).
70 Duby (1980: 302).
71 Dronke (1968: 7). The ennobling power of courtly love 'lies in the cost to oneself, its beauty and value lie in the lover's giving all he has, in his enduring pain and sacrifice for love's sake, in looking constantly to a more-than-human love . . . without distraction, without calculation of success, even if necessary without hope of gaining his desire on earth'. See also Dronke (1986: 13).
72 Bernard of Clairvaux, in Herlihy (1970: 292).
73 Llull, in Herlihy (1970: 318).
74 Schutz (1976, vol. 2: 135–58).
75 Sociologists have been slow to exploit the works of literary critics. The works of Bakhtin, Curtius, and Auerbach are rich in valuable insights into the cultural categories of feudalism.
76 Bakhtin (1968).
77 Ibid., p. 6.
78 Ibid., p. 7.
79 Ibid.
80 Ibid.
81 For a detailed description, see Ladurie (1981: 190–1).
82 Bakhtin (1968: 10).
83 Millenarian and Utopian social movements in the late Middle Ages existed independently of the carnival tradition. See Cohn (1970), Lambert (1979).
84 Bakhtin (1968: 19).
85 Ibid., p. 67. The 'grotesque body' is not regulated, as in the 'primary process', by a constancy principle; it is characterized above all by excess of appetite.
86 Ibid., p. 67.
87 Ibid., p. 71.
88 Davis (1975: 97–100) for a discussion of these views in the context of a somewhat later period.
89 Bakhtin (1968: 88).
90 Ibid., p. 83.

91 Ibid., p. 90.
92 Ibid., p. 318.
93 Freud (S.E., vol. 8: 168–70).
94 Ladurie (1981: 289).
95 Ibid., p. 193.
96 Bernard of Clairvaux (1940), p. 225, remarks that 'Only the highest and lowest fly without hindrance . . . . In the former *perfect love*, in the latter consummate wickedness *casteth out fear*.'
97 Huizinga (1965: 9–29). Even as an ideal knighthood was only gradually 'spiritualized', a development notable in the works of Chrétien de Troyes; see Bednar (1974: 149).
98 Auerbach (1984), Dronke (1974: 38).

CHAPTER SIX

1 Kibre (1984). The more adventurous proposed philosophy as a similarly advanced and synthetic study; see, for example, Hugh of St Victor (1961: 51): 'Philosophy is the discipline which investigates comprehensively the ideas of all things, human and divine', though, even for radicals, the aim of the arts was 'to restore within us the divine likeness' (ibid., p. 61). See also Rashdall (1962, vol. 1: 33–6).
2 Masi (1983: 13–14). There thus developed within feudal societies a tradition that held, on the basis of analogy, that 'mathematics was the link between God and the world' (ibid., p. 32).
3 The Cathedral of Chartres has been aptly described as 'frozen music'; quoted in Masi (1983: 33).
4 The commentaries of Albertus Magnus on Aristotle's natural history are notable exceptions; see Weisheipl (1980), Wilms (1933: 20–43).
5 Haskins (1924), (1927) remains the fundamental statement of such a view. For a recent defence, see Stiefel (1985).
6 Another twelfth-century propagandist of the arts, John of Salisbury (1962: 9), makes clear the context of such a renaissance. In spite of the 'burden of his earthly nature and the sluggishness of his physical body, man may still rise to higher things. Borne aloft, so to speak, on wings of reason and speech, he is thus ennobled by this felicitous shortcut, to outstrip all other beings and to attain the crown of true happiness.' See also Southern (1970: 42).
7 Southern (1986: 23).
8 Curtius (1953) dates its composition to between 1145 and 1153. An excellent English translation is available, Wetherbee (1973), and a comprehensive analytical and historical study by Stock (1972).
9 Wetherbee (1973: 2). Curtius (1953: 111) describes it as 'a link in the "golden chain" which connects late Paganism with the Renaissance of the twelfth century'.
10 Wetherbee (1973: 67).
11 Ibid., p. 69. Noys 'carries the exemplary image of all that is to be generated' (Dronke 1978: 29).

12 Wetherbee (1973: 67).

13 Ibid.

14 Ibid. p. 68.

15 Ibid., p. 67. 'Her capaciousness, confined by no boundaries or limitations . . . . And since it was liable to assume any shape, it was not specifically stamped with the seal of a single form of its own' (ibid., p. 71). The order of nature is latent with Silva, but is brought into existence only by a divine act of unconditional benevolence. Alan of Lille (1980: 144–5), deeply influenced by the *Cosmographia*, similarly tempers his 'paganism' or 'humanism' with a strictly orthodox view of Creation:

> He constructed the marvellous form of the kingdom of the world by the command of His deciding will alone, not by the operative aid of any external factor, not by the help of any pre-existing matter, not at the insistent urgings of any need .

16 Wetherbee (1973: 68).

17 Ibid., p. 70.

18 Ibid., p. 71.

19 Ibid., p. 72.

20 Ibid., p. 75.

21 Ibid., p. 89.

22 Ibid., p. 76. See Stock (1972: 28–9) on the astrological implications.

23 Wetherbee (1973: 69). As Dronke (1978: 30), remarks: 'material existence is intrinsically imperfect because it is material'.

24 Bachelard (1964).

25 Wetherbee (1973: 94).

26 Ibid., p. 96.

27 Ibid. p. 97.

28 Alan of Lille (1980: 118–19).

> For the movement of reason, springing from a heavenly origin, escaping the destruction of things on earth, in its process of thought turns back again to the heavens. On the other hand, the movements of sensuality, going planet-like in opposition to the fixed sky of reason, with twisted course slip to the destruction of earthly things. (Ibid., p. 119)

Though we cannot know the fullness of being, reason itself participates in its mystery.

> Our chain of reason extends too far when it dares to lift our discourse to the ineffable secrets of the godhead, although our mind grows faint in sighs for a knowledge of this matter. The image of this perfectly organised state shines forth in man. Wisdom, that gives orders, rests in the citadel of his head and the other powers, like demi-goddesses, obey her as if she were a goddess. (Ibid., p. 121)

29 Wetherbee (1973: 96).

30 Thorndike (1923, vol. 2: 104–5).

31 Thorndike (1923, vol. 1: 640). On the 'radical Aristotelianism' of the Arab commentators, see Booth (1983: 95–152), Haskins (1924: 3–19, 113–29).

32 Grosseteste (Riedl 1942: 10): 'The first corporeal form which some call corporeity is in my opinion light.' For a brief overview of Grosseteste, see Callus (1955: 1–69).

33 Riedl (1942: 10).

34 Ibid., p. 11. This is not only a physical process. Light, as 'radiance', is also the word of God: 'The Word is eternal, spiritual light, physical creation is corporeal light or form' McEvoy (1982: 59). See also Crombie (1953: 109).

35 Riedl (1942: 13).

36 Ibid. p. 15. The distribution of matter is governed by this 'taxiarchic law of subordination' (McEvoy 1982: 98), which is 'A final westernisation of the Neoplatonic hierarchy' (ibid., p. 106).

37 McEvoy (1982: 182).

38 Quoted in Southern (1986: 216).

39 His apparently 'scientific' analysis of colour, for example, contrasts with his insistence that comets are sublunary phenomena because 'change' in the heavens is inconceivable (Southern 1986: 148–9). For the view of Grosseteste as a pioneer of science, see Sharp (1930) and particularly Crombie (1953), and in Callus (1955: 98–120).

40 Grosseteste would not have gone so far as, for example, Moses Maimonides in claiming that 'everything is licit in which any natural cause appears' (quoted in Thorndike 1924, vol. 2: 209). One of the most important of the early translators of Arabic scientific texts, Michael Scot, was careful to distinguish between *mathesis* (knowledge) and *matesis* (divination), and between *mathematics* (legitimate method) and *matematica* (forbidden to Christians). Genuine knowledge could not be exclusively naturalistic. See Thorndike (1924, vol. 2: 316).

41 Quoted in Thorndike (1924, vol. 2: 165).

42 Klibansky, Panofsky, and Saxl (1964: 67–123, 127–33).

43 Talbot (1967: 127), Kealey (1981: 2–5).

44 Quoted in Thorndike (1924, vol. 2: 584).

45 Surpassed only by his pupil, Thomas Aquinas.

46 Quoted in Thorndike (1924, vol. 2: 588). Booth (1983: 166) points out the centrality of the metaphysics of light in the work of Albertus. He presents

> a total account of light as forming things, and meeting the quality which it has already put in them, giving them a continuing radiance, and thereby activating the medium and the sight; light standing for intelligibility, and intelligibility being brought about by an immaterial passage to and from the object whose origin is also in the same intelligibility. (Ibid., p. 169)

47 Wyckoff (1967).
48 Ibid., p. 178.
49 Ibid., pp. 63, 192: 'For (the stars) above, by their substance and light, position, motion, and configuration pour down into things below all the noble powers they possess.' See also Talbot (1967: 127).
50 Wyckoff (1967: 192); Kibre (1984: 191).
51 See, for example, Albertus Magnus – Best and Brightman (1973), Weisheipl (1980), Arber (1953), White (1954), Bianciotto (1980).
52 Best and Brightman (1973: 53–4). And a typical example from White (1954: 132–3):

> The Pelican is excessively devoted to its children. But when these have been born and begin to grow up, they flap their parents in the face with their wings, and the parents, striking back, kill them. Three days afterward the mother pierces her breast, opens her side, and lays herself across the young, pouring out her blood over the dead bodies. This brings them to life again. In the same way, Our Lord Jesus Christ, who is the originator and maker of all created things, begets us and calls us into being out of nothing. We, on the contrary, strike him in the face.

53 Grant (1981: 5–8), Koyré (1978: 4–8), Duhem (1985: 139–43).
54 Clagett (1959: 65). 'Heaviness' is defined as 'the force with which a heavy body moves towards the centre of the world ... the force is inherent in the body and is not extrinsic. It acts so long as the heavy thing is away from the centre of the world.' Albertus Magnus, giving a non-kinematic example, observes that things 'kept a long time away from the place where they were produced, are destroyed, and no longer rightly deserve their specific name' (Wyckoff 1967: 65).
55 Clagett (1968: 17–18, 45, 237, 299).
56 Hence the traditional argument against the possibility of a real vacuum; resistance reduced to nothing, a body would move from one place to another instantaneously, which is absurd (Koyré 1978: 7–8).
57 Ibid., p. 9.
58 Ibid., p. 7; Butterfield (1968: 6).
59 Social order, that is, required the continuous application of force, a force guided by the intelligence of a superior being. Nature is here an idealized version of the feudal social relation. Nature 'behaves itself' more perfectly than does society. In the natural world, therefore, force is unnecessary and tends to remove things from their 'proper' place. Within nature, 'all motion implies a cosmic disorder' (Koyré 1978: 5).
60 Duhem (1985); foreword by Stanley Jaki and preface by Roger Ariew.
61 Clagett (1959), Wallace (1981: 37–40), Moody (1975: 189–201), Grant (1974).
62 Clagett (1959: xxv).
63 The so-called Merton or mean speed theorem (Grant 1974: 237–52).
64 Thus, Moody (1975: 197): 'What is perhaps most striking about the

work of the Oxford calculators is their complete indifference to the empirical interpretations of which their highly developed kinematic analyses were susceptible'. See also Clagett (1959: 218).

65 Grant (1974: 247).
66 Clagett (1968: 17).
67 Ibid. pp. 38–9. For the Pythagorean tradition in arthmetic, see Masi (1983).
68 Clagett (1968: 41).
69 Ibid.
70 Ibid., p. 245.
71 Ibid.
72 Ibid., p. 299.
73 The orthodox view is expressed, for example, by Moody (1975: 201): 'The medieval interest in physics was dialectical and abstract, rather than experimental.'
74 Grant (1974: 275–84), Clagett (1959: 509–35), Wallace (1981: 50–7).
75 Clagett (1959: 534).
76 Similar arguments were put forward by Francesco Bonamico; see Koyré (1978: 17–18).
77 Murray (1978: 25–58). 'Money, rather than being a solvent of medieval society as it might at first appear, was a prerequisite for its most characteristic achievements – such as cathedrals, pilgrimages, and crusades' (ibid., p. 59).
78 Oresme, ed. and trans. Charles Johnson (1956).
79 Ibid., p. 4. A coin can operate as a *symbol* only because it has an inherent value. It is not here a conventional *sign* and need not have the radical implications which, for example, Sohn-Rethel claims on its behalf. There is a typically feudal use of money, just as there is a typically feudal 'science' and a typically feudal 'state'.
80 The political implications of 'adulteration' were therefore highly significant. 'To take or augment profit by adulteration of the coinage is fraudulent, tyrannical and unjust, and moreover it cannot be persisted in without the kingdom being, in many other respects also, changed to a tyranny' (ibid., p. 47).
81 Murray (1978: 27, 60).

THE MIRROR OF GOD

1 The first approach ('historical') runs the risk of failing to distinguish feudalism as a 'type' of society and thus allows back into bourgeois society those very elements of discarded culture we wish to regard as 'other'. The second approach ('sociological'), in positing 'feudalism' and 'capitalism' as mutually exclusive categories, runs the risk of being unable to account for the 'transition' between them. As this is not an historical work, the second approach has seemed preferable.
2 Fun is therefore a temporary condition. Sooner or later it must hit upon self-negation as one of its possibilities – something completely

bewildering in childhood.

3 A typical form of *Wertrationalität*; see Weber (1978: 99).

4 The medieval cosmos, claims Lewis (1964: 99), 'presents us with an object in which the mind can rest, overwhelming in its greatness but satisfying in its harmony'.

5 Exchange, in other words, is by no means direct and unmediated. Its medium, however, is not money and the 'rationality' of material needs so much as the personal relations of the hierarchic order.

6 Thus the apologetic tone with which Lewis begins his popular exposition.

CHAPTER SEVEN

1 For an overview, see Holton (1985).

2 Burckhardt's *The Civilisation of the Renaissance in Italy* was originally published in 1860. Hay, in Chastel *et al.* (1982: 1), remarks the 'tenacity with which the categories established by Jacob Burckhardt have survived criticism'. See also Ferguson, in Werkmeister (1963).

3 See above, 'The Mirror of God', n.1.

4 Burckhardt (1921: 130).

5 Heller (1978: 200).

6 A conception often attributed, wrongly, to Freud.

7 Manuel and Manuel (1979: 16–17).

8 Burckhardt (1921: 175) makes clear his view that 'The Renaissance is not a mere fragmentary imitation or compilation, but a new birth'. Modern scholars have defined this new birth in terms of a revival in the systematic study of grammar and rhetoric (see Kristeller 1979: 91–2).

9 Of course many Renaissance writers and artists remained self-consciously religious, but their religiosity increasingly expressed itself in forms separated and kept apart from secular culture. See Trinkaus (1970) and Heller (1978) for contrasting interpretations.

10 Cassirer (1963: 7) claims Cusanus provides a 'simple focal point' of Renaissance thought.

11 Cusanus (1954: 7). 'Distinctions, therefore, are only found to exist among things which are susceptible of "more" and "less"' (ibid., p. 13).

12 Ibid., p. 8.

13 Cassirer (1963: 20). And Cusanus (1954: 70): 'it is absolutely impossible to arrive either at a maximum or a minimum'.

14 Cusanus (1954: 9).

15 Ibid., p. 78.

16 Ibid., p. 107. See also Poulet (1966), who views the Renaissance as a turning point in the imagery of the circle and sphere. Then: 'It is man who, equally with God, discovers himself to be centre and circumference of an infinite sphere' (ibid., p. xxvii).

17 Cusanus (1954: 108). The 'earth' here represents the extended sphere immediately beneath the orbit of the moon.

18 Ibid., p. 108.

19 Thus Manetti; 'to act and to know is the proper function of man alone', quoted in Trinkaus (1970: 250).

20 Quoted in Trinkaus (1970: 117).

21 Ibid. And Garin (1965: 45), also discussing Lorenzo Valla, epitomizes his view as a 'suffusion of pleasure with morality', and more directly that '*Voluptas* ought to be desired for its own sake'. Garin also quotes Poggio Bracciolini, 'I cannot understand how one can forget the body, for man is by no means pure soul' (ibid., p. 47).

22 Trinkaus (1970: 117).

23 Valla again: 'the pleasure that you derive observing the heavens and the stars is not greater than the pleasure that I derive looking at a beautiful face' (ibid., p. 123). Nor is pleasure amoral. Alberti (1969: 98) insists that 'pleasure leads to benevolence'.

24 Cassirer (1963: 79). Pleasure, however, as an indeterminate internal state does not prompt the ego into the world just to have it swallowed up by matter. Man must be 'completely *turned towards* the world and yet completely *distinguish* themselves from it' (ibid., p. 86).

25 Cassirer, Kristeller, and Randall (1958: 36–46). A revaluation of the isolated ego he justifies by opening St Augustine's *Confessions* at random and chancing upon an appropriate text: 'And men go to admire the high mountains, the vast floods of the sea, the huge streams of the rivers, the circumference of the ocean, and the revolutions of the stars – and desert themselves' (ibid., p. 44). He closed the book, 'angry with myself that I still admired earthly things'. As Garin (1965: 21), points out, this is quite unlike a monastic retreat. Augustinian inwardness, increasingly devoid of religious content, becomes 'an exaltation of the world of man'; it is 'the rediscovery of the whole richness of the inner life'.

26 Distinctions which in any event have little meaning for the Renaissance.

27 See particularly Alberti (1972), Gadol (1969), Hersey (1976), Wittkower (1962). Pythagoreanism was also significant in Renaissance literary movements – see Heninger (1974) – and played a part in more explicitly cosmological works – see Heninger (1977) and Barkan (1975).

28 Panofsky (1970: 29).

29 Ibid., p. 120. Gadol (1969: 27–35).

30 Alberti (1972: 49): 'A painting will be the intersection of a visual pyramid at a given distance, with a fixed centre and certain position of lights, represented artistically with lines and colours on a given surface.'

31 Ibid., p. 33. Gadol (1969: 103-4): 'by the fifteenth century, the primary artistic aim was no longer to refer to something that transcends experience, but to represent visual experience itself'. The Flemish masters, emerging from the medieval metaphysics of light tradition,

exploited luminosity rather than linear perspective, to this end.

32 Alberti, quoted in Panofsky (1970: 120). And it was individuals rather than symbolic figures that were to be depicted. See generally Abercrombie, Hill, and Turner (1986: 57–68).

33 Gadol (1969: 22).

34 Alberti provides a secular definition of beauty, for example, as 'the harmony and concord of all the parts achieved in such a manner that could be added or taken away or altered except for the worse' (quoted in Wittkower (1962: 33).

35 Barkan (1975: 130-1), Koenigsberger (1979: 7–9). Pacioli in *Divinia proportione*, for which Leonardo provided illustrations, writes: 'from the human body derive all the measures and their demonstrations and in it is to be found all and every ratio and proportion by which God reveals the innermost secrets of nature' (quoted in Wittkower 1962: 15). This conception of arithmetical relations must be distinguished from the 'absolutely impersonal, abstract, interchangeable and measurable quantities', of classical science. See, for example, the confusion in Alfred von Martin (1944), from which the above quotation is taken (ibid., p. 21).

36 Its very title, 'Revolution of the Heavenly *Spheres*' (not Heavenly *Bodies*), indicates his acceptance of much of traditional celestial physics. See Kuhn (1966: 134–5, 146–7, 154). In spite of studying canon law at Bologna and medicine at Ferrara, Copernicus seems to have been less influenced by Cusanus than was, for example, Alberti. Scholastic assumptions are evident at many places in *De revolutionibus*. The universe, for example, is held to be spherical because it is 'the most perfect shape' (Duncan 1976: 36).

37 And 'the movement of a sphere is a revolution in a circle, *expressing its shape by the very action*, in the simplest figures, where neither beginning nor end is to be found' (Duncan 1976: 38).

38 Ibid. Less radical in their conclusions, these formulations none the less display a Pythagorean influence. See Birkenmajer, in Bienkowska (1973: 38–46).

39 Ibid., p. 20.

40 Stressed particularly by Ravetz (1965: 47–55). 'Trepidation' was a measure of the long-term changes in the orientation of the earth *and* planets to the backdrop of 'fixed stars'. It was to account for this that Copernicus introduced a third motion of the earth, geometrically equivalent to the traditional astronomer's 'eighth sphere'.

41 In the *Commentariolus*, a simplified epitome of his new astronomy published before *De revolutionibus*, Copernicus had claimed that his system could be completed 'with fewer and much simpler constructions than were formerly used' (Rosen 1971: 58). In fact the Copernican system when fully developed was geometrically as complex as the Ptolemaic but was much more highly systematized and integrated (Koyré 1973: 43, 49).

42 'But nothing is more repugnant to the whole pattern and form of the universe than for something to be out of its own place' (Duncan 1976:

299

45). In spite of the scholastic allusion to 'place' this is not unlike Alberti's conception of beauty; see above, n. 34.

43 Quoted by Birkenmajer, in Bienkowska (1973: 42).

44 Rosen (1971: 20).

45 Ibid., p. 58.

46 The reception of Copernican ideas did not depend primarily upon technical arguments. See Johnson (1937: 161–210), Burtt (1932: 45), Bienkowska, in Dobrzycki (1972: 80). 'To be for or against Copernicus, then was to *pro* or *contra* on the question of human freedom, greatness, and dignity' (Heller 1978: 375).

47 Febvre and Martin (1976: 109–58).

48 Mandrou (1978: 30–2), Febvre (1982: 30).

49 Mandrou (1978: 121).

50 Randall (1961, 1962), Schmitt (1983: 16-17), Kristeller (1979: 99).

51 Randall (1962, vol. 1: 75).

52 Quoted in ibid., p. 75.

53 Ibid.

54 Ibid.

55 Ibid., p. 84.

56 Ibid., p. 204; Grant (1981: 190–210).

57 Randall (1962, vol. 1: 210).

58 Ibid.: 'Extension, far from being abstracted from things, it is presumed by them.'

59 Pleasure, that is to say, can be viewed *historically* as the secularization of happiness, or *psychologically* as the individuation of fun.

60 Kristeller, (1964: 37–53).

61 Walker (1958: 6). And in a letter to Antonio Canigiani, Ficino writes: 'the universal soul and body, as well as each living being, conform to musical proportion . . . that nearly all living beings are made captive by harmony' (Ficino 1975, vol. 1: 143).

62 Walker (1958: 15). Thus, 'the heavenly spheres and their orbits make a marvellous harmony', (Ficino 1975, vol. 1: 45), a conviction fundamental to Copernicus, and even more explicitly to Kepler.

63 Ficino (1975, vol. 1: 45).

64 Cassirer, Kristeller, and Randall (1956: 223–56), Kristeller (1979: 174).

65 Cassirer (1963: 118). Garin (1976: 27) summarizes the Renaissance problem clearly: 'if human science is to be valid, iron laws of nature are necessary; but if universal and necessary laws of nature exist, how is free and creative human activity possible?'

66 Cassirer (1963: 84).

67 Ibid., p. 59. 'Man could not raise himself without ennobling the world' (ibid., p. 66).

68 Ibid., p. 56.

69 Heller (1978) makes much of this distinction, but note Garin's remarks (1965: 3–5).

70 Febvre (1982), Part IV.

71 Manuel and Manuel (1979: 238–40).

72 Yates (1964: 1-10). An English translation of the *Corpus Hermeticum* is available in Scott (1924), and in part in Grese (1979: 1-33).

73 A view clearly expressed by Weber; see, for example, Weber (1978: 399-401).

74 Indeed movement *is* life, again reminiscent of the five-year-old. 'We know that there is nought but one sky, one immense ethereal region where those magnificent lights keep their proper distances in order to participate in perpetual life' (Bruno 1977: 91). Bruno was not the first to extend the Copernican revolution to a picture of the created cosmos as actually infinite. Thomas Digges, less eloquently, had made a similar proposal. See Johnson (1937: 161-70).

75 Garin (1976: 70).

76 Ibid., pp. 72-3, quotes A. Chastel: 'The universe appears as a gigantic organism in perennial vibration because the stars are the origin of the active forces in insensible matter, in plants and also in animals.'

77 Yates (1964: 6).

78 Ibid., p. 26.

79 Quoted ibid., p. 136.

80 See particularly Yates (1964, 1966), French (1972), Pagel (1982: 203-300), Rossi (1968), Walker (1958), Debus (1977, 1987). Its significance cannot be assessed simply in terms of its 'influence' on early modern scientists, a fascinating if controversial question which regrettably narrows the interpretive context. See, for example, the cautious remarks of Hall, in Chastel (1982) and Heilbron (1979: 26-30).

81 It is union with the whole of creation, rather than any limited practical end, which is the goal of the hermeticist. In Bruno's eroticized cosmos the frenzied operator expands his ego to fill the world. 'By intellectual contact with that godlike object he becomes a god' (Bruno 1964: 109). Paracelsus proclaimed a similar ambition: 'If man, the climax of creation, unites in himself all the constituents of the world surrounding him – minerals, plants, animals and celestial bodies – he can acquire knowledge of nature in a much more direct and "internal" way than the "external" consideration of outside objects by the rational mind' (Pagel 1982: 50). See also Debus (1977: 65-71). Knowledge is for both a function of the personality as a receptive microcosm rather than an intellectual operation.

82 Yates (1969: 197-309).

83 For Bruno, reviving Llull's art of memory, recollection is a process of mediation between himself and the cosmos. To adopt a modern analogy, the cosmos constitutes the universal 'hardware' and memory is the 'machine language' or software. The 'operator' can, by finding the appropriate 'code', enter himself into the system. The 'start-up' procedure for anyone unfamiliar with the system is bound to appear 'magical'.

84 Yates (1969: 239-59) on Bruno's 'memory objects', and his 'unending search for a really operative organisation of the psyche'.

85 Yates (1964: 432-55).

86 Yates (1972), Rossi (1968).

87 Heller (1978: 50–61).

88 The 'calculative' rationality of the Renaissance was a form of political cynicism rather than a generalized 'rationality'. The Prince could not tolerate any potential centre of power to develop; 'no other individuality could be suffered to live and thrive but their own and their nearest dependents' (Burckhardt 1921: 12).

89 For a detailed examination of the English case see Corrigan and Sayer (1985).

CHAPTER EIGHT

1 Merton (1970); for subsequent literature see particularly Webster (1974 and 1975: 484-520).

2 For example, Zilsel (1942: 544–62). Borkenau (1987), and from a different point of view Jacob (1976 and 1981). Compare the work of Freudenthal (1986) and Shapin and Schaffer (1985).

3 Koyré (1965: 6:)

> I am convinced that the rise and growth of experimental science is not the source but, on the contrary, the result of the new *theoretical*, that is, the new *metaphysical* approach to nature that forms the content of the scientific revolution of the seventeenth century, a content which we have to understand before we can attempt an explanation (whatever this may be) of its historical occurrence.

This view is also expressed forcefully in Burtt (1924).

4 See particularly the work of Yates (1964, 1966) and Rossi (1968).

5 Koyré (1965: 6).

6 Ibid., p. 7.

7 Ibid. Koyré notes the sharp distinction between the classical scientific world view and that of the Renaissance, as well as that of the medieval period. It 'implies the disappearance – or the violent expulsion – from the scientific thought of all considerations based on value, perfection, harmony, meaning, and aim, because these concepts, from now on *merely subjective*, cannot have a place in the new ontology' (ibid.).

8 Ibid., p. 8. Emphasis added.

9 Marx (1976: 125).

10 Ibid.

11 Ibid., p. 128. Thus, 'the exchange relation of commodities is characterised precisely by its abstraction from their use-values' (ibid., p. 127).

12 Ibid., p. 138. In this 'it is the direct opposite of the coarsely sensuous objectivity of commodities as physical objects'.

13 Ibid., p. 142. Marx quotes the Newtonian experimentalist Benjamin Franklin approvingly, 'Trade in general being nothing else but the exchange of labour for labour the value of all things . . . is most justly

measured by labour.'

14 Ibid., p. 128.
15 Ibid., p. 126.
16 Ibid., p. 163.
17 Thus the 'mystical character of the commodity' has nothing to do with its use-value, but consists 'simply in the fact that the commodity reflects the social characteristics of men's own labour as objective characteristics of the products of labour themselves' (ibid., pp. 164-5). See Schmidt (1971: 63–75).

Lukács (1968: 87) expresses Marx's idea as follows: 'a world of objects and relations between things springs into being (the world of commodities and their movements on the market)'. It is a world which organizes itself into a system of forces: 'they confront him as invisible forces that generate their own power'. And it is a situation in which 'the personality can do more than look on helplessly while its own existence is reduced to an isolated particle and fed into an alien system' (ibid., p. 90).

18 Marx (1973: 143–4).
19 Marx (1976: 205).
20 For example, Toënnies (1955: 81): 'In the scientific system [concepts] come together in much the same way as commodities do on the market. A supreme scientific concept that no longer denotes something real, e.g. the concept of the atom or the concept of energy, is similar to the concept of money.'
21 Marx (1973: 146).
22 Ibid., p. 149.
23 Koyré (1973: 122).
24 Ibid., Holton (1973: 71).
25 Koyré (1973: 128). Kepler notes in the same passage that even at this early date 'I was already trying to ascribe the Sun's motion to the Earth by virtue of *physical*, or if preferred metaphysical reasons, as Copernicus had done by virtue of mathematical reasons.'
26 Koyré (1973: 146) quotes Kepler as follows:

> The Earth [the sphere of the Earth] is the measure for all the other spheres. Circumscribe a Dodecahedron about it, then the surrounding sphere will be that of Mars; circumscribe a Tetrahedron about the sphere of Mars, then the surrounding sphere will be that of Jupiter; circumscribe a Cube about the sphere of Jupiter, then the surrounding sphere will be that of Saturn. Now place an Icosahedron within the sphere of the Earth, then the sphere which is inscribed is that of Venus; place an Octahedron within the sphere of Venus, and the sphere which is inscribed is that of Mercury. There you have the reason for the number of the planets (emphasis added).

See figure from Abel (1971).
27 And 'The soul is the incorporeal image of God', quoted by Caspar

(1959: 378).

28 Koyré (1973: 133), Holton (1973: 72). On Kepler's scientific realism more generally, see Jardine (1984).

29 Yates (1964: 440–3). W. Pauli, in Jung and Pauli (1955: 196): 'For it is for the vulgar mathematicians to concern themselves with *quantitative shadows*; the alchemist and Hermetic philosophers, however, comprehend the true core of the natural bodies.' Kepler, Fludd claimed, 'excogitates the exterior movements of the created thing' whereas he (Kepler) contemplated 'the internal and essential impulses that issue from nature herself' (ibid., pp. 196-7, emphasis added). On the Fludd–Kepler debate see also Debus (1978: 123–6).

30 Holton (1973: 72).

31 Caspar (1959: 136–7. His rejection was less radical but more specific than Bruno's (ibid., p. 385).

32 Holton (1973: 71).

33 Quoted by Holton (1973: 73).

34 On Brahe, see particularly Dryer (1963).

35 Koyré (1973: 192).

36 Kepler was impressed by William Gilbert's pioneering book *De Magnete* published in 1600, in which it had been claimed that the earth was itself a large magnet and its motion was a consequence of this force; see Gilbert (1900: 211–25). Kepler remarks: 'I build my whole astronomy upon Copernicus' hypothesis concerning the world, upon the observations of Tycho Brahe, and upon the Englishman, William Gilbert's philosophy of magnetism.' Quoted by Wallace (1972: 169).

37 Koyré (1973: 215).

38 Quoted in Wallace (1972: 172). See also Jammer (1957: 85):

> Here is the true doctrine of gravity. Gravity is a mutual affection among related bodies which tends to unite and conjoin them . . . . Thus, no matter whereto the earth is transported, it is always toward it that heavy bodies are carried, thanks to the faculty animating it.

39 A feature of his work brought out with particular vividness by Koestler (1964: 317–48).

40 Koyré (1973: 334–42). And Caspar (1959: 92–3), brings out Kepler's deeply rooted Pythagoreanism: 'By sensuous means, music reveals a transcendental world in which everything is as it should be'; musical harmony is truly a 'world shaping relationship'.

41 Wallace (1977), particularly pp. 63-4, 71, 84. Shea (1972: 85) even refers to his 'decadent Aristotelianism'.

42 Drake (1957: 29): 'I succeeded in constructing for myself so excellent an instrument that objects seen by means of it appeared nearly one thousand times larger and over thirty times closer than when regarded with our natural vision.' The date of publication of Kepler's *Astronomia Nova* was 1609.

43 Drake (1957: 32–7).

44 His estimate of 4 miles for the height of some lunar mountains was in fact more accurate than the maximum of 1 mile he allowed for terrestrial mountains (Drake 1957: 40–1).

45 Ibid., p. 47: 'a host of other stars are perceived through the telescope which escape the naked eye; these are so numerous as almost to surpass belief' – though in fact the cosmological significance of the resolution of the Milky Way into individual stars was little discussed. See Jaki (1973: 105).

46 Drake (1957: 89–103). Galileo's observational discoveries should be seen in the context of a lively discussion over the nature of comets and 'new stars' which Tycho Brahe had shown lay beyond the moon's orbit and therefore contradicted the orthodox assumption of the heaven's incorruptibility. See Drake and O'Malley (1960), Dryer (1963: 38–69, 186–97).

47 Oresme clearly stated the independence of optical and physical arguments over the earth's rotation: Oresme, Menut, and Denomy (1968: 521), Clavelin (1974: 68–9).

48 Wallace (1972, vol. 1: 154).

49 Clavelin (1974: 224–67).

50 Koyré (1978: 145).

51 Koyré in particular stresses the role of 'thought experiments' in Galileo's work: 'The explanation of the real by reference to the impossible' (ibid., p. 155). Stillman Drake has vigorously defended the view that genuine experimentation played a vital role in the development of Galileo's thought; see Drake (1970), in which he suggestively traces the origin of Galileo's experimental attitude to his father's practical and experimental search for a *physical* theory of harmony (ibid., p. 54). See also McMullin (1967: 295–404), Wisan, in Butts and Pitt (1978), and for an attempted compromise in the view of Galileo as pursuing a programme of practical 'reasoning' rather than Platonic reason, Finocchiaro (1980).

52 Galilei (1957: 160–2).

53 Ibid., p. 162.

54 First clearly stated by Descartes, and formalized by Newton; see Westfall (1971: 56–7).

55 Salviati can risk a rhetorical question on the subject, obviously believing that his demonstration has been convincing:

> And that stone which is on the round top, does it not move as being, together with the ship, carried about by the circumference of a circle about the centre; and therefore consequently by a motion indelible in it, if all external obstacles are removed? And is not this motion as swift as that of the ship? (Galilei 1957: 162)

56 Clavelin (1974: 220–1). Circularity remained, however, a unique expression of the underlying orderliness of the cosmos. Salviati claims that 'if we grant the excellent disposition and perfect order of the parts of the universe, then there is nothing but circular motion and

rest' (Galilei 1957: 54).

57 Clavelin (1974: 235–44).

58 Galilei (1957: 34): 'Hence I think it may be rationally concluded that, for the maintenance of perfect order among the parts of the Universe, it is necessary to say that bodies are movable only circularly.'

59 See particularly Wallace (1981: 35–57).

60 A correspondence noted by Sohn-Rethel (1978: 123–31).

61 In opposition, it should be noted, to Descartes' more obviously 'mechanistic' cosmology. See Westfall (1971: 56–98) and Westfall (1977: 120–6).

62 Perhaps regrettably, all history tends to the 'Whig interpretation'. It is impossible to forget who 'wins'. For an account that makes it much less obvious, see Shapin and Schaffer (1985).

63 See Cohen (1980: 61–8).

64 Newton (1946: 398–400). For interesting discussions of Newton's 'Rule III' (concerning the 'universal qualities' of bodies) of his 'Rules of Reasoning in Philosophy' see particularly Thackray (1970), McMullin (1978), and Freudenthal (1986).

65 Newton (1946: 6), a connection brought out clearly by Freudenthal (1986: 14).

66 Newton (1946: 420–8), Cohen (1980: 47–54).

67 Westfall (1971: 146–94) offers a clear account of one of his most sophisticated *scientific* critics, Huyghens. Purely philosophical complaints were, of course, more general.

68 In the 'General Scholium' appended to the second and subsequent editions of the *Principia*, Newton states, 'But hitherto I have not been able to discover the cause of those properties of gravity from phenomena, and I frame no hypotheses.' This is not, of course, a declaration of empiricism, only an insistence that hypotheses must not be introduced on an arbitrary basis. In the same passage Newton himself introduces the 'hypothesis' that 'absolute space' can be conceived as God's sensorium. Cohen (1966: 66) suggests, 'I feign no hypotheses' as a better translation of *Hypotheses non fingo*.

69 Quoted by Cohen (1980:131). Newton's rejection of 'action at a distance' was obscured by Cote's widely read 'Preface' to the second edition (1713) of the *Principia*. See Newton (1946: xx-xxxiii).

70 Ibid. p. 131.

71 Koyré (1965: 39).

72 See further, below Chapter 10. The 'rationality' of exchange-relations, that is, still depends upon a tacit acceptance of the 'irrationality' of use-value.

73 Newton (1931: 400). Variation in the 'size and figure' of corpuscles in fact left considerable scope for differentiating various 'types' of matter, and provided a theoretical programme for chemistry.

74 Thackray (1970: 13–16). Hence Newton's early alchemical speculations. His atomism provided the basis upon which a mechanical account of transmutation could be advanced. If all matter was a more or less 'porous' structure of the same fundamental 'stuff,' then it was

perfectly conceivable that 'every body can be transformed into a body of another kind' (quoted in ibid., p. 14).

75 As favoured, for example, by Robert Boyle. See Shapin and Schaffer (1985: 65–7), Cohen (1966: 95–103); and for more detailed illustrations of this tradition, Schofield (1970), Allan and Schofield (1980), Donovan (1975), Hiebert, Ihde, and Schofield (1980).

76 Westfall (1971: 222–7).

77 See Dijksterhuis (1961).

78 Kepler, for example, expressed a wish 'to show that the celestial machine is to be likened not to a divine organism but rather to a clockwork'. Quoted by Holton (1973: 72). Clockwork is in fact suggestive of a 'pure' mechanism in the sense that it does not 'do' anything. Its 'purpose' depends wholly upon a subjective act of perception. More generally 'machines' transmit forces for some more specific purpose – a purpose, additionally, which controls the construction of the machine in the first instance. The cosmos as clockwork is in fact an ambiguous image. It allows for the possibility that God had some subjective purpose in its creation and structured it accordingly, but equally, could suggest the notion of an autonomous and 'blind' mechanism.

79 Newton (1931: 401).

80 Ibid., p. 339, 'Do not Bodies act upon Light at a distance'?

CHAPTER NINE

1 Unger (1975: 38–46), makes the point with particular clarity but insists on the conventional 'Enlightenment' interpretation, that reason as intellect is opposed by passion so that rational action is always a form of 'self-control', a 'suppression' of pleasure. See also Macpherson (1962 and 1973: 185–94).

2 Reason was the most ideally distributed good. Thus the opening passage of Descartes' *Discourse on Method*: 'Good sense is, of all things among men, the most equally distributed ... the power of judging aright and of distinguishing truth from error, which is properly what is called good sense or reason, is by nature equal in all men.' As the distinguishing criterion of human nature it must be genuinely universal.

3 Hobbes (1840, vol. iv: 1–76), in which the perceptual world is held to be an '*apparition* into us of the *motion*, agitation, or alteration, which the *object* worketh in the *brain*, or spirits, or some internal substance of the head' (ibid., p. 4). Affect can be conceptualized similarly: '*pleasure*, which is nothing really but motion about the heart, as conception is nothing but motion in the head'. See Raphael (1969, vol 1: 4). For critical reactions to Hobbes, see particularly Mintz (1962).

4 Boyle (1979: 18–53).

5 Hartley (1810, vol.1: i). and for an explicit statement of his indebtedness to Newton, ibid., pp. 5 and 115.

6 Ibid., p. 21.
7 Thus he claims that 'all the most complex ideas arise from sensation, and that reflection is not a distinct source, as Mr Locke makes it' (ibid., p. 373).
8 Ibid., p. 75.
9 Ibid., p. 281.
10 Ibid., p. 35.
11 Ibid.
12 But now as a *consequence* of their both being subordinated to the same mechanical laws. Hartley makes this assumption quite explicit – e.g., p. 515: 'By the mechanism of human actions I mean, that each action results from the previous circumstances of body and mind, in the same manner, and with the same certainty, as other effects do from their mechanical causes.'
13 Ibid., p. 492.
14 Ibid., p. 513.
15 Ibid., p. 518.
16 Ibid., p. 515.
17 A paradox central to the tradition of academic psychology to the present.
18 La Mettrie (1912) and Vartanian (1953, 1960).
19 Vartanian (1960: 15).
20 La Mettrie (1912: 109). He continues: 'Only through nature do we have any good qualities; to her we owe all that we are.'
21 Ibid., p. 93.
22 Vartanian (1960: 19).
23 La Mettrie (1912: 128). A principle clearly expressed also by Diderot, particularly in *D'Alembert's Dream*, which can be read as a materialist rejoinder to Cartesian dualism. See Vartanian (1953: 13), who traces this line of thought, 'I think therefore matter thinks', back to Hobbes.
24 La Mettrie (1912: 143).
25 Vartanian (1960: 19).
26 D'Holbach (1966, vol. 1: 130).
27 La Mettrie (1912: 143).
28 Condillac's objection to 'systems' was to metaphysical systems; see Knight (1968: 52), Frankel (1969: 44). 'Nature' and 'man' were, however, constituted as 'real' systems and consequently could be understood geometrically.
29 Quoted in Knight (1968: 28).
30 Ibid., p. 52. We must therefore renounce 'fictitious systems'.
31 Ibid. Sentiments echoed by D'Alembert: 'The art of reducing, as far as possible, a great number of phenomena to a single one which can be regarded as the principle of them . . . constitutes the true *esprit systématique*, which one must be careful not to take for the *esprit de système* with which it does not always coincide' (quoted in Frankel 1969: 44).
32 Particularly in Berkeley's *A New Theory of Vision* first published in 1707, and Diderot's *Lettre sur les aveugles* which appeared in 1749. On

Condillac's 'statue-man', see Knight (1968: 79–108). It is sensation that 'gives birth to the whole system of man, a complete system all of whose parts are linked and mutually sustaining' (ibid., p. 79). Condillac's 'statue-man' was indebted to Buffon's account of the 'first' sensory experiences of archaic man (see Fellows and Milliken 1972).

33 Knight (1968: 86).

34 Quoted in Knight (1968: 80).

35 Perkins (1982: 44).

36 D'Holbach (1966: 59):

> Les physiciens ont nommé cette tendance ou direction *gravitation sur soi*; Newton l'appelle *force d'inertie*, les moralistes l'ont appellée dans l'homme *amour de soi*, qui n'est que la tendance à se conserver, le désir du bonheur, l'amour du bien-être et du plaisir ... cette *gravitation sur soi* est donc une disposition nécessaire dans l'homme et dans les êtres.

37 Parsons (1949: 51–60) remains the most penetrating analysis.

38 Hobbes (1962: 114).

39 Mintz (1962), Macpherson (1962).

40 Mandeville (1957, vol. 1: 231).

41 It is only since Durkheim's attack on utilitarianism and more particularly since Parsons' brilliant analysis of it that we have come to regard functionalism and utilitarianism as incompatible perspectives on social life. See Parsons (1937: 51–60, 344–50).

42 Mandeville (1957, vol. 1: 51).

43 Ibid.

44 Ibid., p. 68.

45 Ibid., p. 72.

46 Bentham (1970: 33).

47 Sidgwick (1907: 117), the first, that is, in England. In many ways, however, Shaftesbury appears as a conventionalization of themes explored more resolutely by Montaigne.

48 Grean (1967: 2).

49 Shaftesbury (1964: 62). And fails therefore to observe one of the key problems of the age, the threat of 'enthusiasm' which is a kind of 'superstition' of the moral sense, when 'the rage of the people ... has put them beyond themselves' in a way unsanctioned by conventional religion. Shaftesbury offers a 'mechanistic' account of such unreasonableness which once again, with its reference to 'fermentation' reminds us of Query 31 of the *Opticks*, 'strange ferments of the blood, which in many bodies occasion an extraordinary discharge; so in reason, too, there are heterogeneous particles which must be thrown off by fermentation' (ibid., p. 12).

50 Shaftesbury (1964: 66). He complained of Mandeville's view in which 'An honest heart is only a more cunning one; and honesty and good-nature, a more deliberate or better-regulated self-love' (ibid., p. 79).

51 Ibid., p. 183.
52 Ibid., p. 184. The 'self' required education and moral regulation but its prior existence was accepted as the 'given' of human nature, a view that was central to the development of the 'Scottish Moralists'. Later, for example, Thomas Reid was to write:

> Every man of sound mind finds himself under a necessity of believing his own identity and continued existence. The conviction of this is immediate and irresistible; and, if he should lose this conviction, it would be a certain proof of insanity, which is not to be remedied by reasoning. (Reid 1961: 26)

53 Newton's formulation of a law of gravity is likened to Euclid's geometry: 'how beautiful is the *Theorem*, and how are we ravished with its first discovery' (Hutcheson 1971, vol. 1: 29). Hutcheson claimed that a similar aesthetic pleasure in rightness sustained the diversity of moral duties 'deduced in the various relations of human life' (ibid., p. 30).
54 Ibid., p. 31. 'It is easy to see how men are charmed with the beauty of such knowledge' (ibid., p. 30).
55 See particularly his *The Original of our Ideas of Virtue* (Hutcheson 1971, vol. 2).
56 Smith (1976: 8).
57 This against Hutcheson's tautological moral theory. Significantly, Hutcheson defended a theology of design, whereas Smith was drawn to a Stoic philosophy within which the Creator 'contrived and conducted the immense *machine* of the universe'.
58 Smith (1976: 9).
59 Ibid., p. 10.
60 Indeed, it is only because we do not directly imitate his *feelings* that we grasp the 'most dreadful . . . of all the calamities to which the condition of mortality exposes mankind' (ibid., p. 12).
61 Ibid., p. 13.
62 There is, in other words, an 'economy of feeling' which is one important aspect of an internal 'market' in sentiments; a market structured by the universal dimensions of self-identity (the 'equivalent' of time in the cosmic sphere and labour in the social world) and sympathy (which is the 'equivalent' of space in the cosmos and money in the social sphere).
63 There is therefore a powerful 'social motive' in the sharing of sorrow, and we are 'not half so anxious that our friends should adopt our friendships, as that they should enter into our resentments' (ibid., p. 15).
64 The Newtonian ambition of his treatise should again be noted. James Woodrow remarked of his theory of moral sentiments that it was 'a very ingenious attempt to account for the principal phenomena in the moral world from this one general principle, like that of gravity in the natural world' (ibid., 'Introduction', p. 3).

65 As opposed to the frivolity of sensuousness it is pleasure in acting virtuously that lends 'weight' to the personality. Luhmann (1986) provides an interesting analysis of developing codes of intimacy as an important aspect of the process of psychic individuation.

66 Smith (1976: 50).

67 Ibid.

68 Ibid.

69 Ibid., pp. 50–1. For a reformulation of this view see Douglas and Isherwood (1979: 56–71).

70 'The man of rank and distinction . . . is observed by all the world . . . to conceive, at least by sympathy, that joy, and exultation with which his circumstances naturally inspire him' (ibid., p. 51). Furthermore, 'upon this disposition of mankind, to go along with the passions of the rich and the powerful, is founded the distinction of ranks, and the order of society' (ibid., p. 52). This subtle regulation of sentiment, however, can be degraded into an envy and a vulgar love of wealth, 'the great and most universal cause of the corruption of our moral sentiments'.

71 For example, in *Soliloquy*, 'One would think there was nothing easier for us than to know our own minds. . . . But our thoughts have generally such an obscure implicit language, that 'tis hardest thing in the world to make them speak out distinctly' (Shaftesbury 1964: 113).

72 Hume's piercing scepticism in the matter, in any event quite exceptional, was not directed at the immediacy of the experience of the self (which was undeniable) but towards the chain of faulty deductions drawn from it: 'The idea of ourselves is always intimately present to us, and conveys a sensible degree of veracity to the idea of any other object, to which we are related' (Hume 1975: 354). See also reference to Reid, above, n. 52.

73 See particularly Löwith (1964).

74 It is constructed through a process of 'negation'. 'This tarrying with the negative is the magical power that converts it into being' (Hegel 1977: 19).

75 Ibid. p. 115.

76 Ibid. p. 125 – the 'dizziness' which is here termed *fun.*

77 Ibid., pp. 111-19. See also Kojeve (1969: 3–30), Lauer (1982: 116–20), and more generally Butler (1987: 17–100).

78 Self-consciousness 'is this whole process itself, of passing out of itself as simple category into a singular individual, into the object, and of contemplating this process in the object, nullifying the object as distinct [from it], *appropriating* it as its own, and proclaiming itself as this certainty of being all reality, of being both itself and its object' (Hegel 1977: 144).

79 Unger (1975).

CHAPTER TEN

1 From, say, Newton's *Principia* (1686), to the invention of non-Euclidean geometry and the development of critical psychology in the 1840s.

2 The use of the word 'contains' here is not intended to convey any sense either of Hegelian immanence or of conscious purposiveness.

3 'We are all convention' Montaigne (1958: 190) boldly declares. For a brilliant exposition of his 'modernism', see Starobinski (1985).

4 Characteristically he lacks *memory*: 'I find scarcely a trace of it in myself', he boasts (Montaigne 1958: 28). See, more generally, Starobinski (1985: 9–14).

5 Or that Cervantes' novel can appear as an 'exemplary story' in the work of Foucault (1970: 46–50) and Schutz (1976, vol. 2: 135–58).

6 As suggested by Paterson (1970: 41).

7 Cole (1971) provides a systematic comparison with Freud.

8 Kierkegaard in general addresses his works to the 'single individual'; for example (1941: 109), and writes, *Journals and Papers* (1970, vol. 2: 411) that 'He can be in kinship only with "the single individual", and only "the single individual" can be in kinship with God'. As the highest 'potentiation' of the self, the 'single individual' is a particular personality, and completely unlike its 'Hegelianized' abstraction, Max Stirner's 'unique one', or absolute ego, which is wholly indeterminate and 'free'. See Stirner (1971: 257–61) and Löwith (1964) for comparison.

9 Not that he did not make many brilliant comments on the contemporary world and his relation to it. In saying (1978: 97), 'The present age is essentially a sensible age, devoid of passion', he is surely correct, but adds 'and therefore it *has nullified the principle of contradiction*' (emphasis in original), he misses the 'passionless' contradiction of hypocrisy.

10 The biographical material is not simply a useful 'background' to understanding his writings. See Kierkegaard (1962), Lowrie (1962), Thompson (1974).

11 See, generally, Thulstrup (1980), particularly pp. 320–82; Taylor (1980: 105–40), Malantschuk (1971: 9–101).

12 Lukács (1974: 28–41).

13 For example, a *Journal* entry (1970, vol. 2: 225): '*The System* "goes forward by necessity," so it is said. And look, it never for a moment is able to advance as much as half an inch ahead of existence, which goes forward in freedom.' But the 'present age' (1978: 68) 'is essentially a sensible, reflecting age devoid of passion, flaring up in superficial, short-lived enthusiasm and prudentially relaxing in indolence'.

14 Hence the 'disjointed' structure of *Either* as a description of 'immediacy'. On the differences between Kierkegaard's and Marx's rejection of Hegel, see Löwith (1964).

15 Kierkegaard (1959, vol. 2: 217).

16 Ibid.

17 Philosophical reflection empties the self of all content; by reflection 'I can circumnavigate myself, but I cannot erect myself above myself, I cannot find the Archimedean point' which lies in existence itself (*Repetition*, p. 90). It is felt by the intellect as paradox, 'for the paradox is the source of the thinker's passion' (*Philosophical Fragments*, p. 46). It is therefore to 'repetition' rather than to Socratic 'recollection' that the passionate thinker must turn. 'Recollection is the pagan life-view, repetition is the modern life-view; repetition is the *interest* of metaphysics, and at the same time the interest upon which metaphysics founders' (*Repetition*, p. 53).

18 Kierkegaard (1959, vol. 1: 37). For details of his relation with Regine, see Lowrie (1962: 191–231) and Thompson (1974: 101–16).

19 In turn representing a conflict between his own withdrawn melancholy and an urge to speak 'directly'. A 'quiet despair' (*Journals*, vol. 1: 345–6), and a 'passionate coldness' (*Letters*: 133–8). In short, lack of 'faith' (Lowrie 1962: 226).

20 In fact she married Fritz Schlegel six years later, in 1847.

21 The 'aesthetic' self 'loses itself in the multifarious' (Kierkegaard 1959, vol. 2: 171); Judge William comments of the 'young man', 'you are constantly only in the moment, and therefore your life dissolves' (ibid., p. 183).

22 His soul had been 'anaesthetised by despair' (ibid., p. 226).

23 Particularly *Either/Or, Repetition, The Sickness unto Death* and *The Concept of Anxiety*.

24 And attributes his failure to attain it to his own 'melancholy', 'the most faithful mistress I have known' (Kierkegaard 1959, vol.1: 20).

25 'Pleasure is in itself a multiplicity' (ibid., vol. 2: 188). And in another sense is the *boredom* of immediacy, 'the nothingness which pervades reality' (ibid., vol. 1: 287).

26 Accidental, that is to say, from an 'inward' viewpoint. 'The aesthetic choice is either entirely immediate and to that extent no choice, or it loses itself in the multifarious' (ibid., vol. 2: 171).

27 He has become 'abstracted' and literally philosophical, 'you are united with the philosophers. What unites you is that life comes to a stop'; the Hegelian method becomes a reality. 'You mediate contradictions in a higher madness, philosophy mediates them in a higher unity' (ibid., vol. 2: 175–6).

28 Kierkegaard (1962: 76–7).

29 Kierkegaard (1959, vol. 2: 229–35).

30 Kierkegaard (1965).

31 Kierkegaard (1959). 'Doubt is a despair of thought, despair is a doubt of the personality' (ibid., vol. 2: 215).

32 For example, marriage, 'is the true transfiguration of romantic love' (ibid, vol. 2: 31).

33 Ibid. (vol. 2: 181–2).

34 Kierkegaard (1941: 468–93). 'So the essential consciousness of guilt is the first deep plunge into existence' (ibid., p. 473).

35 Kierkegaard (1962: 44–63).

313

36 *Journals* (1978, vol. 6: 223–7).

37 'The thing of being a Christian is not determined by the *what* of Christianity but by the *how* of the Christian. This *how* can only correspond to one thing, the absolute paradox' (Kierkegaard 1941: 540).

38 Ibid., p. 457.

39 For a sympathetic assessment, see Carroll (1974).

40 Kierkegaard (1957).

41 Kierkegaard (1980); the archaic term 'dread' is still much richer in associations.

42 Cole (1971).

43 'Dread is a *sympathetic antipathy and an antipathetic sympathy*. . . .This dread belongs to the child so essentially that it cannot do without it; even though it alarms him, it captivates him nevertheless by its sweet feeling of apprehension' (Kierkegaard 1957: 38).

44 Carroll (1974).

45 'Underground Man' is a somewhat cynical version of the 'young man' of the *Either*, and Raskolnikov might be considered as his 'demonic' brother. Prince Myshkin has the comical incognito of the authentic Christian.

46 Bakhtin (1973: 4–7), in particular, has drawn attention to the 'astonishing inner independence of Dostoevsky's heroes' and to the 'plurality of independent and unmerged voices and consciousnesses and the genuine polyphony of full-valued voices' characteristic of his novels.

47 Mochulsky (1967: 463).

48 Thus, 'in Dostoevsky's world even *agreement* (*soglasie*) retains its *dialogical* character, i.e. it never leads to a *merging* of voices and truths in a single *impersonal* truth, as is the case in the monological world' (Bakhtin 1973: 78).

49 In *The Sickness unto Death* Kierkegaard might have been describing the paradoxically self-conscious 'Underground Man'; 'Yet despair is precisely *self*-consuming, but it is an impotent self-consumption which is not able to do what it wills; and this impotence is a new form of self-consumption, in which again, however, the despairer is not able to do what he wills, namely, to consume himself' (ibid., p. 151). Mochulsky (1967: 244) also draws attention to the parallels between 'Underground Man's' striking insights and those of Kierkegaard and Nietzsche.

50 But rather than complete the 'abstract' development of the Kierkegaardian 'young man', this is accompanied by a characteristic 'fall' from the ethical to the aesthetical, to cynical pleasure (Mochulsky 1967: 249).

51 As had been expressed particularly by Chernyshevsky in his novel *What is to be Done*, and symbolized by the Crystal Palace (Dostoevsky 1972: 34–43) (see Mochulsky 1967: 253–4).

52 'They had so little understanding of the most essential things, so little interest in the most inspiring subjects, that I could not help looking

on them as my inferiors. . . . I abominated them, although I was perhaps worse than they were' (Dostoevsky 1972: 68–9).

53 Kierkegaard (1957: 103-5) speaks eloquently of the 'sophistry of remorse'; cf. Freud's (1971, vol. 20) later view of symptom formation as a means of dealing with 'existential' anxiety.

54 Mochulsky (1967: 272–3).

55 Porfiry understands this perfectly well, 'We're dealing with quite a fantastic affair here, a sombre affair, a modern one, a case characteristic of our time . . . here we are faced with a determination to take the first step' (Dostoevsky 1966: 467).

56 A theme deepened in the figure of Ivan Karamazov, in whom 'all is permitted' turns into 'metaphysical rebellion'. See Camus (1971: 50–6).

57 A second 'unplanned' murder is introduced to ensure conventional moral culpability.

58 Mochulsky (1967: 312). In the 'Epilogue' Raskolnikov states quite clearly, 'My conscience is clear' (Dostoevsky 1966: 552).

59 Ibid., pp. 432-3: 'I just did it; I did it for myself alone . . .the devil had dragged me there, and it was only afterwards that he explained to me that I had no right to go there because I was the same kind of louse as the rest.'

60 This does not, of course, resurrect some 'normal' feeling in Raskolnikov; she follows him to Siberia where she is tormented 'by his rude and contemptuous attitude towards her' (ibid., p. 551).

61 Ibid., pp. 468-74.

62 Dostoevsky had previously dealt with the idea of the 'double', a theme he had taken from Gogol (Dostoevsky 1972).

63 Mochulsky (1967: 307–11).

64 Bakhtin (1973: 87–113). Interestingly, John Cowper Powys (1974) also points, through the notion of *play*, to carnival imagery in Dostoevsky.

65 Time and space in his novels are broken into incommensurable fragments (Bakhtin 1973: 124).

66 'The individual is in the evil and is in dread of the good. The bondage of sin is an unfree relation to the evil, but the demoniacal is an unfree relation to the good' (Kierkegaard 1957: 106); see also Kierkegaard (1968: 200–7).

67 Kierkegaard (1967: 110–14).

68 His less exalted 'positive' characters – Razumikhin, for example, or Sonia – are also less striking than their correspondingly minor 'devils'.

69 Kierkegaard (1944); Nietzsche (1969, 1973).

70 He rejects, of course, any unconditional moral *code*. His morality, like Kierkegaard's religiosity, begins with a 'teleological suspension of the ethical', with the 'stern earnestness' of 'self-overcoming' (Jaspers 1965: 154–6).

71 The really striking fact in this, Nietzsche points out, is the extent to which science has none the less vindicated itself. 'Indeed, we are so convinced of the uncertainty and fantasies of our judgments and of the eternal change of all human laws and concepts that we are really

amazed how *well* the results of science stand up' (Nietzsche 1974: 111).

72 In a typical passage we read:

> We operate only with things that do not exist: lines, planes, bodies, atoms, divisible time spans, divisible spaces. How should explanations be at all possible when we first turn everything into an *image*, our image! It will do to consider science as an attempt to humanise things as faithfully as possible: as we describe things and their one-after-another, we learn how to describe ourselves more and more precisely . . . in truth we are confronted by a continuum out of which we isolate a couple of pieces. . . . The suddenness with which many effects stand out misleads us; actually, it is sudden only for us. In this moment of suddenness there is an infinite number of processes that elude us. (Nietzsche 1974: 172–3)

73 Nietzsche (1969: 24–56).

74 Ibid., pp. 97–163).

75 A view denied by Scheler (1972), who none the less took over the category of *ressentiment.*

76 Of course we need not suppose that a spiritual liberation will destroy the world.

> What at first was appearance becomes in the end, almost invariably, the essence and is effective as such. How foolish it would be to suppose that one only needs to point out this origin and this misty shroud of delusion in order to *destroy* the world that counts for real, so-called 'reality'. (Nietzsche 1974: 122)

77 The 'self' betrays an outmoded metaphysical prejudice in favour of a unified and 'deep' reality. But 'would it not be rather probable that, conversely, precisely the most superficial and external aspect of existence – what is most apparent, its skin and sensualisation – would be grasped first – and might even be the only thing that allowed itself to be grasped?' (ibid., p. 335).

78 'Forgetting is no mere *vis inertiae* as the superficial imagine; it is rather an active and in the strictest sense positive faculty of repression' (Nietzsche 1967: 57).

79 For a somewhat different reconstruction, see Carroll (1974).

CHAPTER ELEVEN

1 'There is no "reality" for us' Nietzsche (1974: 121). Cf. n. 72, Chapter 10.

2 Nietzsche's madman who declares that 'God is dead. God remains dead. And we have killed him' is filled with authentic spirituality (Nietzsche 1974: 181). But as the absolute reality expressed in God's

existence is human personality, it is the 'rational self' which is dead.

3 Robert Musil was one of the few really important modern writers with a deep understanding of contemporary scientific movements. Significantly, he opens his thesis on Mach – which he wrote before turning to literature – not simply with an orthodox expression of *philosophical* scepticism but with the confident assertion that '*Exact science* has shown that there are no such things as causal connections' (Musil 1982: 15) (emphasis added).

4 See Freudenthal (1986). A Leibnizian relational physics, rather like a Hegelian relational social theory, would tend however to a *systematic* abstraction at odds with any form of subversive modernism.

5 Boscovich (1961) in the mid-eighteenth century proposed a conceptual reduction of matter to identical 'point-masses' connected by an oscillatory force whose strength and direction varied by distance. These 'perfectly indivisible and non-extended points' (ibid., p. 105) have some of the elusive properties of modern quanta. It is hardly surprising that his work was ignored by his contemporaries.

6 Amrine, Zucker, and Wheeler (1987).

7 Carnot (1960) and, for a clear description of his work, Segré (1984: 192–200).

8 Kline (1972: 861–81) and Kline (1954: 410–31).

9 Hendry (1986: 6–45) calls the first approach 'mechanistic' and the second 'dynamistic'. Both, however, are part of the 'classical' scientific picture of the world; one begins with individuated 'objects' and treats 'forces' as secondary; the other regards 'forces' as the more fundamental. Both aimed at a complete, logically coherent 'system of the world', and both could claim, with some justification, the authority of Newton. For the underlying equivalence of both, see particularly Meyerson (1930: 63–103).

10 Meyerson (1930: 215–18) was among the first to draw the attention of a non-scientific audience to the significance of this point. See also Grünbaum (1973: 213-14). Max Planck (1925: 19) also makes it clear that all reversible processes are idealizations of nature.

11 Ibid., p. 31: 'Thus the law cannot directly express reality.'

12 Ibid., p. 265. An idea developed within the classical framework most powerfully by Boltzmann. See Boltzmann (1974: 13–32), Segré (1984: 237–45).

13 Carnot (1960: 9).

14 Meyerson (1930: 265).

15 Particularly as expounded by Grünbaum (1973: 209–19).

16 This is the physical meaning of Boltzmann's celebrated 'H-Theorem': one is amazed that, on the assumption that the world is a large system with a finite number of bodies . . . not even the world as a whole can be a perpetuum mobile' (Boltzmann 1974: 30).

17 Westfall (1971: 150–5), for some significant attempts. Impact physics was no more comprehensible, in spite of its 'intuitive' appeal, than 'action at a distance' (Myerson 1930: 97).

18 '[I]n spite of constant mutual influence each molecule pursues its

own independent path, appearing as it were as an autonomously acting individual' (Boltzmann 1974: 19).

19 Though only after the kinetic theory of heat had been well established. Thomas Brown made the observation in 1828, but its significance only became 'obvious' on the publication of Einstein's paper in 1905.

20 Hendry (1986: 72–82).

21 Segré (1984: 84–97), Schaffner (1972: 11–19), Hesse (1961: 189–98), Whittaker (1951, vol. 1: 94–127).

22 Whittaker (1951, vol. 1: 128–70), Schaffner (1972: 40–75).

23 A view championed by Whittaker in opposition to Einstein's revival of corpuscularism. See, e.g., Whittaker (1951, vol. 1: 303).

24 Hendry (1986: 79–82).

25 Maxwell (1890, vol. 1: 452): 'We are dissatisfied with the explanation founded on the hypothesis of attractive and repellent forces.' Also p. 455: 'Let us now suppose that the phenomena of magnetism depend on the existence of a tension in the direction of the lines of force,' and p. 527: '. . . produced by actions which go on in the surrounding medium as well as in the excited bodies'. Einstein and Infeld (1947: 148) regard Maxwell's formulation of the field equations as 'the most important event in physics since Newton's time'. And de Broglie (1962: 22), concurring, particularly stresses the non-classical time asymmetry in Maxwell's theory. For a general and biographical introduction, see Domb (1963) and Tolstoy (1981). Hendry (1986) provides a much-needed scholarly treatment.

26 A viewpoint already clearly established in an essay on 'Analogy' delivered as an undergraduate at Cambridge to the 'Apostles'. See Campbell and Garnett 1882: 235–44).

27 Not only mechanism but also space and time are structures of analogy linking nature and intellect (Campbell and Garnett 1882: 238).

28 Quoted in Campbell and Garnett (1882: 243). Hence the significance of self-conscious analogy, or theory: 'The dimmed outline of phenomenal things all merge into one another unless we put on the focussing glass of theory and screw it up sometimes to one pitch of definition, and sometimes to another, so as to see down into different depths through the great millstone of the world' (ibid., p. 237). This, Maxwell is quite willing to concede, implies the notion that 'causes' are simply 'reasons' 'analogically referred to objects instead of thoughts' (ibid. p. 238). It is tempting to see in Maxwell, and in his near-contemporary at Cambridge, William Clifford, a presciently 'modern' conception of science as a system of 'representations'. Duhem (1954, original French 1914: 39) might have drawn his inspiration from Maxwell rather than Mach when he argued that 'physical theories should be regarded as condensed representations and not as explanations'.

29 In this Maxwell was more original than Thompson (Lord Kelvin), who always strove to provide a mechanical picture of the 'underlying process' of nature (Hendry 1986: 144).

30 Quoted in Jammer (1954: 138). Stallo (1848: 24) was among the first of the modern 'relativists': 'There is nowhere absolute rest, but motion everywhere.'

31 Mach (1893). Einstein admitted that 'Ernst Mach's *The Science of Mechanics* shook this dogmatic faith in mechanics as the final basis of all physical thinking' (quoted in Holton 1973: 223). Poincaré, in Čapek (1976: 317–27).

32 Mach (1893: 231–3).

33 He thus argues that 'the motions of the universe are the same whether we adopt the Ptolemaic or the Copernican mode of view. Both views are indeed equally *correct*; only the latter is more simple and more *practical* (ibid., p. 232).

34 Mach himself became more interested in psychology than physics; see, e.g., Mach (1976). Duhem turned to the history of science, Musil to literature.

35 FitzGerald's account of the Michelson–Morley result, for example, had recourse to an *ad hoc physical* hypothesis. See Whittaker (1949: 52–3), Holton (1973: 268).

36 A. A. Michelson and E. W. Morley, 'On the Relative Motion of the Earth and the Luminiferous Aether'(1887), reprinted in Schaffner (1972).

37 For the immediate background to Einstein's paper, see Holton (1973: 197–218).

38 Lorentz *et al.* (1952: 37).

39 Ibid., p. 117.

40 Ibid., p. 38 – a proposition derived from Maxwell, and already utilized by Lorentz to account for the FitzGerald 'contraction'. See Einstein (1920: 30–4).

41 Namely, the *universality* of nature's laws. Methodologically, however, Einstein shifted away from an early 'Machian' phenomenalism to a rational realism, from which viewpoint he resisted the new quantum physics (Holton 1973: 219–59).

42 An example drawn from Davies (1977: 33–5).

43 Thus, 'we cannot attach any *absolute* signification to the concept of simultaneity, but that two events which, viewed from a system of co-ordinates are simultaneous, can no longer be looked upon as simultaneous events when envisaged from a system which is in motion relatively to that system' (Einstein 1952: 42–3).

44 Minkowski, in Lorentz *et al.* (1952: 75).

45 A 'mediation' that is characteristic of the collapsing bourgeois ego. In Dostoevsky's novels desire (which is always and by definition desire of and for the self) is felt only through a series of unstable relationships. See Girard (1965), particularly, pp. 83–95.

46 It is *generality* rather than simplicity that is Einstein's theoretical ambition. Einstein (1952: 117). And as we can only directly experience a limited range of metrical variations, this generality can be achieved only at the expense of immediate 'picturability'. The tendency to a Lévi-Straussian 'structuralism' can again be detected. In

relativity theory, 'the structure of reason expresses itself in the arbitrariness of admissible systems' (Reichenbach 1965: 90). The experimental 'tests' of the theory cannot rest therefore with 'everyday' observations. For an account of the standard 'verifications' of Relativity Theory, see Shapiro in Woolf (1980: 115–36).

47 Minkowski, in Einstein (1952: 76).

48 Ibid, p. 76.

49 A physical interpretation of Riemann's geometry. Stein, in Earman, Glymour, and Stachel (1977), Jammer (1954: 145–60). Kline (1972: 889–96) notes in passing that 'the history of non-Euclidean geometry reveals in a striking manner how much mathematicians are influenced not by the reasoning they perform but by the spirit of the times'.

50 Clifford (1885: 224). He also anticipated much of the physical and logical 'weirdness' of quantum discontinuities (see Clifford 1879: 114–18).

51 Einstein and Infeld (1947: 227–8).

52 Einstein (1920: 65–70): 'The *same* quality of a body manifests itself according to circumstances as "inertia" or as "weight".'

53 An observational test of Einstein's theory carried out in 1919 by Eddington, who, during an eclipse, was able to measure the bending of starlight by the gravitational field of the sun (Eddington 1920: 115–16).

54 Einstein and Infeld (1947: 251–7), Tonnelat, in Taton (1966: 89–93). For non-technical expositions, see Narlikar (1982), Geroch (1978). Reichenbach (1965: 3) points out that the General Theory 'asserted nothing less than that Euclidean geometry is not applicable to physics'.

55 Einstein and Infeld (1947: 251).

56 Ibid., p. 275.

57 Ibid., p. 257.

58 Ibid., p. 275.

59 This is not to say that the nature revealed by twentieth-century science might not in time become a 'commonsense' reality, only that it found itself in contradiction to the commonsense world of classical science and the systematic reason it had justified.

60 Kuhn (1978); Hermann (1971) for a detailed historical reconstruction. It is little consolation to discover, Jammer (1966: 2), that quantum physics might well have developed in relation to a less complex problem!

61 A solution to which he was driven, he claimed, 'as simply an act of desperation' (quoted in Hermann 1971: 1).

62 Jammer (1966, 30–6); Gibbins (1987: 21–2).

63 On the wave analogy it would be as if large waves threw floating debris no farther up on to the beach than could tiny ripples (de Broglie 1939: 27–8).

64 Not only did light appear to be, in some circumstances, particulate, but matter could, in other circumstances, appear wave-like; a duality

which became particularly prominent (de Broglie 1939). See Hey and Walters (1987: 1–12) for a clear account of the classic 'two-slit' experiment.

65 Jammer (1966: 65–80).

66 It meant 'a complete renunciation as regards a time description' (Bohr 1934: 80).

67 Murdoch (1987: 18–19).

68 Quoted in Hund (1974: 34).

69 Quoted in Murdoch (1987: 21). Since Einstein's early papers, 'free from contradiction' meant consistent with Maxwell's electrodynamics rather than Newton's mechanics.

70 Quoted in Murdoch (1987: 55). Meyerson (1985, original French 1925) was also quick to point to Planck's constant of action 'as a new kind of irrationality'. It was an irrationality that Einstein, having played an important role in its inception, could not accept. Interestingly, quantum 'weirdness' seems to have been anticipated to some extent, not only by Clifford but also by Riemann, who speculated on the breakdown of 'normal' metric relations at infinitely small distances. See Akhundov (1986: 140).

71 De Broglie (1939: 47).

72 Schroedinger (1935: 149–53) regarded any physical 'picture' of nature as an unfortunate limitation. His formal approach he regarded as a species of *play*. Genuine science, like art and play, were not 'determined by the aims imposed by the necessities of life' (ibid., p. 28). The difficulties remain unresolved, see, e.g., Gibbins (1987), Krips (1987), Polkinghorne (1984), Reichenbach (1948), Herbert (1985). And for interesting discussions of the relationship between modern movements in art and science, Richter (1985), Richardson (1971).

73 Heisenberg (1959: 96).

74 De Broglie, in Taton (1966: 83) contrasts the view that such imprecision was 'merely an expression of our ignorance', with the conviction that 'transition probabilities became the expression of a kind of pure chance'.

75 Quantum theory, therefore, 'does not allow a completely objective description of nature' (Heisenberg 1959: 96). The quantum formalism was not simply imprecise. It was not that a 'particle' might be either 'here' or 'there'; the superposition principle was a way of saying that it was *both* 'here' and 'there'. See Polkinghorne (1984: 19–21).

76 Murdoch (1987), Folse (1985), Holton (1973: 115–64).

77 Quoted in Murdoch (1987: 15).

78 Too much should not be made of a direct connection with Kierkegaard, as Folse (1985: 47) has rightly stressed. The similarity in their philosophical outlook is, consequently, all the more striking.

79 A principle of complementarity applicable to human knowledge in general and not just to knowledge of nature. See Holton (1973: 154–5). Einstein always resisted such arguments and remained

faithful to the classical ideal of descriptive *completeness*. 'I still believe in the possibility of a model of reality – that is to say, of a theory which represents things themselves and not merely the probability of their occurrence' Schilpp (1949: 256).
80 As revealed particularly in his replies to Einstein's critical thought experiments (Murdoch 1987: 155–78).
81 Heisenberg (1959: 42) – again a statement which bears comparison with Kierkegaard's *Philosophical Fragments*.
82 Heisenberg (1959: 96).
83 Thus Pagels (1982: 144): 'The electron seems to spring into existence as a real object only when we observe it!'
84 See, e.g., Johnston (1972), Schorske (1979), Frisby (1985).

CHAPTER TWELVE

1 Mach, whose writings on mechanics were among the severest criticisms of classical science, was oddly reluctant to consider newer 'relativistic' or phenomenological psychologies. His insistence on the 'relational' aspects of experience could not conceal that he was at heart a sensationalist. See Mach (1976: 6); Hiebert, in Machamer and Turnbull (1976); Musil (1982).
2 Increasingly the classical bourgeois conception of rationality has been allied with technological systems, rather than to scientific theories, as pointed out, from a somewhat different perspective, by, e.g., Marcuse (1964: 19–32) and Habermas (1971: 81–123).
3 The Complementarity Principle is by no means restricted to a particular interpretation of quantum physics (Folse 1985: 27–31). It has become quite characteristic of modern thought to regard empirical reality as only one, usually arbitrary, selection from a range of theoretical possibilities, the generality of which cannot be rationally reconstructed.
4 'Bourgeois' is here used as a broader concept than 'capitalism'. The latter refers primarily to the rational order of the commodity, the former to the forms of life in fact sustained by such an order.
5 There is no suggestion here that a particular social relation is the 'cause' of a new cosmological view; particularly as the language of 'causality' is foreign to such a view.
6 Again, it is not that *either* production *or* exchange holds the key to a process of rationalization. Rather, each sphere of activity is marked by a typical opposition between rational and non-rational forms.
7 Cf. above, Chapter 3.
8 It is within just such a framework that Freud's work is still frequently read; thus Rieff (1965) makes of him an orthodox bourgeois moralist, and Hartmann *et al.* (1964) claims him for 'ego psychology'.
9 Capek (1961: 135–40). Freud's initial attempt at a psychological synthesis was couched in mechanistic terms and was quickly abandoned. Freud (S.E., vol. 1: 283–399).

10 There is, then, a considerable irony in Freud being regarded as a 'developmental' psychologist.

11 See Freud (S.E., vol. 1: 312–21).

12 Especially in important theoretical works – for example, in *The Interpretation of Dreams* (S.E., vol. 5: 537–44, 573–82, 610–17); and the *Introductory Lectures* (S.E., vol. 16: 356–7).

13 Breuer and Freud (S.E., vol. 2: 180).

14 In 1912; Freud (S.E., vol. 12: 99–108, 159–71).

15 Murdoch (1987: 94–103).

16 An approach which does not, of course, borrow specific concepts from psychoanalysis. For an approach which does, see Bachelard (1964).

17 Thus, in dreams, for example. 'Thoughts which are mutually contradictory make no attempt to do away with each other, but persist side by side' (Freud, S.E., vol. 5: 596).

18 Approximately what Freud referred to as 'condensation'; see Freud (S.E., vol. 4: 279–82).

19 And 'displacement', see Freud (S.E., vol. 4: 305–9).

20 Thus Feynman's warning remark: 'I think I can safely say that nobody understands quantum mechanics' (quoted in Hey and Walters 1987: 1). Understanding, that is to say, is still identified with classical scientific rationality. Bergson's attempt to shift the philosophical foundation of natural knowledge to active *intuition* was largely ignored. See Čapek (1971: 30–51).

21 Feinberg (1977: 59–63); Hey and Walters (1987: 76–8); Feynman *et al.* (1965, vol. 3: 4–13) remark that: 'In fact, almost all the peculiarities of the material world hinge on this wonderful fact.'

22 On the 'particle honeymoon' of the very early universe, see Rowan-Robinson (1985: 230–88); Barrow and Silk (1984); Weinberg (1977).

23 In fact, Hubble's account of the observed red shift of distant objects was accepted almost at once and in the absence of reliable data. See Wagoner and Goldsmith (1982: 103–4); Harrison (1981: 208–9); Silk (1980: 43–9); Peebles (1971: 6–7).

24 See, for example, Veblen (1925: 29): 'It becomes indispensable to accumulate, to acquire property, in order to retain one's good name.' And Douglas and Isherwood (1978: 25–55).

25 A distance common to the natural sciences and to bourgeois psychology.

26 Cf. above, Chapter 3, for specific reference to the writings of Bergson and Proust in this regard. Within a more academic tradition, William James (1980, vol. 1: 224–400), and particularly pp. 373–9, is most fully aware of the instability inherent in the classical model.

27 This much at least can be granted to the critics of 'mass society'; see particularly Riesman (1965).

28 Kierkegaard referred to his pseudonymous works as a 'maieutic art', meaning literally 'obstetric'. They were intended as an 'indirect communication' and, in an extreme version, a means of deceiving

people into the truth (Kierkegaard 1962: 147–50).

29 Simmel, in Wolff (1965: 409–24).

30 Consumption, that is to say, is a *gamble*, and while the commodity cannot be guaranteed to induce a state of excitement, the risk involved in its acquisition can. It thus remains 'rational' to consume, whatever the outcome.

31 A superficial paradox, to be sure. Behaviourist psychologists have long been aware that intermittent reward is the most secure foundation of habit.

32 The 'failure rate', never fully determined, is more of a known quantity for old commodities. Given that it is always high, new commodities are always 'worth a try'. Additionally, as familiarity and excitement seem to be inversely related, the probability of excitement decreases with repeated use.

33 From this perspective Freud can be viewed as continuing the tradition of the Enlightenment rather than of nineteenth-century medicine. See above, Chapter 1.

34 An association explored by Thompson (1961) and Sohn-Rethel (1978) but made evident first, and most strikingly, by Simmel.

35 Characteristics of money analysed in the *Grundrisse*, Marx (1973: 146–51), and summed up in his aphorism, 'All commodities are perishable money; money is the imperishable commodity' (ibid., p. 149). See, from a different perspective, Douglas in Firth (1967).

36 Thus, 'Money is the purest form of the tool' (Simmel 1978: 210); 'money is detached from all specific contents and exists only as a quantity' (ibid., p. 216).

37 Ibid., p. 79. Though even before it is objectified in exchange, 'subjective' value exists as a *form* that is never 'merely capricious' but 'exists in our consciousness as a fact that can no more be altered than can reality itself' (ibid., p. 63).

38 Ibid., p. 108.

39 Ibid., p. 119.

40 Simmel points out that 'Money is the *representative* of abstract value' (emphasis added), and explicitly draws attention to the linguistic distinction between signifier and signified (ibid., p. 120).

41 Ibid., p. 128 (emphasis added).

42 Ibid., p. 129. He continues: 'Thus money is the adequate expression of the relationship of man to the world.'

43 Ibid., p. 231.

44 See, for example, Weber (1978: 107–9), a calculability which simultaneously depends upon a number of other formal and historical conditions.

45 Because of the length, complexity and objectivity of the 'sequence of purposes' 'given' in the condition of modern society.

46 Ibid., p. 235. One is reminded of Kierkegaard's depiction of 'aesthetic' existence; the 'young man' of *Either* 'hovers above existence'.

47 Compare ibid., pp. 326–31, on the role of money in the development of the sense of individual worth with, for example, pp. 389-94, on money and the ideal of distinction. Thus 'Money thoroughly destroys that self-respect that characterises the distinguished person' (ibid., p. 394).

48 A process Marx noted as a general characteristic of capitalism; see, for example, Marx (1976: 1052–3), only after he had considered it in relation to money (Marx 1973: 196–9).

49 Simmel (1978: 429).

50 See particularly ibid., pp. 238-47, on greed and avarice. Simmel is clearly aware of the unpredictability inherent in the wishful relation to a commodity; 'the relation of wish to its fulfilment is an infinitely diverse one, because the wish almost never allows for all aspects of the object and its effect upon us' (ibid., p. 243). If wishes are 'transferred' to money, as a 'thing absolutely lacking in qualities', this particular source of uncertainty is diminished.

51 Ibid., p. 248.

52 Simmel's discussion is reminiscent not only of Kierkegaard's description of aesthetic despair but also of Dostoevsky's celebration of uncertainty in *The Gambler*.

53 Simmel (1978: 255).

54 Ibid., p. 296.

55 For an interesting discussion of the implications of incompleteness in a number of fields, see Hofstadter (1979).

56 An idea central also to Max Weber's view of modern society; for a brief discussion, see Gerth and Mills (1964: 77–128).

57 This is usually referred to as the Cosmological Principle or Copernican Principle, and describes the *large-scale* structure of the universe. In its modern form it is usually attributed to Milne or Einstein, who put it succinctly: 'all places in the universe are alike' (quoted in Harrison 1981: 88). See also Peebles (1971: 31–42); Silk (1980: 51–5); Rees, Ruffini, and Wheeler (1974: 270–5); and, for a fuller historical discussion, Barrow and Tipler (1986: 367–444).

58 In the classical sciences not only are all movements ideally reversible, the cosmos as a whole is stable and 'complete'. The 'self' similarly is composed of internal 'exchanges' and constitutes in itself a 'finished' structure.

59 A view recently championed by Braudel and his school, rather than by conventional Marxists. Wallerstein, for example, used the revealing title *Historical Capitalism* as an introduction to his work.

60 Though not of course without a good deal of bloodshed.

61 And most directly by Laplace, see Čapek (1961: 122).

62 Far less of these developments spelling the end of capitalism, as periodically, since Schumpeter (1943), is suggested. See, for example, Galbraith (1967).

63 A typical comment, for example, from *On The Genealogy of Morals*: 'Europe is rich and inventive today above all in means of *excitation*; it seems to need nothing so much as stimulants and brandy' (Nietzsche

1969: 159) (emphasis added). And there is nothing so exciting as novelty; 'what distinguishes truly original minds' is the ability to see '*as new* what is old, long familiar, seen and overlooked by everybody' (ibid., p. 176). Of course, as most people are not original in this sense, they require the artificial stimulus of novel commodities, or at least a good illness, as, 'being sick can even become an energetic *stimulus* for life' (ibid., p. 224).

64 A view persuasively espoused by Hawking (1988).

65 Čapek (1971) argues that Bergson, in introducing duration into nature, is one of the most important and neglected sources of modern cosmology. Even in Weyl and Minkowski, he claims, there is no genuine succession and time is again reduced to a spatial order as 'blindfolded consciousness creeps along its world line to discover the future' (Čapek 1961: 165).

66 Bohm suggests that if genuine duration is introduced into nature then no fixed mode of being can remain permanent or any conceptual distinction remain inviolate. Nature then becomes a 'qualitative infinity', implying that 'the development of the universe in time will lead to an inexhaustible diversity of new things' (Bohm, in Čapek 1976: 559).

67 Ideally, as with motor-car licence plates, by dating.

68 Classical equilibrium on the large scale was a tenacious assumption. Einstein in 1917, finding no stable solution to his new field equations, felt obliged to introduce an arbitrary cosmological constant; see Narlikar (1977: 111–12). Yet, ten years later, Hubble's discovery of a direct correlation between distance and apparent recession velocity of distant objects was very quickly accepted (Peebles 1971: 7).

69 Bondi (1961).

70 This proved socially as well as scientifically wide of the mark. Matter is again treated as something 'different' from space-time, and 'creation' is really a device to maintain a 'timeless' equilibrium.

71 There are a number of variants, open or closed, finite or infinite, depending on the choice of constant; Silk (1980: 94–9); Harrison (1981: 293–307).

72 By Penzias and Wilson in 1965 (Silk 1980: 75–9).

73 Several non-technical accounts of modern cosmology take this approach; for example, Wald (1977), Weinberg (1977), Goldsmith and Levy (1974); and for interesting variations, Rowan-Robinson (1985) and Barrow and Silk (1984).

74 Rowan-Robinson (1985: 236–41, 288); Weinberg (1977: 101–6), Barrow and Silk (1984: ix) describe this primordial state of nature as a period 'when all the laws of physics were on an equal footing, all nature's elementary constituents, heavy and light, interacted freely and democratically. The most exotic particles known, or even dreamt of, by man were liberated to participate in this unrestrained interchange'.

75 The universe retains clues to its original playfulness. Weinberg (1977: 149), points out that 'The present universe is so cold that the

symmetries among the different particles and interactions have been obscured by a kind of freezing.' Thus there is in the presently observable universe an 'imbalance' of matter over anti-matter, and of photons over protons, as well as a 'differentiation' of different forces. See particularly Barrow and Silk (1984) on these and related observations. One is forcefully reminded of Freud's theory of sexual 'development', a kind of differentiation and 'cooling', from an original 'hot' state of 'polymorph perversity'.

76 And cosmologists formulate their accounts of the universe in terms of the specific initial conditions of the universe, rather than in terms of continually acting 'laws of nature'; and aim from these conditions at 'nothing less than a complete reconstruction of the past history of the universe' (Barrow and Silk 1984: 208).

77 In Weber's famous argument a *spirit* of capitalism is one of the 'initial conditions' of capitalism itself. See also Abercrombie, Hill, and Turner (1980: 95–127).

78 This is the case even where 'new technologies' make use of effects, such as lasers, which remain classically incomprehensible.

79 Barrow and Silk (1984), Feinberg (1977), Weinberg (1983), Polkinghorne (1979), and Pagels (1982) for these and many other modern exoticisms.

80 See above, Chapter 3.

THE SYMMETRY OF CREATION

1 Husserl (1970: 60).

2 Meyerson's *Identity and Reality* dates from 1908 – prior, that is, to much of the turmoil of modern science, but not of course prior to the general cultural transition of which science was to form such a significant part.

3 A view powerfully expressed by Bergson and Proust as well as by Freud; see above, Chapter 3.

4 Symmetry, that is to say, has largely replaced 'force' as the most fundamental of physical concepts. See Gal-Or (1981: 30–1), Davies and Brown (1988: 33–47), and more generally (and technically), Shubnikov and Koptsik (1974).

5 As represented, for example, by Lukács (1980).

CONCLUSION

1 Calvino (1983), Mann (1958). Kuckuck is of course a critical and comic portrait. His worldly naïvety does not impinge, however, on his scientific authority. His portrait of the cosmos rivals Humboldt's in its classical self-confidence.

BIBLIOGRAPHY

Publication date refers to edition consulted and not necessarily to the first edition.

Abel, F. (1971) *Johannes Kepler 1571/1971*, Bonn.

Abercrombie, N., Hill, S., and Turner, Bryan S. (1980) *The Dominant Ideology Thesis*, London.

(1986) *Sovereign Individuals of Capitalism*, London.

Abro, A. d' (1951) *The Rise of the New Physics: Its Mathematical and Physical Theories*, New York.

Adams, Jeremy duQuesnay (1969) *Patterns of Medieval Society*, Englewood Cliffs, NJ.

Aelred, of Rievaulx (1974) *Spiritual Friendship*, Trans. Mary Eugenia Laker, Washington DC.

(1981) *Dialogue on the Soul*, Trans. C. H. Talbot, Kalamazoo, MI.

Akhundov, Murad D. (1986) *Conceptions of Space and Time*, Cambridge, MA.

Alan of Lille (1980) *The Plaint of Nature*, Trans. James J. Sheridan, Toronto.

⚹ Albertus Magnus (1967) *Book of Minerals*, Trans. and ed. Dorothy Wyckoff, Oxford.

⚹ (1973) *The Book of Secrets of Albertus Magnus*, Michael Best and Frank H. Brightman (eds), Oxford.

Alberti, Leon Battista (1955) *Ten Books on Architecture*, Trans. James Leoni, London.

(1969) *The Family in Renaissance Florence*, Trans. Renée Watkins, New York.

(1972) *On Painting and Sculpture*, Trans. Cecil Grayson, London.

Alexander, H. G. (ed.) (1956) *The Leibniz–Clarke Correspondence*, Manchester.

Alexander, Peter (1985) *Ideas, Qualities and Corpuscles; Locke and Boyle on the Extended World*, Cambridge.

328

Allan, D. G. C. and Schofield, R. E. (1980) *Stephen Hales: Scientist and Philosopher*, London.

Allier, Raoul (1929) *The Mind of the Savage*, London.

Amrine, F., Zucker, F. J., and Wheeler, H (eds) (1987) *Goethe and the Sciences: a Reappraisal*, Dordrecht/Boston.

✳ Aquinas, St Thomas (1911–25) *Summa Theologiae*, London.

✳ (1955) *On the Truth of the Catholic Faith*, New York.

✳ (1964-81) *Summa Theologiae*, Latin/English, London.

✳ Anselm, St (1974) *Anselm of Canterbury*, J. Hopkins and H. Richardson, eds and trans., 4 vols, London.

Anzieu, Didier (1986) *Freud's Self-Analysis*, London.

Arber, Agnes (1953) *Herbals: Their Origin and Evolution, a Chapter in the History of Botany 1470-1670*, Cambridge.

Ariès, Philippe (1973) *Centuries of Childhood*, Harmondsworth, Middx.

(1983) *The Hour of Our Death*, Trans. H. Weaver, London.

Arieti, Silvano (1974) *Interpretations of Schizophrenia*, London.

Artin, Tom (1974) *The Allegory of Adventure: Reading Chrétien's Erec and Yvain*, Lewisburg, WV.

Ashtor, Eliyahu (1983) *Levant Trade in the Later Middle Ages*, Princeton, NJ.

Auerbach, Erich (1961) *Dante: Poet of the Secular World*, Chicago.

(1968) *Mimesis: the Representation of Reality in Western Literature*, Trans. Willard R. Trask, Princeton, N.J.

(1984) *Scenes from the Drama of European Literature*, Manchester.

Augé, Marc (1982) *The Anthropological Circle: Symbol, Function, History*, Cambridge.

✳ Augustine, St (1972) *The City of God*, Harmondsworth, Middx.

Babb, L. (1951) *The Elizabethan Malady: a Study of Melancholia in Elizabethan Literature 1580-1640*, East Lansing, MI.

(1959) *Sanity in Bedlam: a Study of Robert Burton's Anatomy of Melancholy*, East Lansing, MI.

Bachelard, Gaston (1964) *The Psychoanalysis of Fire*, Trans. Alan Ross, Boston.

Bachofen, J. J. (1967) *Myth, Religion and Mother Right*, Trans. Ralph Mannheim, Princeton, NJ.

Bakhtin, Mikhail (1968) *Rabelais and His World Mass*, Trans. H. Iswolsky, Cambridge, MA.

(1973) *Problems of Dostoevsky's Poetics*, Trans. R. W. Rotsel, New York.

Baldwin, J. M. (1895) *Mental Development in the Child and the Race*, New York and London.

Bantock, G. H. (1984) *Studies in the History of Educational Theory*, 2 vols, London.

Barkan, Edward (1975) *Nature's Work of Art: the Human Body as Image of the World*, New Haven, CT, and London.

Barrow, John D. and Silk, Joseph (1984) *The Left Hand of Creation: the Origin and Evolution of the Expanding Universe*, London.

Barrow, John D. and Tipler, Frank J. (1986) *The Anthropic Cosmological Principle*, Oxford.

Barthes, Roland (1967) *Elements of Semiology*, Trans. Annette Lavers and Colin Smith, London.

(1976) *Sade, Fourier, Loyola*, Trans. Richard Miller, London.

Baudrillard, Jean (1975) *The Mirror of Production*, Trans. Mark Poster, St Louis, MO.

(1981) *For a Critique of the Political Economy of the Sign*, Trans. Charles Levin, St. Louis, MO.

Beattie, James (1974, (1783)), *Dissertations Moral and Critical*, Hildesheim.

Bell, Daniel (1960), *The End of Ideology*, New York.

Bell, Daniel N. (1984), *The Image and Likeness: the Augustinian Spirituality of William of St Thierry*, Kalamazoo, MI.

Bednar, John (1974) *La Spiritualité et le Symbolisme dans les Oeuvres de Chrétien de Troyes*, Paris.

Benedict, St (1960) *The Rule of Saint Benedict*, Justin McCann (ed. and trans.), London.

Benson, Robert L. and Constable, Giles (1982) *Renaissance and Renewal in the Twelfth Century*, Oxford.

✴ Bentham, Jeremy (1970 (1789, 1823)) *An Introduction to the Principles of Morals and Legislation*, J. H. Burns and H. L. A. Hart (eds), London.

Benveniste, Emile (1973) *Indo-European Languages and Society*, Trans. Elizabeth Palmer, London.

Bergson, Henri (1911) *Matter and Memory*, New York.

(1935) *Two Sources of Religion and Morality*, London.

✳ Berkeley, George (1910, 1709)) *A New Theory of Vision and Other Selected Philosophical Writings*, London and Toronto.

Berman, Marshall (1982) *All That is Solid Melts into Air*, New York.

Bernard of Clairvaux (1940) *The Steps of Humility*, G. B. Burch (ed.), Cambridge, MA.

(1971, 1976, 1979, 1980) *The Works of Bernard of Clairvaux*, Shannon and Kalamazoo, MI.

Bianciotto, Gabriel (1980) *Bestiaires du Moyen Age*, Paris.

Bienkowska, Barbara (ed.) (1973) *The Scientific World of Copernicus*, Dordrecht and Boston.

Bleuler, Eugene (n.d.; German edn 1916) *Textbook of Psychiatry*, Trans. A. A. Brill, London.

(1950 (1911)) *Dementia Praecox: or the Group of Schizophrenias*, Trans. Joseph Zirkin, New York.

Bloch, Marc (1965) *Feudal Society*, Trans. L. A. Manyon, London.

Boase, T. S. R. (1972) *Death in the Middle Ages*, London.

Boscovich, Roger Joseph (1961) *Roger Joseph Boscovich: Studies of His Life and Work*, Lancelot Law Whyte (ed.), London.

Bohr, Niels (1934) *Atomic Theory and the Description of Nature*, London.

✳ Boltzmann, Ludwig (1974) *Theoretical Physics and Philosophical Problems*, Brian McGuinness (ed.), Dordrecht and Boston.

Bondi, H. (1961) *Cosmology*, Cambridge.

Booth, Edward (1983) *Aristotelian Aporetic Ontology in Islamic and Christian Thinkers*, Cambridge.

Bolton, Brenda (1983) *The Medieval Reformation*, London.

Borkenau, F. (1987) 'The sociology of the mechanistic world-picture', *Science in Context* 1 (1).

Bowie, Theodore (ed.) (1959) *The Sketchbook of Villard de Honnecourt*, Bloomington, Ind., and London.

Boxer, C. R. (1965) *The Dutch Seaborne Empire 1600-1800*, London.

Boyle, Robert (1979) *Selected Philosophical Papers* M. A. Stewart (ed.), Manchester.

Braudel (Fernand) (1979) *The Structure of Everyday Life*, Trans. Sian Reynolds, London.

(1983) *The Wheels of Commerce*, Trans. Sian Reynolds, London.

(1984) *The Perspective of the World*, Trans. Sian Reynolds, London.

Broglie, Louis de (1939) *Matter and Light: the New Physics*, London.

(1962) *New Perspectives in Physics*, Edinburgh and London.

Brooke, Rosalind and Brook, Christopher (1984) *Popular Religion in the Middle Ages*, London.

Bruno, Giordano (1964) *Giordano Bruno's The Heroic Frenzies*, Trans. Paul Eugene Memmo, Jr, Chapel Hill, NC.

(1977) *The Ash Wednesday Supper*, Trans. Edward A. Gosselin and Lawrence S. Lerner, Hamden, CT.

Buchdahl, Gerd (1969) *Metaphysics and the Philosophy of Science: the Classical Origins, Descartes to Kant*, Oxford.

Bühler, K. (1930) *The Mental Development of the Child*, London.

Burckhardt, Jacob (1921) *The Civilisation of the Renaissance in Italy*, London.

Burrow, J. W. (1966) *Evolution and Society: a Study in Victorian Social Theory*, Cambridge.

Burton, Robert (1932 (1621)) *The Anatomy of Melancholy*, 3 vols, London.

Burtt, E. A. (1924) *The Metaphysical Foundations of Modern Physical Science*, London.

Butler, Cuthbert (1919) *Benedictine Monachism: Studies in Benedictine Life and Rule*, London.

Butler, Judith P. (1987) *Subjects of Desire: Hegelian Reflections in Twentieth-Century France*, New York.

Butterfield, H. (1968) *The Origins of Modern Science 1300-1800*, London.

331

Butts, Robert E. and Davis, John W. (1970) *The Methodological Heritage of Newton*, Oxford.

Butts, Robert E. and Pitt, Joseph C. (1978) *New Perspectives on Galileo*, Dordrecht and Boston.

Buytendijk, F. T. J. (1935) *The Mind of the Dog*, London.

Bynum. W. F., Porter, Roy, and Shepherd, Michael (1985) *The Anatomy of Madness*, 3 vols, London.

Callus, D. A. (1955) *Robert Grosseteste: Scholar and Bishop*, Oxford.

Calvino, Italo (1982) *Cosmicomics*, Trans. William Weaver, London.

(1983) *Mr Palomar*, Trans. William Weaver, London.

Campbell, Lewis and Garnett, William (1882) *The Life of James Clerk Maxwell*, London.

Camus, Albert (1971) *The Rebel*, Trans. Anthony Bower, Harmondsworth, Middx.

Canetti, Elias (1962) *Crowds and Power*, Trans. Carol Stewart, London.

Čapek, Milic (1961) *The Philosophical Impact of Modern Physics*, Princeton, NJ.

(1971) *Bergson and Modern Physics*, Dordrecht.

(ed.) (1976) *The Concepts of Space and Time*, Dordrecht and Boston.

Capellanus, Andreas (1959) *The Art of Courtly Love*, Trans. John Jay Parry, New York.

Carnot, Sadi (1960) *Reflections on the Motive Power of Fire*, E. Mendoza (ed.), New York.

Carroll, John (1974) *Break-out from the Crystal Palace*, London.

Caspar, Max (1959) *Kepler*, London and New York.

Cassirer, Ernst (1953) *The Platonic Renaissance in England*, Trans. James P. Pettegrove, London.

(1953, 1955, 1957) *The Philosophy of Symbolic Forms*, Trans. Ralph Mannheim, 3 vols, New Haven, CT, and London.

(1963) *The Individual and the Cosmos in Renaissance Philosophy*, Oxford.

(1966) *The Philosophy of the Enlightenment*, Trans. Fritz C. A. Koelln and James P. Pettegrove, Boston.

Cassirer, Ernst, Kristeller, Paul Oskar, and Randall, John Herman (1956) *The Renaissance Philosophy of Man*, Chicago.

Cazeneuve, Jean (1972) *Lucien Lévy-Bruhl*, Trans. Peter Riviere, Oxford.

Charlton, D. G. (1959) *Positivist Thought in France during the Second Empire, 1852-1870*, Oxford.

Chastel, A., Grayson, C., Hall, M. B., Hay, D., Kristeller, P. O., Rubinstein, N., Schmitt, C. B., Trinkaus, C., and Ullman, W. (1982) *The Renaissance: Essays in Interpretation*, London and New York.

Cheyette, Frederic L. (ed.) (1968) *Lordship and Community in Medieval Europe*, New York.

Chrétien de Troyes (1966) *Cligés: a Romance*, Trans. L. J. Gardiner, New York.

(1982) *Perceval: the Story of the Grail*, Trans. Nigel Bryant, Cambridge.

Cipolla, Carlo (1967) *Clocks and Culture 1300-1900*, London.

Clifford, William Kingdom (1879) *Lectures and Essays*, Leslie Stephen and Frederick Pollock (eds), London.

(1885) *The Common Sense of the Exact Sciences*, London.

Clagett, Marshall (1959) *The Science of Mechanics in the Middle Ages*, Madison, Wis., and London.

* (1968) *Nicole Oresme and the Medieval Geometry of Qualities and Motions*, Wisconsin.

Clark, G. N. (1970) *Science and Social Welfare in the Age of Newton*, Oxford.

Clavelin, Maurice (1974) *The Natural Philosophy of Galileo*, Trans. A. J. Pomerans, Cambridge, MA.

Codrington, Robert Henry (1891) *The Melanesians: Studies in Their Anthropology and Folklore*, Oxford.

Cohen, I. Bernard (1966) *Franklin and Newton*, Cambridge, MA.

(1980) *The Newtonian Revolution*, Cambridge.

Cohn, Norman (1970) *The Pursuit of the Millennium*, London.

Cole, J. Preston (1971) *The Problematic Self in Kierkegaard and Freud*, New Haven, CT., and London.

Cole, Luella (1950) *A History of Education*, New York.

Collingwood, R. G. (1945) *The Idea of Nature*, Oxford.

(1946) *The Idea of History*, Oxford.

Collins, Ardis B. (1974) *The Secular is Sacred: Platonism and Thomism in Marsilio Ficino's Platonic Theology*, The Hague.

Compayré, G. (1893) *L'Evolution intellectuelle et morale de l'enfant*, Paris.

Comte, Auguste (1883) *The Positive Philosophy*, Trans. Harriet Martineau, 2 vols, London.

* Condillac, E. B. de (1973) *Essais sur l'origine de connaissances humaines*, Auvers-sur-Oise.

Condorcet, Marie Jean Antoine, marquis de (1955) *Sketch for a Historical Picture of the Progress of the Human Mind*, New York.

Copernicus, Nicolaus (1971), *Three Copernican Treatises*, Edward Rosen (ed.), New York.

* (1976) *Copernicus: On The Revolution of the Heavenly Spheres* A. M. Duncan (ed. and trans.), Newton Abbot.

Corrigan, Philip and Sayer, Derek (1985) *The Great Arch: English State Formation as Cultural Revolution*, Oxford.

Coulton, G. G. (1923, 1927, 1936) *Five Centuries of Religion*, 3 vols, Cambridge.

Cowdrey, H. E. J. (1970) *The Cluniacs and the Gregorian Reform*, Oxford.

Critchley, J. S. (1978) *Feudalism*, London.

Crocker, J. C. (1985) *Vital Souls: Bororo Cosmology, Natural Symbolism and Shamanism*, Tucson, Arizona.

Crombie, A. C. (1953) *Robert Grosseteste and the Origin of Experimental*

Science, Oxford.

(1961) *Augustine to Galileo*, 2 vols, London.

Curtius, Ernst Robert (1953) *European Literature and the Latin Middle Ages*, London.

Cusanus, Nicolas (1954) *Of Learned Ignorance*, London.

Davidoff, L. and Hall, C. (1987) *Family Fortunes: Men and Women of the English Middle Class* 1780–1850, London.

Davies, P. C. W. (1977) *Space and Time in the Modern Universe*, Cambridge.

Davies, P. C. W. and Brown, J. (eds.) (1988) *Superstrings: a Theory of Everything?* Cambridge.

Davis, D. B. (1970) *The Problem of Slavery in Western Culture*, Harmondsworth, Middx.

Davis, N. Z. (1975) *Society and Culture in Early Modern France*, London.

Debus, Allen G. (1977) *The Chemical Philosophy*, 2 vols, New York.

(1978) *Man and Nature in the Renaissance*, Cambridge.

(1987) *Chemistry, Alchemy and the New Philosophy, 1550-1700*, London.

Déchanet, Jean Marie (1972) *William of St Thierry: the Man and his Work*, Trans. Richard Strachan, Spencer, MA.

Delacroix, H. (1934) *Les Grandes Formes de la Vie Mentale*, Paris.

Deleuze, Gilles (1972) *Proust and Signs*, Trans. Richard Howard, London.

(1983) *Nietzsche and Philosophy*, Trans. Hugh Tomlinson, London.

Deleuze, Gilles and Guattari, Felix (1977) *Anti-Oedipus: Capitalism and Schizophrenia*, Trans. Robert Hurley, Mark Seem, and Helen R. Lane, New York.

Descartes, René (1912) *A Discourse on Method*, London.

Diderot, Denis (1935) *Supplément au Voyage de Bougainville*, Paris and Oxford.

(1966) *Rameau's Nephew / D'Alembert's Dream*, Trans. Leonard Tancock, Harmondsworth, Middx.

Dijksterhuis, E. J. (1961) *The Mechanization of the World Picture*, Trans. C. Dikshoorn, Oxford.

Dobrzycki, Jerzy (ed.) (1972) *The Reception of Copernicus's Heliocentric Theory*, Dordrecht and Boston.

Doerner, Klaus (1981) *Madmen and the Bourgeoisie*, Oxford.

Domb, C. (ed.) (1963) *Clerk Maxwell and Modern Science*, London.

Donkin, R. A. (1978) *The Cistercians: Studies in the Geography of Medieval England and Wales*, Toronto.

Donovan, A. L. (1975) *Philosophical Chemistry in the Scottish Enlightenment*, Edinburgh.

Donzelot, Jacques (1980) *The Policing of Families: Welfare versus the State*, London.

Dostoevsky, Fyodor (1953) *The Devils*, Trans. David Magarshack,

Harmondsworth, Middx.

(1955) *The Idiot*, Trans. David Magarshack, Harmondsworth, Middx.

(1958) *The Brothers Karamazov*, Trans. David Magarshack, 2 vols, Harmondsworth, Middx.

(1962) *The House of the Dead*, Trans. H. Sutherland Edwards, London.

(1966) *Crime and Punishment*, Trans. David Magarshack, Harmondsworth, Middx.

(1966) *The Gambler/Bobok/A Nasty Story*, Trans. Jessie Coulson, Harmondsworth, Middx.

(1972) *Notes from Underground/The Double*, Trans. Jessie Coulson, Harmondsworth, Middx.

Douglas, Andrew Halliday (1910) *The Philosophy and Psychology of Pietro Pomponazzi*, Cambridge.

Douglas, Mary (1970) *Purity and Danger*, Harmondsworth, Middx.

(1972) *Natural Symbols*, Harmondsworth, Middx.

(ed.) (1973) *Rules and Meanings*, Harmondsworth, Middx.

Douglas, Mary and Isherwood, Baron (1979) *The World of Goods: Towards an Anthropology of Consumption*, Harmondsworth, Middx.

Drake, Stillman (1970) *Galileo Studies*, Ann Arbor, MI.

(1978) *Galileo at Work: His Scientific Biography*, Chicago.

(1980), *Galileo*, Oxford.

Drake, Stillman (ed.) (1957) *Discoveries and Opinions of Galileo*, New York.

Drake, Stillman and Drabkin, I. E. (1969) *Mechanics in Sixteenth-Century Italy*, Madison, Milwaukee.

Drake, Stillman and O'Malley, C. D. (1960) *The Controversy on the Comets of 1618*, Philadelphia, PA.

Dronke, Peter (1968) *Medieval Latin and the Rise of the European Love-Lyric*, vol.1, Oxford.

(1974) *Fabula: Explorations into the Uses of Myth in Medieval Platonism*, Leiden and Cologne.

(ed.) (1978) *Cosmographia*, Leiden.

(1984) *The Medieval Poet and His World*, Rome.

(1986) *Poetic Individuality in the Middle Ages*, London.

Dryer, J. L. E. (1963) *Tycho Brahe*, New York.

Duby, Georges (1977) *The Chivalrous Society*, London.

(1980) *The Three Orders: Feudal Society Imagined*, Chicago.

Duerr, Hans Peter (1985) *Dreamtime: Concerning the Boundary between Wilderness and Civilisation*, Trans. Felicitas Goodman, Oxford.

Duhem, Pierre (1954) *The Aim and Structure of Physical Theory*, Princeton, NJ.

(1985) *Medieval Cosmology: Theories of Infinity, Place, Time, Void, and the Plurality of Worlds* Roger Ariew (ed. and trans.), Chicago and London.

Duncan, A. M. (ed.) (1976) *Copernicus: On the Revolution of the Heavenly Spheres*, Newton Abbot.

Dunlop, D. M. (1958) *Arabic Science in the West*, Karachi.

Durkheim, Emile (1964) *The Elementary Forms of the Religious Life*, Trans. Joseph Ward Swain, London.

Durkheim, Emile and Mauss, Marcel (1963) *Primitive Classification*, Trans. Rodney Needham, London.

Eadmer (1962) *The Life of St Anselm*, ed. and trans., with notes and introduction, by R. W. Southern, Oxford.

Earman, J., Glymour, C., and Stachel, J. (1977) *Foundations of Space-Time Theories*, Minneapolis, MN.

Eddington, A. S. (1920) *Space, Time and Gravitation*, Cambridge.

Einstein, Albert (1920) *Relativity: The Special and the General Theory*, London.

—— (1956) *Investigations on the Theory of the Brownian Movement*, R. Furth (ed.), (A. D. Cowper trans.), New York.

Einstein, Albert and Infeld, Leopold (1947) *The Evolution of Physics*, Cambridge.

Elias, N. (1978) *The Civilizing Process*, vol. 1, *The History of Manners*, Trans. E. Jephcott, Oxford.

Ellenberger, Henri F. (1970) *The Discovery of the Unconscious: The History and Evolution of Dynamic Psychiatry*, London.

Emmerson, Richard Kenneth (1981) *Antichrist in the Middle Ages*, Manchester.

Evans, G. R. (1978) *Anselm and Talking About God*, Oxford.

Evans, G. R. (1980) *Anselm and a New Generation*, Oxford.

Evans-Pritchard, E. E. (1937) *Witchcraft, Oracles, and Magic among the Azande*, Oxford.

—— (1965) *Theories of Primitive Religion*, Oxford.

—— (1981) *A History of Anthropological Thought*, London.

Febvre, Lucien (1982) *The Problem of Unbelief in the Sixteenth Century: the Religion of Rabelais*, Trans. Beatrice Gottlieb, Cambridge, MA.

Febvre, Lucien and Martin, Henri-Jean (1976) *The Coming of the Book*, Trans. David Gerard, London.

Feinberg, Gerald (1977) *What is the World Made Of?* New York.

Fellows, Otis E. and Milliken, Stephen F. (1972) *Buffon*, New York.

Ferguson, Adam (1966 (1767)) *An Essay on the History of Civil Society*, Duncan Forbes (ed.), Edinburgh.

—— (1973 (1798)), *Principles of Moral and Political Science*, 2 vols, New York.

Ferguson, Harvie (1983) *Essays in Experimental Psychology*, London.

Feuchtersleben, Baron Ernst von (1847) *The Principles of Medical Psychology*, London.

Feuer, L. S. (ed.) (1959) *Marx and Engels: Basic Writings on Politics and Philosophy*, New York.

Feyerabend, Paul (1975) *Against Method: Outline of an anarchistic theory of knowledge*, London.

Feynman, Richard P., Leighton, Robert B., and Sands, Matthew (1965) *The Feynman Lectures on Physics*, 3 vols, New York and London.

Ficino, Marsilio (1944) *Marsilio Ficino's Commentary on Plato's Symposium*, Sears Reynolds Jayne (ed.), New York.

(1975) *The Letters of Marsilio Ficino*, 3 vols, London.

Finocchiaro, Maurice A. (1980) *Galileo and the Art of Reasoning*, Dordrecht and Boston.

Firth, Raymond (ed.) (1967) *Themes in Economic Anthropology*, London.

(1973) *Symbols: Public and Private*, London.

Folse, Henry J. (1985) *The Philosophy of Niels Bohr: the Framework of Complementarity*, Amsterdam.

Foucault, Michel (1965) *Madness and Civilization: a History of Insanity in the Age of Reason*, Trans. Richard Howard, London.

(1970) *The order of things: an Archaeology of the Human Sciences*, London.

(1977) *Discipline and Punish: the Birth of the Prison*, Trans. Alan Sheridan, London.

(1979) *The History of Sexuality*, vol. 1, Trans. Robert Hurley, London.

Fourquin, Guy (1976) *Lordship and Feudalism in the Middle Ages*, London.

France, Peter (1988) *The 'Confessions' of Rousseau*, Oxford.

Frankel, Charles (1969) *The Faith of Reason: the Idea of Progress in the French Enlightenment*, New York.

Frazer, Sir James (1900) *The Golden Bough: a Study in Magic and Religion*, London.

(1922), *Belief in Immortality*, 3 vols, London.

French, A. P. (1972) *John Dee: the World of an Elizabethan Magus*, London.

Freud, Sigmund (1953-74) *The Standard Edition of the Complete Psychological Works of Sigmund Freud*, Trans. and ed. James Strachey, Alix Strachey, and Alan Tyson, 24 vols, London.

(1954) *The Origins of Psycho-Analysis: Letters to Wilhelm Fliess, Drafts and Notes: 1887-1902*, Eds Marie Bonaparte, Anna Freud, Ernst Kris; Trans. Eric Mosbacher and James Strachey, London.

Freudenthal, Gideon (1986) *Atom and Individual in the Age of Newton: On the Genesis of the Mechanistic World View*, Dordrecht.

Frisby, David (1984) *Georg Simmel*, London and New York.

(1985) *Fragments of Modernity: Theories of Modernity in the Work of Simmel, Kracauer and Benjamin*, Cambridge.

Froebel, (1897) *Pedagogics of the Kindergarten*, Trans. Josephine Jarvis, London.

Gabel, J. (1975) *False Consciousness: an Essay on Reification*, Oxford.

Gadol, Joan (1969) *Leon Battista Alberti: Universal Man of the Early Renaissance*, Chicago and London.

Galbraith, John Kenneth (1967) *The New Industrial State*, London.

Galilei, Galileo (1914) *Dialogue Concerning Two New Sciences*, Trans. Henry Crow and Alfonso de Salvio, New York.

—— (1953), *Dialogue on the Great World Systems*, Trans. T. Salusbury, revised and annotated by Giorgio de Santillana, Chicago.

—— (1957) *Discoveries and Opinions of Galileo*, Stillman Drake (ed. and trans.), New York.

—— (1977), *Galileo's Early Notebooks: the Physical Questions*, Notre Dame and London.

—— (1981) *Cause, Experiment and Science*, Stillman Drake (ed.), Chicago and London.

Gal-Or, Benjamin (1981) *Cosmology, Physics and Philosophy*, New York.

Ganshof, F. L. (1952) *Feudalism*, London.

Garin, Eugenio (1965) *Italian Humanism: Philosophy and Civic Life in the Renaissance*, Trans. Peter Munz, Oxford.

—— (1976) *Astrology in the Renaissance: the Zodiac of Life*, London.

—— (1978) *Science and Civic Life in the Italian Renaissance*, Trans. Peter Munz, Gloucester, MA.

Gay, Peter (1966, 1969) *The Enlightenment: an Interpretation*, 2 vols, London.

—— (ed.) (1973) *The Enlightenment: a Comprehensive Anthology*, New York.

Gennep, A. van (1910-14) *Religions, moeurs et legendes*, Paris.

Genovese, E. D. (1971) *The World the Slaveholders Made*, New York.

Geroch, Robert (1978) *General Relativity: from A to B*, Chicago.

Gerth, H. H. and Mills, C. W. (eds) (1964) *From Max Weber: Essays in Sociology*, London.

Gibbins, Peter (1987) *Particles and Paradoxes: the Limits of Quantum Logic*, Cambridge.

Gilbert of Hoyland (1978, 1979) *Sermons on the Song of Songs*, Trans. Lawrence Braceland, 2 vols, Kalamazoo, MI.

Gilbert, William (1900) *On the Magnet*, London.

Gilson, Etienne (1924) *The Philosophy of St Thomas Aquinas*, Cambridge.

—— (1936) *The Spirit of Medieval Philosophy*, London.

—— (1955) *History of Christian Philosophy in the Middle Ages*, London.

Ginzburg, Carlo (1980) *The Cheese and the Worms: the Cosmos of a Sixteenth-Century Miller*, Trans. John and Anne Tedeschi, London.

Girard, René (1965) *Deceit, Desire and the Novel: Self and Other in Literary Structure*, Trans. Yvonne Freccero, Baltimore, MD.

Goldmann, Lucien (1973) *The Philosophy of the Enlightenment*, London.

Goldsmith, Donald and Levy, Donald (1974) *From the Black Hole to the Infinite Universe*, San Francisco.

Gould, Stephen Jay (1977) *Ontogeny and Phylogeny*, London and Cambridge, MA.

Grant, Edward (ed.) (1974) *A Source Book in Medieval Science*, Cambridge, MA.

(1977) *Physical Science in the Middle Ages*, Cambridge.

(1981) *Much Ado about Nothing: Theories of Space and Vacuum from the Middle Ages to the Scientific Revolution*, Cambridge.

Grean, Stanley (1967) *Shaftesbury's Philosophy of Religion and Ethics: a Study in Enthusiasm*, New York.

Grese, William C. (1982) *Corpus Hermeticum XII and Early Christian Literature*, Leiden.

Griaule, Marcel (1965) *Conversations with Ogotemmeli: an Introduction to Dogon Religious Ideas*, Oxford.

Groos, Karl (1898) *The Play of Animals*, London.

(1901) *The Play of Man*, New York.

Gruber, Howard E. (1974) *Darwin on Man: a Psychological Study of Scientific Creativity*, London.

Grünbaum, Adolf (1973) *Philosophical Problems of Space and Time*, Dordrecht and Boston.

Gurevich, A. J. (1985) *Categories of Medieval Culture*, London.

Gusdorf, Georges (1967) *Les sciences humaines et la pensée occidentale*, vol. 2, Paris.

Habermas, Jurgen (1971) *Toward a Rational Society*, London.

Haeckel, Ernst (1879) *The Evolution of Man: a Popular Exposition of the Principal Points of Human Ontogeny and Phylogeny*, London.

Hall, A. Rupert (1983) *The Revolution in Science 1500-1750*, Harlow.

Hall, G. S. (1904) *Adolescence: Its Psychology and Its Relation to Physiology, Anthropology, Sociology, Sex, Crime, Religion and Education*, 2 vols, New York.

Hallier, Amedée (1969) *The Monastic Theology of Aelred of Rievaulx*, Shannon.

Hamilton, Bernard (1986) *Religion in the Medieval West*, London.

Haren, M. (1985) *Medieval Thought: The Western Intellectual Tradition from Antiquity to the Thirteenth Century*, London.

Harris, William T. (1898) *Psychological Foundations of Education*, New York.

Harrison, Edward R. (1981) *Cosmology*, Cambridge.

Hartley, David (1810 (1749)) *Observations on Man: His Frame, His Duty, and His Expectations*, London.

Hartmann, H., Kris, E., and Lowenstein, R. (1964) *Papers on Psychoanalytic Psychology*, New York.

Hartshorne, Charles (1962) *The Logic of Perfection*, La Salle, IL.

Haskins, Charles Homer (1924) *Studies in the History of Medieval Science*, Cambridge, MA.

(1927) *The Renaissance of the Twelfth Century*, Cambridge, MA.

Hawking, S. (1988) *Brief History of Time*, London.

Hazard, P. (1965) *European Thought in the Eighteenth Century*, Harmondsworth, Middx.

✳ Hegel, G. W. F. (1977 (1807)) *Hegel's Phenomenology of Spirit*, Trans. A. V. Miller, Oxford.

Heilbron, J. L. (1979) *Electricity in the Seventeenth and Eighteenth Centuries*, Berkeley, CA.

Heisenberg, Werner (1959) *Physics and Philosophy: The Revolution in Modern Science*, London.

Heller, Agnes (1978) *Renaissance Man*, London.

Hendry, John (1986) *James Clerk Maxwell and the Theory of Electromagnetic Field*, Bristol and Boston.

Heninger, S. K. (1974) *Touches of Sweet Harmony: Pythagorean Cosmology and Renaissance Poetics*, San Marino, CA.

(1977) *The Cosmological Glass: Renaissance Diagrams of the Universe*, San Marino, CA.

Herbert, Nick (1985) *Quantum Reality: Beyond the New Physics*, London.

Herlihy, David (ed.) (1970) *The History of Feudalism*, London.

Hermann, Armin (1971) *The Genesis of Quantum Theory 1899-1913*, Trans. Claude W. Nash, Cambridge, MA.

Hersey, G. L. (1976) *Pythagorean Palaces: Magic and Architecture in the Italian Renaissance*, Ithaca, NY.

Hertz, R. (1960) *Death and the Right Hand*, Trans. R. Needham and C. Needham, with an Introduction by E. E. Evans-Pritchard, London.

Hesse, Mary (1961) *Forces and Fields*, London.

Hey, Tony and Walters, Patrick (1987) *The Quantum Universe*, Cambridge.

Hiebert, E. N., Ihde, A. J., and Schofield, R. (1980) *Joseph Priestley: Scientist, Theologican and Metaphysician*, Lewisburg, WV, and London.

Hilton, R. H. (1975) *The English Peasantry in the Later Middle Ages*, Oxford.

(ed.) (1979) *The Transition from Feudalism to Capitalism*, London.

(1985) *Class Conflict and the Crisis of Feudalism*, London.

✳ Hobbes, Thomas (1840) *The English Works of Thomas Hobbes*, Sir William Molesworth (ed.), London.

✳ (1962 (1651)) *Leviathan*, John Plamenatz (ed.), London.

Hodgen, Margaret T. (1964) *Early Anthropology in the Sixteenth and Seventeenth Centuries*, Philadelphia, PA.

Hofstadter, Douglas R. (1979) *Godel, Escher, Bach: an Eternal Golden Braid*, Hassocks.

Holbach, Paul-Henry Th. d' (1966 (1770)) *Système de la nature*, 2 vols, Hildesheim.

Holt, J. (1974) *Escape from Childhood: The Needs and Rights of Children*, Harmondsworth, Middx.

Holton, Gerald (1973) *Thematic Origins of Scientific Thought: Kepler to Einstein*, Cambridge, MA.

Holton, R. J. (1985) *The Transition from Feudalism to Capitalism*, London.

Horton, Robin and Finnegan, Ruth (eds) (1973) *Modes of Thought: Essays in Thinking in Western and Non-Western Societies*, London.

Hubert, H. and Mauss, M. (1964) *Sacrifice: Its Nature and Function*, Trans. W. D. Halls, with a Foreword by E. E. Evans-Pritchard, London.

Hugh of St Victor (1961) *Didascalion*, Trans. Jerome Taylor, London and New York.

Hugh-Jones, Christine (1979) *From the Milk River: Spatial and Temporal Processes in Northwest Amazonia*, Cambridge.

Hugh-Jones, Stephen P. (1979) *The Palm and the Pleiades: Initiation and Cosmology in Northwest Amazonia*, Cambridge.

Hughes, G. E. (1982) *John Buridan on Self-Reference*, Cambridge.

Huizinga, J. (1965) *The Waning of the Middle Ages*, Trans F. Hopman, London.

(1971) *Homo Ludens: a Study of the Play Element in Culture*, London.

Humboldt, Alexander von (1856) *Cosmos: Sketch of a Physical Description of the Universe*, 4 vols, London.

Hume, David (1854) *The Philosophical Works of David Hume*, Edinburgh.

(1963) *Hume on Religion*, London.

(1975) *A Treatise on Human Nature*, L. A. Selby-Bigge (ed.), Oxford.

Hund, Friedrich (1974) *The History of Quantum Theory*, London.

Hunt, David (1970) *Parents and Children in History: the Psychology of Family Life in Early Modern France*, New York.

Hunt, Noreen (1967) *Cluny under Saint Hugh 1049-1109*, London.

(ed.) (1971) *Cluniac Monasticism in the Central Middle Ages*, London.

Hunter, Richard and Macalpine, Ida (eds) (1963) *Three Hundred Years of Psychiatry*, London.

Husserl, Edmund (1970) *The Crisis of European Sciences and Transcendental Phenomenology*, Trans. David Carr, Evanston, IL.

Hutcheson, Francis (1971 (1725)) *Collected Works*, vols 1 and 2, Hildesheim.

Idung of Prüfening (1977) *Cistercians and Cluniacs: the Case for Cîteaux*, Kalamazoo, MI.

Jacob, J. R. (1977) *Robert Boyle and the English Revolution*, New York.

Jacob, Margaret C. (1976) *The Newtonians and the English Revolution 1689-1720*, Hassocks.

(1981) *The Radical Enlightenment: Pantheists, Freemasons and Republicans*, London.

Jaki, Stanley L. (1973) *The Milky Way*, Newton Abbot.

James, William (1980 (1918)) *The Principles of Psychology*, 2 vols, New York.

Jammer, Max (1954) *Concepts of Space: the History of Theories of Space in Physics*, Cambridge, MA.

(1957) *Concepts of Force: a Study in the Foundations of Dynamics*,

341

Cambridge, MA.

(1961) *Concepts of Mass*, Cambridge, MA.

(1966) *The Conceptual Development of Quantum Mechanics*, New York.

(1974) *The Philosophy of Quantum Mechanics*, London and New York.

Jardine, N. (1984) *The Birth of History and Philosophy of Science: Kepler's A Defence of Tycho against Ursus*, Cambridge.

Jaspers, Karl (1963 (1913, 1923)) *General Psychopathology*, Trans. J. Hoenig and Marion W. Hamilton, Manchester.

(1965) *Nietzsche*, Trans. Charles F. Wallraff and Frederick J. Schmitz, Tucson, AZ.

Jevons, Frank Byron (1896) *An Introduction to the History of Religion*, London.

John of Salisbury (1962) *The Metalogicon of John of Salisbury*, Trans. Daniel D. McGarvy, Berkeley, CA, and Los Angeles.

Johnson, Francis R. (1937) *Astronomical Thought in Renaissance England*, London.

Johnson, Samuel (1963) *The Idler and the Adventurer*, New Haven, CT.

(1976) *Johnson on Johnson*, John Wain (ed.), London.

Johnston, Mark D. (1987) *The Spiritual Logic of Ramon Llull*, Oxford.

Johnston, William M. (1972) *The Austrian Mind*, Berkeley, CA, and Los Angeles.

Jones, Kathleen (1955) *Lunacy, Law and Conscience 1744-1845*, London.

Jung, C. G. and Pauli, W. (1955) *The Interpretation of Nature and the Psyche*, London.

Kames, Lord (Henry Home) (1761) *Elements of Criticism*, Edinburgh.

(1766) *Principles of Equity*, 2 vols, Edinburgh.

(1813) *Sketches of the History of Man*, Edinburgh.

Kandinsky, Wasily (1914) *The Art of Spiritual Harmony*, Trans. M. T. H. Sadler, London.

Kant, Immanuel (1963) *On History*, Lewis White Beck (ed.), Indianapolis, IN.

Kantorowicz, Ernst H. (1957) *The King's Two Bodies: a Study in Medieval Political Theology*, Princeton, NJ.

Kargon, Robert Hugh (1966) *Atomism in England from Hariot to Newton*, Oxford.

Kaufmann, William J. (1979) *The Cosmic Frontiers of General Relativity*, Harmondsworth, Middx.

Kealey, Edward J. (1981) *Medieval Medicine: a Social History of Anglo-Norman Medicine*, London and Baltimore, MD.

Kelsen, Hans (1946) *Nature and Society*, London.

Kettler, David (1965) *The Social and Political Thought of Adam Ferguson*, Ohio.

Kibre, Pearl (1984) *Studies in Medieval Science: Alchemy, Astrology, Mathematics and Medicine*, London.

Kierkegaard, Søren (1941) *Concluding Unscientific Postscript*, Trans. David F. Swenson and Walter Lowrie, Princeton, NJ.

— (1944) *Attack upon 'Christendom'*, Trans. and ed. Walter Lowrie, Princeton, NJ.

— (1944) *For Self-Examination and Judge for Yourselves!*, Trans. Walter Lowrie, Princeton, NJ.

— (1954) *Fear and Trembling and The Sickness Unto Death*, Trans. Walter Lowrie, Princeton, NJ.

— (1957) *The Concept of Dread*, Trans. Walter Lowrie, Princeton, NJ.

— (1959) *Either/Or*, Trans. David F. Swenson and Lillian Marvin Swenson, 2 vols, Princeton, NJ.

— (1962) *The Point of View of My Work as an Author: a Report to History*, Trans. Walter Lowrie; ed. Benjamin Nelson, New York.

— (1962a) *Philosophical Fragments*, Trans. and ed. David Swenson, Niels Thulstrup, and Howard V. Hong, Princeton, NJ.

— (1964) *Repetition: an Essay in Experimental Psychology*, Trans. Walter Lowrie, New York and London.

— (1965) *The Concept of Irony*, Trans. Lee M. Capel, Bloomington, IN., and London.

— (1967) *Stages on Life's Way*, Trans. Walter Lowrie, New York.

— (1972) *Training in Christianity*, Trans. Walter Lowrie, Princeton, NJ.

— (1978) *Two Ages*, ed. and trans. Howard V. Hong and Edna H. Hong, Princeton, NJ.

— (1980) *The Concept of Anxiety*, Trans. and ed. Reidar Thomte and Albert B. Anderson, Princeton, NJ.

— (1967-78) *Søren Kierkegaard's Journals and Papers*, ed. and trans. Howard V. Hong and Edna H. Hong, 6 vols, Bloomington, IN., and London.

King, John H. (1892) *The Supernatural: Its Origin, Nature and Evolution*, London.

Kleczek, Josip (1976) *The Universe*, Dordrecht.

Klein, Melanie (1932) *The Psycho-Analysis of Children*, London.

Klibansky, Raymond (ed.) (1979) *Abbot Suger on The Abbey Church of St-Denis*, Princeton, NJ.

Klibansky, Raymond (1982) *The Continuity of the Platonic Tradition During the Middle Ages*, London and New York.

Klibansky, R., Panofsky, E., and Saxl, F. (1964) *Saturn and Melancholy*, London.

Kline, Morris (1954) *Mathematics in Western Culture*, London.

— (1972) *Mathematical Thought from Ancient to Modern Times*, New York.

Knight, Isabel F. (1968) *The Geometric Spirit*, New Haven, CT.

Knowles, David (1949) *The Monastic Order in England: a History of its Development from the Times of St Dunstan to the Fourth Lateran Council 943-1216*, Cambridge.

— (1969) *Christian Monasticism*, London.

Knowles, David and Obolensky, Dimitri (1972) *The Christian Centuries – the Middle Ages*, London and New York.

Koenigsberger, Dorothy (1979) *Renaissance Man and Creative Thinking: a History of Concepts of Harmony 1400-1700*, Hassocks.

Koerner, E. F. K. (1973) *Ferdinand de Saussure: Origin and Development of his Linguistic Thought in Western Studies of Language*, Braunschweig and Oxford.

Koestler, A. (1964) *The Sleepwalkers: a History of Man's Changing View of the Universe*, Harmondsworth, Middx.

Kohler, Wolfgang (1927) *The Mentality of Apes*, London.

Kojeve, Alexandre (1969) *Introduction to the Reading of Hegel*, Trans. James H. Nichols, Jr, New York.

Kolakowski, L. (1972) *Positivist Philosophy: From Hume to the Vienna Circle*, Harmondsworth, Middx.

Komesaroff, Paul A. (1986) *Objectivity, Science and Society*, London.

Koyré, Alexandre (1957) *From the Closed World to the Infinite Universe*, Baltimore.

(1965) *Newtonian Studies*, London.

(1973) *The Astronomical Revolution*, London.

(1978) *Galileo Studies*, Trans. John Mepham, Hassocks.

Krajewski, Wladyslaw (1977) *Correspondence Principle and Growth of Science*, Dordrecht and Boston.

Kretschmer, Ernst (1936) *Physique and Character: an Investigation of the Nature of Constitution and of the Theory of Temperament*, Trans. W. J. H. Sprott, London.

Kretzmann, Norman (1982) *Infinity and Continuity in Ancient and Medieval Thought*, Ithaca, NY, and London.

✗ Krips, Henry (1987) *The Mathematics of Quantum Theory*, Oxford.

Kristeller, Paul Oskar (1961) *Renaissance Thought*, New York.

(1964) *Eight Philosophers of the Renaissance*, Stamford.

(1972) *Renaissance Concepts of Man and Other Essays*, New York.

(1979) *Renaissance Thought and Its Sources*, New York.

✗ Kuhn, Thomas (1957) *The Copernican Revolution: Planetary Astronomy and the Development of Western Thought*, Cambridge, MA.

✗ (1962) *The Structure of Scientific Revolutions*, Chicago.

✗ (1978) *Black-Body Theory and the Quantum Discontinuity 1894-1912*, Oxford.

Ladd, George Turnbull (1897) *Philosophy of Knowledge*, London and New York.

Ladurie, Emmanuel Le Roy (1981) *Carnival in Romans: a People's Uprising at Romans 1579-1580*, Trans. Mary Feeney, Harmondsworth, Middx.

Lambert, M. D. (1961) *Franciscan Poverty: the Doctrine of the Absolute Poverty of Christ and the Apostles in the Franciscan Order 1210-1323*,

Hayek & Lévi-Strauss: what ties?

London.

(1977) *Medieval Heresy: Popular Movements from Bogomil to Hus*, London.

Landes, David S. (1983) *Revolution in Time: Clocks and the Making of the Modern World*, London and Cambridge, MA.

Lang, Andrew (1887) *Myth, Ritual and Religion*, 2 vols, London.

La Salle, J.-B. de (n.d. (1720)) *Cahiers Lasalliens*, Rome.

Lassus, J. B. A. (1968) *Album de Villard de Honnecourt*, Paris.

Lauer, Quentin (1982) *A Reading of Hegel's Phenomenology of Spirit*, New York.

Lawrence, C. H. (1984) *Medieval Monasticism: Forms of religious life in Western Europe in the Middle Ages*, London and New York.

Lea, Henry Charles (1888) *A History of the Inquisition of the Middle Ages*, 4 vols, London.

Leach, E. (1982) *Social Anthropology*, Glasgow.

Leclerq, J. (1978) *The Love of Learning and the Desire for God*, New York.

(1979) *Monks and Love in Twelfth-Century France*, Oxford.

Le Goff, Jacques (1988) *Medieval Civilization, 400-1500*, Oxford.

(1980) *Time, Work and Culture in the Middle Ages*, Chicago and London.

Lehmann, W. C. (1930) *Adam Ferguson and the Beginnings of Modern Sociology*, New York.

Lévi-Strauss, Claude (1966) *The Savage Mind*, London.

(1968) *Structural Anthropology*, Trans. Claire Jacobson and Brooke Grundfest Schoepf, London.

(1969) *The Elementary Structures of Kinship*, Trans. James Harle Bell, John Richard von Sturmer, and Rodney Needham, London.

(1970, 1973, 1978, 1981) *Introduction to a Science of Mythology*, Trans. John and Doreen Weightman, 4 vols, London.

(1973) *Tristes Tropiques*, Trans. John and Doreen Weightman, London.

(1977) *Structural Anthropology*, vol. II, Trans. Monique Layton, London.

Lévy-Bruhl, Lucien (1905) *Ethics and Moral Science*, London.

(1928) *The Soul of the Primitive*, London.

(1936) *Primitives and the Supernatural*, London.

(1975) *The Notebooks on Primitive Mentality*, Oxford.

Lewis, C. S. (1958) *The Allegory of Love*, New York.

(1964) *The Discarded Image*, Cambridge.

Lindberg, David C. (1976) *Theories of Vision from Al-Kindi to Kepler*, Chicago and London.

Little, Lester K. (1978) *Religious Poverty and the Profit Economy in Medieval Europe*, London.

Llull, Ramon (1985) *Selected Works of Ramon Llull 1232-1316*, Anthony Bonner (ed.), 2 vols, Princeton, NJ.

Locke, John (1924) *An Essay Concerning Human Understanding*, A. S. Pringle-Pattison (ed.), Oxford.

(1968 (1693)) *The Educational Writings of John Locke*, James L. Axtell (ed.), Oxford.

Loewenberg, J. (1965) *Hegel's Phenomenology: Dialogues on the Life of the Mind*, La Salle, IL.

Lorentz, H. A., Einstein, A., Minkowski, H., and Weyl, H. (1952) *The Principle of Relativity*, New York.

Lovejoy, Arthur O. (1960) *The Great Chain of Being: a Study of the History of an Idea*, New York.

Lowie, Robert H. (1924) *Primitive Religion*, London.

Löwith, Karl (1964) *From Hegel to Nietzsche: the Revolution in Nineteenth-Century Thought*, Trans. David E. Green, London.

Lowrie, Walter (1962) *Kierkegaard*, 2 vols, New York.

Lubbock, Sir John (1870) *The Origin of Civilisation and the Primitive Condition of Man*, London and Edinburgh.

Luhmann, Niklas (1986) *Love as Passion: the Codification of Intimacy*, Trans. Jeremy Gains and Doris L. Jones, Cambridge.

Lukács, Georg (1968) *History and Class Consciousness: Studies in Marxist Dialectics* Trans. Rodney Livingstone, London.

— (1974) *Soul and Form*, Trans. Anna Bostock, London.

— (1968) *History and Class Consciousness: Studies in Marxist Dialectics*, Trans. Rodney Livingstone, London.

— (1980) *The Destruction of Reason*, Trans. Peter Palmer, London.

McCormmach, R. (ed.) (1971) *Historical Studies in the Physical Sciences*, 3 vols, Philadelphia, PA.

McEvoy, James (1982) *The Philosophy of Robert Grosseteste*, Oxford.

Mach, Ernst (1893) *The Science of Mechanics*, Chicago.

— (1976) *Knowledge and Error: Sketches on the Psychology of Enquiry*, Dordrecht and Boston.

Machamer, Peter K. and Turnbull, Robert G. (1976) *Motion and Time Space and Matter*, Ohio.

Mackie, J. C. (1982) *The Miracle of Theism*, London.

McLean, Antonia (1972) *Humanism and the Rise of Science in Tudor England*, London.

McMullin, Ernan (1978) *Newton on Matter and Activity*, Notre Dame and London.

— (ed.) (1967) *Galileo: Man of Science*, New York.

Macpherson, C. B. (1962) *The Political Theory of Possessive Individualism*, Oxford.

— (1973) *Democratic Theory: Essays in Retrieval*, Oxford.

Maffei, Paolo (1980) *Monsters in the Sky*, Cambridge, MA.

Malantschuk, Gregor (1971) *Kierkegaard's Thought*, Howard V. Hong and Edna H. Hong (ed. and trans.), Princeton, NJ.

Mâle, Emile (1949) *Religious Art: from the Twelfth to the Eighteenth Century*, London.

— (1958) *The Gothic Image: Religious Art in France of the Thirteenth Century*,

New York.

Mandeville, Bernard (1957 (1732)) *The Fable of the Bees: Or, Private Vices, Publick Benefits*, F.B. Kaye (ed.), 2 vols, Oxford.

Mandrou, Robert (1978) *From Humanism to Science 1480-1700*, Hassocks.

Mann, Thomas (1958) *The Confessions of Felix Krull, Confidence Man*, Trans. Denver Lindley, Harmondsworth, Middx.

Manuel, Frank E. (1962) *The Prophets of Paris*, Cambridge.

(ed.) (1965) *The Enlightenment*, Englewood Cliffs, NJ.

Manuel, Frank E. and Manuel, Fritzie P. (1979) *Utopian Thought in the Western World*, Oxford.

Marcus, Steven (1974) *The Other Victorians: a Study in Sexuality and Pornography in Mid-Nineteenth-Century England*, New York.

Marcuse, Herbert (1964) *One Dimensional Man*, London.

(1969) *Eros and Civilization*, London.

Marenbon, John (1983) *Early Medieval Philosophy 480-1150*, London.

Marett, R. R. (1932) *Faith, Hope and Charity in Primitive Religion*, Oxford.

(1936) *Tylor*, London.

Martin, Alfred von (1944) *Sociology of the Renaissance*, London.

Martinet, A. (1969) *Elements of General Linguistics*, London.

Marx, Karl (1973) *Grundrisse: Foundations of the Critique of Political Economy*, Trans. Martin Nicolaus, Harmondsworth, Middx.

(1975) *Early Writings*, Trans. Rodney Livingstone and Gregor Benton; Introduction by Lucio Colletti, Harmondsworth, Middx.

(1976) *Capital*, vol. 1, Trans. Ben Fowkes, Harmondsworth, Middx.

Masi, M. (1983) *Boethian Number Theory*, Amsterdam.

Mason, John Hope (1982) *The Irresistible Diderot*, London.

Mause, Lloyd de (1976) *The History of Childhood*, London.

Mauss, Marcel (1970) *The Gift: Forms and Functions of Exchange in Archaic Society*, Trans. Ian Cunnison, London.

(1972) *A General Theory of Magic*, Trans. Robert Brain, London.

May, R., Angel, E., Ellenberger, H. F. (eds) (1958) *Existence: a New Dimension in Psychiatry and Psychology*, New York.

Maxwell, James Clerk (1890) *The Scientific Papers of James Clerk Maxwell*, 2 vols, Cambridge.

Menut, A. D. and Denamy, A. J. (eds) (1968) *Le Livre du Ciel et du Monde*, Michigan and London.

Merton, Robert K. (1970) *Science, Technology and Society in Seventeenth Century England*, New York.

Mészáros, I. (1970) *Marx's Theory of Alienation*, London.

Mettrie, Julien Offray de la (1912 (1747)) *Man a Machine*, La Salle, IL.

(1960) *L'Homme machine*, Introduction and Notes by Aram Vartanian, Princeton, NJ.

Meyerson, Emile (1930) *Identity and Reality*, Trans. Kate Loewenberg, London and New York.

(1985) *The Relativist Deduction: Epistemological Implications of the Theory of Relativity*, Dordrecht and Boston.

Miles, Robert (1987) *Capitalism and Unfree Labour: Anomaly or Necessity*, London and New York.

Mintz, Samuel (1962) *The Hunting of Leviathan*, Cambridge.

Mitteis, Heinrich (1975) *The State in the Middle Ages*, Amsterdam and Oxford.

Mochulsky, Konstantin (1967) *Dostoevsky: His Life and Work*, Trans. Michael A. Minihan, Princeton, NJ.

Monboddo, Lord (James Burnet) (1773) *Origin and Progress of Language*, 4 vols, Edinburgh.

⚹ Montaigne, Michel de (1958) *Essays*, ed. and trans. J. M. Cohen, Harmondsworth, Middx.

Moody, Ernest A. (1975) *Studies in Medieval Philosophy, Science and Logic*, Berkeley, CA.

Moore, J. R. (1979) *The Post-Darwinian Controversies*, Cambridge.

Morgan, Lewis H. (1851) *League of the Iroquois*, London.

(1877) *Ancient Society*, London.

Morris, Colin (1972) *The Discovery of the Individual 1050-1200*, London.

Morton, A. G. (1981) *History of Botanical Science*, London.

Müller, F. Max (1887) *The Science of Thought*, London.

(1898) *Lectures on the Origin and Growth of Religion*, Collected Works, vol. 9, London.

Murdoch, Dugald (1987) *Niels Bohr's Philosophy of Physics*, Cambridge.

Murphy, John (1927) *Primitive Man*, London.

Murray, Alexander (1978) *Reason and Society in the Middle Ages*, Oxford.

Musil, Robert (1968) *The Man Without Qualities*, Trans. Eithne Wilkins and Ernst Kaiser, 3 vols, London.

(1982 (1908)) *On Mach's Theories*, Trans. Kevin Mulligan, Washington DC.

Myers, Henry A. with Wolfram, Herwig (1982) *Medieval Kingship*, Chicago.

Narlikar, Jayant (1977) *The Structure of the Universe*, Oxford.

(1982) *The Lighter Side of Gravity*, San Francisco.

Nebelsick, Harold P. (1985) *Circles of God: Theology and Science from the Greeks to Copernicus*, Edinburgh.

Needham, Rodney (1972) *Belief, Language and Experience*, Oxford.

(ed.) (1973) *Right and Left: Essays in Dual Symbolic Classification*, Chicago and London.

Nelson, John Charles (1958) *Renaissance Theory of Love*, New York.

⚹ Newton, Isaac (1931 (1704)) *Opticks*, Foreword by Einstein, Introduction by E. T. Whittaker, London.

⚹ (1946 (1688, 1713)) *Sir Isaac Newton's Mathematical Principles of Natural Philosophy and His System of the World*, Trans. Andrew Motte (1729), revised by Florian Cajori, Berkeley, CA.

Nicolson, Marjorie (1956) *Science and Imagination*, New York.

✶ Nietzsche, Friedrich (1969) *On the Genealogy of Morals / Ecce Homo*, Trans. Walter Kaufmann and R. J. Hollingdale, New York.

✶ (1973) *Beyond Good and Evil*, Trans. R. J. Hillingdale, Harmondsworth, Middx.

✶ (1974) *The Gay Science*, Trans. Walter Kauffmann, New York.

Norton, David Fate (1982) *David Hume*, Princeton, NJ.

Nygren, Anders (1953) *Agape and Eros*, London.

Ollman, Bertell (1971) *Alienation: Marx's Conception of Man in Capitalist Society*, Cambridge.

Olson, Richard (1975) *Scottish Philosophy and British Physics 1750-1880*, Princeton, NJ.

Onians, Richard Broxton (1951) *The Origin of European Thought about the Body, the Mind, the Soul, the World, Time and Fate*, Cambridge.

✶ Oresme, Nicholas (1956) *The De Moneta of Nicholas Oresme and English Mint Documents*, Trans. Charles Johnson, London.

✶ (1968) *Le Livre du Ciel et du Monde*, Albert D. Menut and Alexander J. Denamy (eds), Michigan and London.

Orrù, Marco (1987) *Anomie: History and Meaning*, Boston.

Otto, Rudolf (1913) *Naturalism and Religion*, New York.

Packard, Sidney R. (1973) *Twelfth Century Europe: an Interpretive Essay*, Amherst, MA.

Pagel, Walter (1982) *Paracelsus*, Basel.

Pagels, Heinz R. (1982) *The Cosmic Code: Quantum Physics as the Language of Nature*, New York.

Painter, Sidney (1961) *Feudalism and Liberty: Articles and Addresses*, Baltimore, MD.

Pannekoek, A. (1961) *A History of Astronomy*, London.

Panofsky, Erwin (1970) *Renaissance and Renascences in Western Art*, London.

(1979) *Abbot Suger: On the Abbey Church of St-Denis and its Art Treasures*, Princeton, NJ.

Parry-Jones, William L. (1972) *The Trade in Lunacy: a Study of Private Madhouses in England in the Eighteenth and Nineteenth Centuries*, London.

Parsons, Talcott (1949) *The Structure of Social Action*, New York and London.

Pascal, R. (1960) *Design and Truth in Autobiography*, London.

Paterson, Antoinette M. (1970) *The Infinite Worlds of Giordano Bruno*, Springfield, IL.

Peebles, P. J. E. (1971) *Physical Cosmology*, Princeton, NJ.

Peel, J. D. Y. (1971) *Herbert Spencer: The Evolution of a Sociologist*, London.

Pelczynski, Z. A. (ed.) (1971) *Hegel's Political Philosophy: Problems and*

Perspectives, Cambridge.

Perella, N. J. (1969) *The Kiss: Sacred and Profane*, Berkeley, CA.

Perkins, Merle L. (1982) *Diderot and the Time-Space Continuum: His Philosophy, Aesthetics and Politics*, Oxford.

Perry, W. J. (1923) *The Origin of Magic and Religion*, London.

Petit-Dutaillis, C. (1983) *The Feudal Monarchy in France and England*, New York.

Piaget, Jean (1956) *The Language and Thought of the Child*, London.

(1973) *The Child's Conception of the World*, London.

(1977) *The Moral Judgment of the Child*, London.

Planck, Max (1925) *A Survey of Physics*, Trans. R. Jones and D. H. Williams, London.

Plantinga, Alvin (1974) *The Nature of Necessity*, Oxford.

Poggi, Gianfranco (1978) *The Development of the Modern State: a Sociological Introduction*, London.

Polkinghorne, J. C. (1979) *The Particle Play: an account of the Ultimate Constituents of Matter*, Oxford and San Francisco.

(1984) *The Quantum World*, London.

Pollock, Linda (1983) *Forgotten Children: Parent–Child Relations 1500-1900*, Cambridge.

Porter, R., (1982) *English Society in the Eighteenth Century*, Harmondsworth, Middx.

Poulet, Georges (1966) *The Metamorphoses of the Circle*, Trans. Carley Dawson and Elliott Coleman, Baltimore, MD.

(1977) *Proustian Space*, Trans. Elliott Coleman, Baltimore, MD, and London.

Powys, John Cowper (1974) *Dostoievsky*, London.

Preyer, Wilhelm (1896) *The Mind of the Child*, 2 vols, New York.

Propp, V. (1968) *Morphology of the Folktale*, Austin, Tex., and London.

Proust, Marcel (1966-70) *Remembrance of Things Past*, Trans. C. K. Scott Moncrieff and Andreas Mayor, London.

Radin, Paul (1957) *Primitive Religion*, New York.

Rand, Edward Kennard (1928) *Founders of the Middle Ages*, Cambridge.

Randall, J. H. (1961) *The School of Padua and the Emergence of Modern Science*, Padua.

(1962) *The Career of Philosophy*, 2 vols, New York.

Raphael, D. D. (1969) *British Moralists*, 2 vols, Oxford.

Rasmussen, Vilhelm (1920) *Child Psychology*, 3 vols, Copenhagen.

Rashdall, J. H. (1962) *The Career of Philosophy*, 2 vols, Columbia.

Ravetz, Jerome R. (1965) *Astronomy and Cosmology in the Achievement of Nicolaus Copernicus*, Warsaw.

Rees, M., Ruffini, R., and Wheeler, J. A. (1974) *Black Holes, Gravitational Waves and Cosmology: an Introduction to Current Research*, New York and London.

Regosin, Richard L. (1977) *The Matter of My Book: Montaigne's Essays as the Book of the Self*, Berkeley, CA.

Reichenbach, Hans (1948) *Philosophic Foundations of Quantum Mechanics*, Berkeley, CA, and Los Angeles.

— (1965) *The Theory of Relativity and A Priori Knowledge*, Berkeley, CA, and Los Angeles.

Reichel-Dolmatoff, Gerardo (1971) *Amazonian Cosmos: the Sexual and Religious Symbolism of the Tukano Indians*, Chicago.

Reid, Thomas (1961 (1785)) *Essays on the Intellectual Powers of Man*, Ed. and abridged by A. D. Woozly, London.

Reidl, C. C. (ed.) (1942) *On Light*, Milwaukee, WI.

Richardson, John Adkins (1971) *Modern Art and Scientific Thought*, Urbana, IL, and London.

Richter, Gottfried (1985) *Art and Human Consciousness*, New York and Edinburgh.

Rieff, Philip (1965) *Freud: the Mind of the Moralist*, London.

Riesman, David (1965) *The Lonely Crowd*, New Haven, CT.

Rittersbuch, Philip C. (1968) *The Art of Organic Forms*, Washington DC.

Rivers, W. H. R. (1924) *Medicine, Magic and Religion*, London.

Robinson, Daniel S. (ed). (1961) *The Story of Scottish Philosophy*, New York.

Robertson, John M. (1910) *Christianity and Mythology*, London.

Robertson Smith, W. (1907) *Lectures on the Religion of the Semites*, London.

Rojek, Chris (1985) *Capitalism and Leisure Theory*, London.

Rorty, Richard (1980) *Philosophy and the Mirror of Nature*, Oxford.

Rosen, E. (ed.) (1971) *Three Copernican Treatises*, New York.

Rosen, George (1968) *Madness and Society*, London.

Rossi, Paolo (1968) *Francis Bacon: from Magic to Science* Trans. Sacha Rabinovitch, London.

— (1970) *Philosophy, Technology and the Arts in the Early Modern Era*, Trans. Salvator Attanasio, New York and London.

Rousseau, G. S. and Porter, Roy (eds) (1980) *The Ferment of Knowledge: Studies in the Historiography of Eighteenth-Century Science*, Cambridge.

Rousseau, Jean-Jacques (1911 (1762)) *Emile: or Education*, Trans. Barbara Foxley, London.

— (1953 (1781)) *The Confessions of Jean-Jacques Rousseau*, Trans. J. M. Cohen, Harmondsworth, Middx.

— (1969) *Essai sur l'origine des langues*, Charles Porset (ed.), Bordeaux.

— (1979) *The Reveries of the Solitary Walker*, Trans. Charles Butterworth, New York.

— (1984) *A Discourse on Inequality*, Trans. with Introduction and Notes by Maurice Cranston, Harmondsworth, Middx.

Rowan-Robinson, Michael (1985) *The Cosmological Distance Ladder*, New

York.

Russell, Jeffrey Burton (1968) *A History of Medieval Christianity*, Arlington, IL.

Saussure, Ferdinand de (1966) *Course in General Linguistics*, Charles Bally and Albert Sechehaye (eds); Trans. Wade Baskin, New York.

Scammel. G. V. (1981) *The World Encompassed: the first European Maritime Empires 800-1650*, London.

Schacht, Richard (1971) *Alienation*, London.

Schaffner, Kenneth F. (1972) *Nineteenth-Century Aether Theories*, Oxford.

Schatzman, Morton (1973) *Soul Murder: Persecution in the Family*, Harmondsworth, Middx.

Scheler, Max (1972) *Ressentiment*, Trans. William W. Holdheim, New York.

Schilder, Paul (1935) *The Image and Appearance of the Human Body: Studies in the Constructive Energies of the Psyche*, London.

Schilpp, Paul Arthur (ed.) (1949) *Albert Einstein: Philosopher–Scientist*, Evanston, IL.

Schlereth, Thomas (1977) *The Cosmopolitan Ideal in Enlightenment Thought*, London.

Schmidt, Alfred (1971) *The Concept of Nature in Marx*, Trans. B. Fowkes, London.

Schmidt, W. (1931) *The Origin and Growth of Religion: Facts and Theories*, London.

Schmitt, Charles B. (1983) *Aristotle and the Renaissance*, Cambridge, MA.

Schneider, L. (ed.) (1967) *The Scottish Moralists: On Human Nature and Society*, Chicago.

Schofield, Robert E. (1970) *Mechanism and Materialism: British Natural Philosophy in an Age of Reason*, Princeton, NJ.

Schorske, Carl E. (1979) *Fin-De-Siècle Vienna*, London.

Schreber, D. P. (1955) *Memoirs of My Nervous Illness*, Trans., edited, with Introduction and Notes by Ida Macalpine and Richard A. Hunter, London.

Schroedinger, Erwin (1935) *Science, Theory and Man*, London.

Schumpeter, J. A. (1943) *Capitalism, Socialism and Democracy*, London.

Schutz, Alfred (1976) *Collected Papers*, vol. 2, The Hague.

Schutz, A. and Luckmann, T. (1974) *The Structures of the Life-World*, Trans. R. M. Zaner and H. T. Englehardt Jr, London.

Sciama, D. W. (1975) *Modern Cosmology*, Cambridge.

Scott, Sir Walter (ed.) (1924) *Hermetica*, 4 vols, Oxford.

Scott, Wilson L. (1970) *The Conflict between Atomism and Conservation Theory 1644-1860*, London.

Scull, Andrew T. (1982) *Museums of Madness: the Social Organisation of Insanity in Nineteenth-Century England*, Harmondsworth, Middx.

(ed.) (1981) *Madhouses, Mad-Doctors and Madmen: the Social History of*

Psychiatry in the Victorian Era, London.

Sedgwick, Peter (1982) *Psycho Politics*, London.

Segré, Emilio (1984) *From Falling Bodies to Radio Waves*, New York.

Shaftesbury, Anthony, Earl of (1964 (1711)) *Characteristics of Men, Manners, Opinions, Times*, John M. Robertson (ed.), Introduction by Stanley Grean, Indianapolis, IN.

Shapin, Stevin and Schaffer, Simon (1985) *Leviathan and the Air-Pump: Hobbes, Boyle and the Experimental Life*, Princeton, NJ.

Sharp, D. E. (1930) *Franciscan Philosophy at Oxford*, Oxford.

Shea, William R. (1972) *Galileo's Intellectual Revolution*, London.

Shubnikov, A. V. and Koptsik, V. A. (1974) *Symmetry in Science and Art*, Trans. G. D. Archard, New York.

Sidgwick, Henry (1907) *The Methods of Ethics*, London.

Silk, Joseph (1980) *The Big Bang: the Creation and Evolution of the Universe*, San Francisco.

Silvestris, Bernard (1973) *The Cosmographia of Bernardus Silvestris*, Trans. Winthrop Wetherbee, New York.

(1978) *Cosmographia*, Peter Dronke (ed.), Leiden.

Simmel, Georg (1964) *The Sociology of Georg Simmel*, Trans. and ed. Kurt H. Wolff, New York.

(1965) *Essays on Sociology, Philosophy and Aesthetics by Georg Simmel et al.*, Kurt H. Wolff (ed.), New York.

(1971) *On Individuality and Social Forms*, D. N. Levine (ed.), Chicago.

(1978) *The Philosophy of Money*, Trans. T. Bottomore and D. Frisby, London.

Simon, W. M. (1963) *European Positivism in the Nineteenth Century: an Essay in Intellectual History*, New York.

Sinclair, Andrew (1977) *The Savage*, London.

Skultans, Vieda (1979) *English Madness: Ideas on Insanity 1550-1890*, London.

Smith, Adam (1976 (1759)) *The Theory of Moral Sentiments*, D. D. Raphael and A. L. MacFie (eds), Oxford.

(1976 (1776)) *An Enquiry into the Nature and Causes of the Wealth of Nations*, R. H. Campbell and A. S. Skinner (eds), Oxford.

(1980) *Essays on Philosophical Subjects*, W. P. D. Wightman and J. C. Bryce (eds), Oxford.

Sohn-Rethel, Alfred (1978) *Intellectual and Manual Labour: a Critique of Epistemology*, London.

Sontag, S. (1983) *Illness as Metaphor*, Harmondsworth, Middx.

Southern, R. W. (1966) *Saint Anselm and his Biographer: A Study of Monastic Life and Thought 1059–c.1130*, Cambridge.

(1970) *Western Society and the Church in the Middle Ages*, Harmondsworth, Middx.

(1970a) *Medieval Humanism and Other Studies*, Oxford.

(1986) *Robert Grosseteste: The Growth of an English Mind in Medieval Europe*, Oxford.

Sperber, Dan (1975) *Rethinking Symbolism*, Cambridge.

Stallo, J. B. (1848) *General Principles of the Philosophy of Nature*, London.

Starobinski, Jean (1985) *Montaigne in Motion*, Trans. Arthur Goldhammer, Chicago.

Steenberghen, Fernand van (1966) *La Philosophie au XIIIe Siècle*, Louvain and Paris.

Steneck, Nicholas H. (1976) *Science and Creation in the Middle Ages*, London.

Stenton, F. M. (1932) *The First Century of English Feudalism*, Oxford.

Stephens, Thomas (ed.) (1905) *The Child and Religion*, New York.

Stern, William (1924) *Psychology of Early Childhood*, London.

Stewart, Dugald (1829) *Elements of the Philosophy of the Human Mind*, Cambridge.

Stiefel, Tina (1985) *The Intellectual Revolution in Twelfth Century Europe*, London.

Stirner, Max (1971) *Max Stirner: The Ego and His Own*, John Carroll (ed.), London.

Stock, Brian (1972) *Myth and Science in the Twelfth Century: a Study of Bernard Silvester*, Princeton, NJ.

Stocking, George W. (1983) *Observers Observed: Essays on Ethnographic Fieldwork*, Wisconsin.

Stone, L. (1979) *The Family, Sex and Marriage in England 1500–1800*, Harmondsworth, Middx.

Sulloway, Frank J. (1980) *Freud Biologist of the Mind: Beyond the Psychoanalytic Legend*, London.

Sully, James (1896) *Studies of Childhood*, London.

Swift, J. (1909) *A Tale of a Tub, the Battle of the Books and Other Satires*, London.

Szasz, T. (1961) *The Myth of Mental Illness*, New York.

Talbot, C. H. (1967) *Medicine in Medieval England*, Oldbourne.

Taton, René (ed.) (1963) *Ancient and Medieval Science*, London.

(ed.) (1965) *Science in the Nineteenth Century*, London.

(ed.) (1966) *Science in the Twentieth Century*, London.

Taylor, Charles (1975) *Hegel*, Cambridge.

Taylor, Mark C. (1980) *Journeys to Selfhood: Hegel and Kierkegaard*, Berkeley, CA.

Thackray, Arnold (1970) *Atoms and Powers: an Essay on Newtonian Matter Theory and the Development of Chemistry*, London.

Thompson, George (1961) *The First Philosophers*, London.

Thompson, Josiah (1974) *Kierkegaard*, London.

✶ Tolstoy, Ivan (1981) *James Clerk Maxwell: a Biography*, Edinburgh.

Thorndike, Lynn (1923-58) *A History of Magic and Experimental Science during the First Thirteen Centuries of Our Era*, 8 vols, New York.

(1949) *The 'Sphere' of Sacrobosco and its Commentators*, Chicago.

Thulstrup, Niels (1980) *Kierkegaard's Relation to Hegel*, Trans. George L. Stengren), Princeton, NJ.

✕ Tönnies, Ferdinand (1955) *Community and Association*, Trans. and supplemented by Charles P. Loomis, London.

Trinkaus, Charles (1970) *In Our Image and Likeness: Humanity and Divinity in Italian Humanist Thought*, 2 vols, London.

Tuke, Daniel H. (1882) *Chapters in the History of the Insane in the British Isles*, London.

✕ Turgot, Anne Robert Jacques baron de l'Aulne (1973) *Turgot on Progress, Sociology and Economics*, London.

Turner, Bryan S.(1984) *The Body and Society: Explorations in Social Theory*, Oxford.

Turner, Frank Miller (1974) *Between Science and Religion: the Reaction to Scientific Naturalism in Late Victorian England*, New Haven, CT, and London.

Turner, V. M. (1969) *The Ritual Process*, Harmondsworth, Middx.

Tylor, Edward B. (1913 (1871)) *Primitive Culture: Researches into the Development of Mythology, Philosophy, Religion, Language, Art and Culture*, 2 vols, London.

Uexküll, J. von (1926) *Theoretical Biology*, London.

✕ Ullmann, Walter (1961) *Principles of Government and Politics in the Middle Ages*, London.

✕ (1966) *The Individual and Society in the Middle Ages*, London.

Unger, Roberto Mangabeira (1975) *Knowledge and Politics*, New York and London.

Vandenbroucke, F. (1972) *Why Monks?*, Washington DC.

Vartanian, Aram (1953) *Diderot and Descartes*, Princeton, NJ.

(ed.) (1960) *L'Homme Machine*, Princeton, NJ.

Venturi, Franco (1971) *Utopia and Reform in the Enlightenment*, Cambridge.

Vereker, Charles (1967) *Eighteenth Century Optimism*, Liverpool.

Vilar, Pierre (1976) *A History of Gold and Money 1450–1920*, London.

✕ Veblen, T. (1925) *The Theory of the Leisure Class*, London.

Wagoner, Robert V. and Goldsmith, Donald W. (1982) *Cosmic Horizons: Understanding the Universe*, San Francisco.

Wald, Robert M. (1977) *Space, Time and Gravity*, Chicago and London.

Walker, D. P. (1958) *Spiritual and Demonic Magic from Ficino to Campanella*, London.

Wallace, William A. (1972) *Causality and Scientific Explanations*, 2 vols, Ann Arbor, MI.

(1977) *Galileo's Early Notebooks: The Physical Questions*, Notre Dame and London.

(1981) *Prelude to Galileo: Essays on Medieval and Sixteenth Century Sources of Galileo's Thought*, Dordrecht.

Wallerstein, Immanuel (1983) *Historical Capitalism*, London.

Weber, Max (1964) *From Max Weber: Essays in Sociology*, H. H. Gerth and C. Wright Mills (eds), London.

(1965) *The Sociology of Religion*, Trans. Ephraim Fischoff, with an Introduction by Talcott Parsons, London.

(1976) *The Protestant Ethic and the Spirit of Capitalism*, Trans. Talcott Parsons, London.

(1978) *Economy and Society*, Guenther Roth and Claus Wittich (eds), 2 vols, Berkeley, CA and Los Angeles.

Webster, Charles (ed.) (1974) *The Intellectual Revolution of the Seventeenth Century*, London.

(1975) *The Great Instauration: Science, Medicine and Reform 1626-1660*, London.

Weinberg, Steven (1977) *The First Three Minutes*, London.

(1983) *The Discovery of Subatomic Particles*, New York.

Weisheipl, James A. (1980) *Albertus Magnus and the Sciences*, Toronto.

Werkmeister, William H. (ed.) (1963) *Facets of the Renaissance*, New York.

Werner, Heinz (1957) *Comparative Psychology of Mental Development*, New York.

Westfall, Richard S. (1971) *Force in Newton's Physics: the Science of Dynamics in the Seventeenth Century*, London and New York.

(1977) *The Construction of Modern Science: Mechanism and Mechanics*, Cambridge.

Westman, Robert S. (ed.) (1975) *The Copernican Achievement*, Berkeley, CA, and London.

Wetherbee, W. (ed.) (1973) *The Cosmographia of Bernardus Silvestris*, New York.

White, T. H. (1954) *The Book of Beasts*, London.

Whittaker, Sir Edmund (1949) *From Euclid to Eddington*, Cambridge.

(1951) *A History of the Theories of Aether and Electricity*, 2 vols, London.

Wilden, Anthony (1972) *System and Structure: Essays in Communication and Exchange*, London.

William of St Thierry (1970) *Exposition on the Song of Songs*, Trans. Mother Columba Hart, Shannon.

(1971) *On Contemplating God*, Trans. Sister Penelope, Shannon.

(1971a) *The Golden Epistle*, Trans. Theodore Berkeley, Spencer, MA.

(1974) *The Enigma of Faith*, Trans. John D. Anderson, Washington DC.

(1979) *The Mirror of Faith*, Trans. Thomas X. Davis, Kalamazoo, MI.

Wilms, Hieronymus (1933) *Albert the Great*, London.

Winnicott, D. W. (1971) *Playing and Reality*, Harmondsworth, Middx.

Wittkower, Rudolf (1962) *Architectural Principles in the Age of Humanism*,

London.

Wood, Charles T. (1970) *The Age of Chivalry*, London.

Wolff, K. H. (ed.) (1965) *Essays on Sociology, Philosophy and Aesthetics by Georg Simmel et al.*, New York.

Woolf, Harry (ed.) (1980) *Some Strangeness in the Proportion*, Reading, MA.

Wrightson, K. (1982) *English Society 1580–1680*, London.

Wyckoff, D. (ed.) (1967) *Book of Minerals*, Oxford.

Yates, Frances (1964) *Giordano Bruno and the Hermetic Tradition*, London.

(1966) *The Art of Memory*, London.

(1972) *The Rosicrucian Enlightenment*, London.

(1979) *The Occult Philosophy in the Elizabethan Age*, London.

(1985) *Llull and Bruno: Collected Essays*, vol. 1, London.

Zaner, Richard M. (1971) *The Problem of Embodiment: Some Contributions to a Phenomenology of the Body*, The Hague.

Zilboorg, Gregory, with Henry, George W. (1941) *A History of Medical Psychology*, New York.

Zilsel, E. (1942) 'The sociological roots of science', *American Journal of Sociology* 47.

NAME INDEX

SUBJECT INDEX

361